Emerging Applications of Nanomaterials

Edited by

N.B. Singh[1], Md. Abu Bin Hasan Susan[2] and Ratiram Gomaji Chaudhary[3]

[1]Department of Chemistry and Biochemistry, SBSR, Sharda University, Greater Noida, India

[2]Department of Chemistry, University of Dhaka
Dhaka 1000, Bangladesh

[3]Department of Chemistry, Seth Kesarimal Porwal College of Arts, Science and Commerce, Kamptee, India

Published by **Materials Research Forum LLC**
Millersville, PA 17551, USA

Published as part of the book series
Materials Research Foundations
Volume 141 (2023)
ISSN 2471-8890 (Print)
ISSN 2471-8904 (Online)

Print ISBN 978-1-64490-228-8
eBook ISBN 978-1-64490-229-5

Distributed worldwide by

Materials Research Forum LLC
105 Springdale Lane
Millersville, PA 17551
USA
https://www.mrforum.com

Manufactured in the United States of America
10 9 8 7 6 5 4 3 2 1

Table of Contents

Preface

The researches on science and engineering of nanomaterials are increasing continuously because of their extensive applications in different sector. Nanomaterials are changing the face of scientific regeneration globally. Today, nanomaterials have applications in diverse areas addressing sustainability to ensure the future of human being. Nanomaterials are small (less than 100nm) but do big things. Keeping the importance of nanomaterials in mind, this book briefly discusses different aspects of nanomaterials, starting from synthesis, characterization, and applications. In this book, there are 14 chapters written by experts in the area. This book will be useful for students and researchers alike.

Prof. N.B. Singh
Prof. Md. Abu Bin Hasan Susan
Dr. Ratiram Gomaji Chaudhary

Chapter 1

Introduction to Nanomaterials

N.B. Singh

Department of Chemistry and Biochemistry, Sharda University, Greater Noida, India

n.b.singh@sharda.ac.in

Abstract

In the last few years, nanoscience and nanotechnology have become a very important branch of science and technology. 10^{-9} stands for nano and particles size less than 100 nm are said to be nanomaterial. Nanoscience connects chemistry, physics, biology, medicine, engineering sciences and others. Nanomaterials are synthesized in a number of ways, but the conventional methods are costly, time consuming and pollute the environment. However, green routes are becoming popular. Synthesized nanomaterials are characterized by using a number of techniques and they are useful in different sectors. A general introduction in brief is given in this chapter.

Keywords

Nanomaterials, Nanotechnology, Green Method, Characterization, Applications

Contents

1.   Introduction

New science and technology sometimes is developed by human imagination. Such imagination and dreams gave birth to 21st-century Nanotechnology. Devices made by matter in the dimensions between 1 and 100 nm, give novel applications known as nanotechnology [1]._Nanomaterials were already available in the atmosphere the name nano was not yet known. The term nanometre was first coined by 'Richard Zsigmondy, the 1925 Nobel Prize Laureate in chemistry'. However, in the modern era, nanotechnology is the brain child of Richard Feynman, who gave a lecture saying ''There's Plenty of Room at the Bottom'', at American Physical Society meeting at Caltech, in 1959 [2]. Feynman hypothesis led to novel ideas of manipulating matter at the atomistic level. He is widely regarded as the father of nanotechnology. Trio scientists Kroto, Smalley and Curl in 1980, discovered fullerene [3]. This further lead to increased interest in the field of nanotechnology. Invention of carbon nanotubes by Japanese scientist Iijima is another development in the history of nanomaterials [4].

Materials play an important role in the development of a nation. Properties of materials depend on size and materials scientist believe that the 21st century is the century of materials particularly nanomaterials (NMs). It is rightly said that no *materials, no progress.* A lot of interests in the field of nanoscience and nanotechnology were seen in the beginning of the 21st Century. Nanoscience has tremendous applications in the field of materials science, mechanical, electrical, and chemical engineering, chemistry, biology, energy, physics, medicine, textile, agrifoods, and many others [5]. Nanomaterials (NMs), being smaller in size, easily get agglomerated. These are then converted into nanocomposites (NCs), which are also used in different sectors.

There are variety of NMs and NCs prepared in different ways and characterized by different techniques. The nanomaterials can be used in different sectors. The present chapter gives a brief account of NMs and NCs.

2.   Classification of NMs

NMs are classified in number of ways such as based on dimensionality, morphology, state, chemical composition, etc. (Fig.1) [6]. There are variety of NMs such as metals, metal oxides, metal chalcogenides, organic, carbon, etc. (Fig.2).

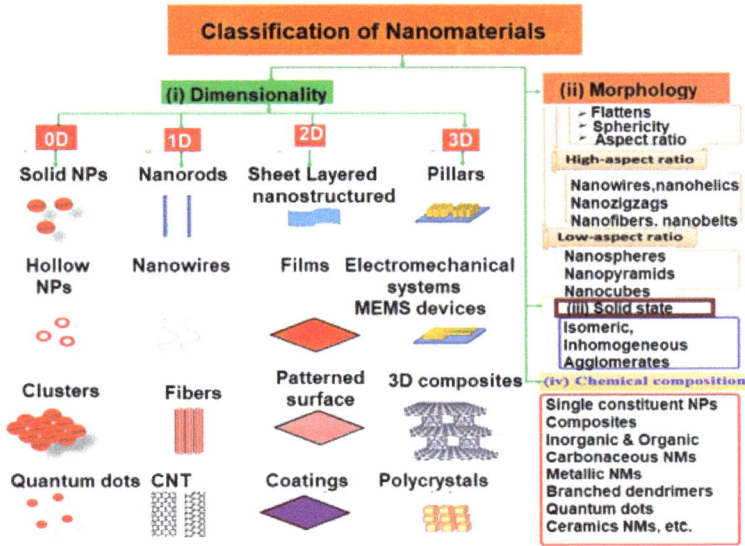

Figure 1: Different types of NMs and their classification [6]

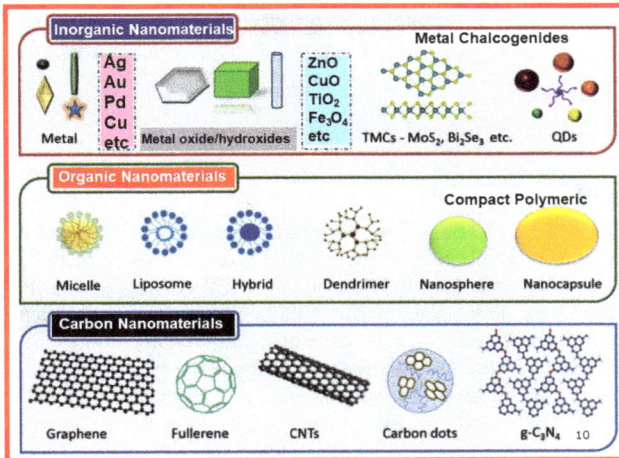

Figure 2: Different NMs

2.1 Nanocomposites

If two materials one of which is a matrix and the other is a reinforcing phase (in the form of fibers, sheets, or particles), are mixed and the properties of the resultant mixture are superior to those of individual components, are called composites. When either of the phases in composite has dimensions less than 100 nanometers (nm), the material is called nanocomposite (NC). The idea of developing NCs started in Japan, in 1980 at Toyota Research Laboratory. In 1998 Chemistry in Britain published an article on nano sandwiches, stating, nature is a master chemist with incredible talent [7]. Nanocomposites have multifunctional properties which depend on various parameters. NCs are basically of two types- polymer based and nonpolymer based (Fig.3). NCs are used in different sectors.

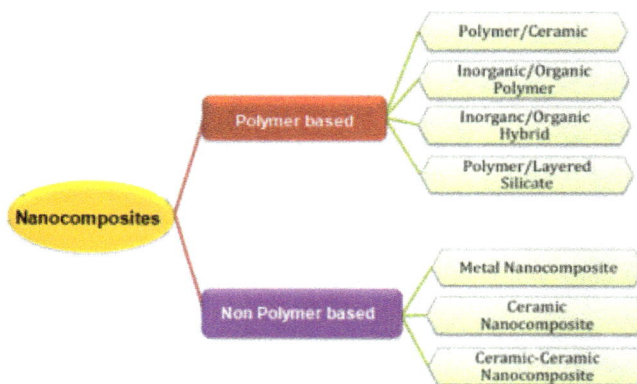

Figure 3: Classification of NCs

3. Synthesis of nanomaterials

NMs are synthesized by top-down and bottom-up approaches. (Fig.4)[8,9]. In top-down approach, number of methods such as mechanical and ball milling techniques are used to reduce the size of bulk materials. In addition, chemical etching, sputtering, evaporation, laser ablation, etc. are also used. In the bottom-up approach, chemical methods are involved. Bottom-up methods include chemical vapour deposition, sol–gel synthesis, chemical reduction, hydrothermal synthesis, and flame spray pyrolysis methods [10,11]. Bottom-up approaches are generally employed in the synthesis of metal NPs and metal oxide NPs [11,12]. These methods use reducing agents and petal precursors [13-17].

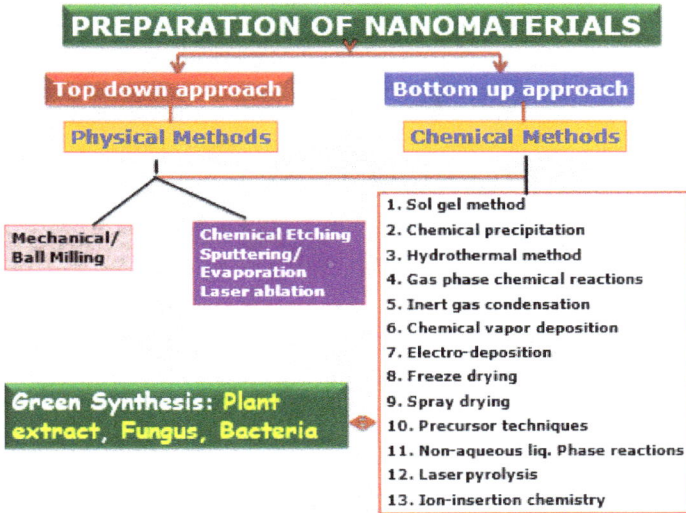

Figure 4: Top down and bottom-up approaches for synthesis of NMs

3.1 Green route for synthesis of NMs

Bottom-up approach is basically chemical method and uses costly and hazardous chemicals. Therefore, althernative ecofriendly methods known as green methods are now being used. There are number of methods as represented by Fig.5. Amongst different methods, plant extracts and microbial compounds are frequently used for the synthesis of metal and metal oxide NPs (Fig.6).

Figure 5: Green routes for the synthesis of NMs

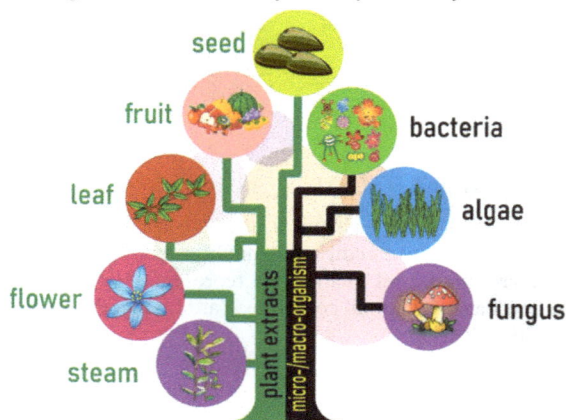

Figure 6: Green methods for the synthesis of NMs using plant extracts and microbial compounds

When salt solutions are mixed with extracts of different parts of plants, reduction of metal ions take place with the formation of NPs. The process depends on pH, concentration, time and temperature. The mechanism can be shown by Fig. 7[18].

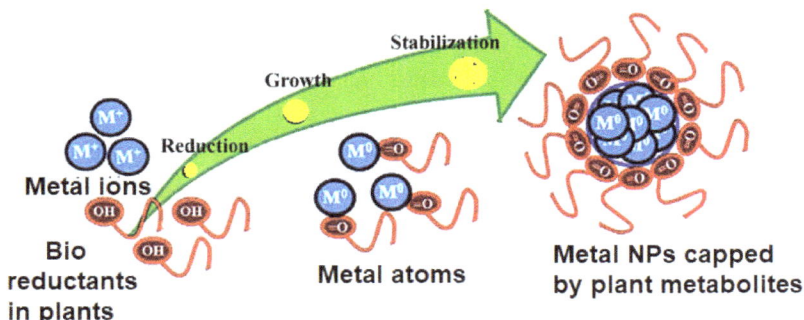

Figure:7 Mechanism of formation of metal NP using plant extracts [18]

4. Characterization of nanoparticles

Nanoparticles (NPs) are characterized by using number of experimental techniques such as Electron microscopy, Transmission electron microscopy, Powder X-ray diffraction, UV-visible spectroscopy, FTIR spectroscopy, Atomic force microscopy and energy dispersive X-ray spectroscopy in order to determine the size and shape, morphology, etc. Details of some of the techniques are given below.

4.1 UV-visible spectroscopy

Metallic nanoparticles, because of surface plasmon resonance (SPR) phenomena, give absorption bands in UV-visible region. Surface plasmon is collective oscillation of conduction electron excited at metal-dielectric interface by electromagnetic radiation. The SPR band is sensitive to shape, size and dielectric constant of the medium. This gives information about the shape and size of the metallic NPs and specifically novel metal NPs [19,20]. In the synthesis of Pd NPs, no peak is obtained but there was a change in colour from light yellow to dark brown [21].

4.2 FTIR spectroscopy

FTIR gives information about functional groups present in the compounds (phytochemicals) useful to identify the presence of reducing and capping agent. During the synthesis of Ag NPs using Nigella sativa extract, the peaks are shifted [22]. Similar results were found in other cases, when NPs were synthesized by using plant extracts.

4.3 Powder X-ray Diffraction

From X-ray diffraction patterns, using Debye-Scherrer equation (Eq.1), the crystallite sizes of NPs are determined.

Materials Research Forum LLC

https://doi.org/10.21741/9781644902295-1

$$d = \frac{K\lambda}{\beta \cos\theta} \ldots\ldots\ldots(1)$$

K= Scherrer constant, d= particle size in nm, β= full width half maximum, λ= wavelength of X-ray, θ= half of Bragg's angle.

4.4 Electron microscopic techniques

Morphologies of NPs and their sizes can be determined by using Scanning Electron Microscopy (SEM) and Transmission Electron Microscopy (TEM). [23,24].

4.5 Energy dispersive X-ray spectroscopy (EDX)

EDX is used to quantify the presence of different elements present in the sample [25].

5. Applications

There are number of applications of NMs in different industrial sectors. Presently there is no industry which can survive without using NMs in one way or the other. NMs are now being used in manufacture of scratchproof eyeglasses, anti-graffitic coatings for walls, stain-repellent fabrics, self-cleaning windows, crack- resistant paints, transparent sunscreens, ceramic coatings for solar cells, cosmetics, agriculture, medicine, food science, food technology, and civil engineering [26–27]. Nanotechnology is used to create substances which are lighter, rustproof, stronger, and stain/fire resistant [28]. A list of different sectors where NMs are used are given in Fig.8.

The incorporation of NM in personal care products seeks to prolong their useful life by increasing the stability of the cosmetic components and improving the delivery of the ingredients through the skin. NM are used to give new colour, transparency, and UV protection. This is the case of titanium dioxide (TiO2NP) and zinc oxide nanoparticles (ZnO NP) that are incorporated in modern sunscreens as an UV filter and to make creams and lotions transparent [29]. TiO_2 and Ag NPs [30, 31] provide antimicrobial properties against Gram positive and Gram-negative bacteria. Nanoelectronics have allowed to reduce power consumption, weight and thickness of the computer and television screens, but also helped to increase the computers response speed and the capacity of hard disk drive storage. Carbon nanotubes (CNT) [32] are incorporated in transistors to magnify the processing power of silicon chips, reducing the use of energy, the heat generated, and the size and weight of the components [33]. Additionally, nanotechnology is applied to conserve power and prolong battery life of small portable electronic gadgets. Sports technology also includes different nanoscale materials, specially CNT, to produce lighter but stiffer equipment and enhance strength and durability of conventional sporting equipment [34]. In the case of textile industry, nanotechnology is used to improve specific functions in clothing, like water repellence, antimicrobial action, conductivity, and antistatic and antiwrinkle characteristics [35]. NM are applied in different areas like drug delivery, imaging, gene therapy, tissue engineering, diagnosis, nanoscale biochips, and alternative therapeutics.

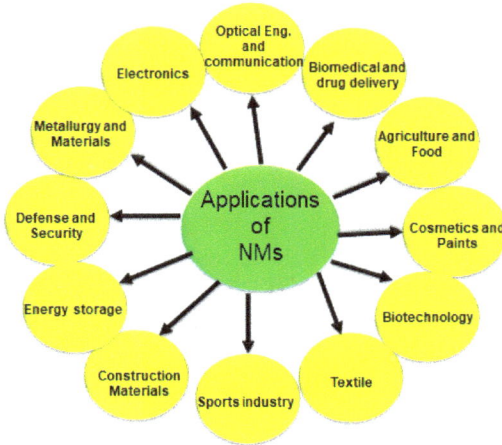

Figure 8: Applications of NMs

NMs have a number of applications in different sectors as shown by Fig.8 but very little is known about their toxicological behaviour [36]. Being small particles, it can easily go into the body and can accumulate. Some of the adverse effect on human health are given in Fig.9 [37,38]. Continuous exposure to NMs is likely to increase their adverse health effects. Therefore, this aspect should be studied in detail.

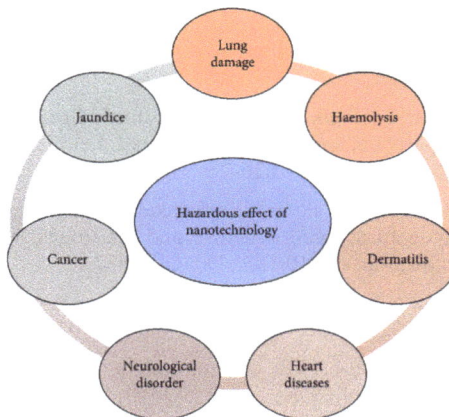

Figure 9 Hazardous effects of nanotechnology

Conclusions

In this chapter, the importance of nanoscience and nanotechnology has been highlighted. Methods used for synthesis of NMs have been discussed in brief. Emphasis is given for synthesis of NMs by using green routes. Different experimental techniques have been referred for characterization of NMs. Applications of NMs in different sectors are highlighted. Research should also be done on toxicological behaviour of NMs.

References

[1] H. Gleiter, Nanostructured materials: basic concepts and microstructure Acta Mater., 48 (2000)1-29 https://doi.org/10.1016/S1359-6454(99)00285-2

[2] R. P. Feynman, There's plenty of room at the bottom, Eng. Sci. 23 (1960) 22-36.

[3] H. W. Kroto, J. R. Heath, S. C. O'Brien, R. F. Curl, R. E. Smalley, C60: Buckminsterfullerene. Nature, 318 (1985) 162-163. https://doi.org/10.1038/318162a0

[4] S. Iijima, Helical microtubules of graphitic carbon. Nature, 354(1991) 56-58. https://doi.org/10.1038/354056a0

[5] S.A. Mazari, E.Ali , R. Abro, F.S. Khan, I. Ahmed, M. Ahmed, S. Nizamuddin, T.H. Siddiqui, N. Hossain, N. M. Mubarak, A. Shah, Nanomaterials: Applications, waste-handling, environmental toxicities, and future challenges-A Review. J. Environ. Chem. Engg., 9(1021)105028 https://doi.org/10.1016/j.jece.2021.105028

[6] T. A. Saleh, Nanomaterials: Classification, properties, and environmental toxicities, Environmental Technology & Innovation, 20 (2020) 101067 https://doi.org/10.1016/j.eti.2020.101067

[7] Khorshidi Mahsa, Asadpour Saeid, Sarmast Narges, Dinari Mohammad. A review of the synthesis methods, properties, and applications of layered double hydroxides/carbon nanocomposites. Journal of Molecular Liquids, 348(2022) 118399. https://doi.org/10.1016/j.molliq.2021.118399

[8] I. Hussain, N.B. Singh, A. Singh, H. Singh, S.C. Singh, Green synthesis of nanoparticles and its potential application, Biotechnol. Lett., 38 (2016) 545-560. https://doi.org/10.1007/s10529-015-2026-7

[9] N. Baig, I. Kammakakam, W. Falath, I. Kammakakam, Nanomaterials: A review of synthesis methods, properties, recent progress, and challenges, Materials, Advances 2 (2021) 1821-1871. https://doi.org/10.1039/D0MA00807A

[10] M. Huston, M. DeBella, M. DiBella, A. Gupta, Green synthesis of nanomaterials, Nanomaterials, 11 (8) (2021) 2130. https://doi.org/10.3390/nano11082130

[11] V. Mody, R. Siwale, A. Singh, H. Mody, Introduction to metallic nanoparticles, Journal of Pharmacy and Bioallied Sciences, 2 (2010) 282. https://doi.org/10.4103/0975-7406.72127

[12] A.A. Yaqoob, K. Umar, M.N.M. Ibrahim, Silver nanoparticles: various methods of synthesis, size affecting factors and their potential applications-a review, Applied Nanoscience (Switzerland), 10 (2020) 1369-1378. https://doi.org/10.1007/s13204-020-01318-w

[13] D. Lima, A. Ribicki, L. Gonçalves, A.C.M. Hacke, L.C. Lopes, R.P. Pereira, K. Wohnrath, S.T. Fujiwara, C.A. Pessôa, Nanoconjugates based on a novel organicinorganic hybrid silsesquioxane and gold nanoparticles as hemocompatible nanomaterials for promising biosensing applications, Colloids Surf., B, 213 (2022). https://doi.org/10.1016/j.colsurfb.2022.112355

[14] L. Jeong, W.H. Park, Preparation and characterization of gelatin nanofibers containing silver nanoparticles, Int. J. Mol. Sci., 15 (2014) 6857-6879. https://doi.org/10.3390/ijms15046857

[15] R. Eluri, B. Paul, Synthesis of nickel nanoparticles by hydrazine reduction: Mechanistic study and continuous flow synthesis, J. Nanopart. Res., 14 (2012)800. https://doi.org/10.1007/s11051-012-0800-1

[16] Q. Wu, W.S. Miao, Y. du Zhang, H.J. Gao, D. Hui, Mechanical properties of nanomaterials: A review, Nanotechnology Reviews, 9 (2020) 259-273. https://doi.org/10.1515/ntrev-2020-0021

[17] J. Yang, J.Y. Lee, T.C. Deivaraj, H.P. Too, A highly efficient phase transfer method for preparing alkylamine- stabilized Ru, Pt, and Au nanoparticles, J. Colloid Interface Sci., 277 (2004) 95-99. https://doi.org/10.1016/j.jcis.2004.03.074

[18] Nasrollahzadeh Mahmoud, Atarod Monireh, Sajjadi Mohaddeseh, Sajadi S Mohammad, Issaabadi Zahra, Chapter 6 - Plant-Mediated Green Synthesis of Nanostructures: Mechanisms, Characterization, and Applications. Interface Science and Technology,28 (2019) 199-322. https://doi.org/10.1016/B978-0-12-813586-0.00006-7

[19] De A, Kumari A, Jain P, Manna AK, Bhattacharjee G (2020) Plasmonic sensing of Hg (II), Cr (III), and Pb (II) ions from aqueous solution by biogenic silver and gold nanoparticles. Inorganic and Nano-Metal Chemistry, 1-12. https://doi.org/10.1080/24701556.2020.1826523

[20] Vijayaraghavan K, Ashokkumar T (2017) Plant-mediated biosynthesis of metallic nanoparticles: a review of literature, factors affecting synthesis, characterization techniques and applications. Journal of environmental chemical engineering, 5(5): 4866-4883. https://doi.org/10.1016/j.jece.2017.09.026

[21] Khodadadi B, Maryam B, Mahmoud N (2017) Green synthesis of Pd nanoparticles at Apricot kernel shell substrate using Salvia hydrangea extract: catalytic activity for reduction of organic dyes. J colloid interf sci 490: 1-10. https://doi.org/10.1016/j.jcis.2016.11.032

Materials Research Forum LLC
https://doi.org/10.21741/9781644902295-1

[22] Amooaghaie R, Saeri, MR, Azizi M (2015) Synthesis, characterization and biocompatibility of silver nanoparticles synthesized from Nigella sativa leaf extract in comparison with chemical silver nanoparticles. Ecotoxicology and Environmental Safety, 120: 400-408. https://doi.org/10.1016/j.ecoenv.2015.06.025

[23] Anand K, Gengan RM, Phulukdaree A, Chuturgoon A (2015) Agroforestry waste Moringa oleifera petals mediated green synthesis of gold nanoparticles and their anti-cancer and catalytic activity. Journal of Industrial and Engineering Chemistry, 21:1105-1111. https://doi.org/10.1016/j.jiec.2014.05.021

[24] Rajiv P, Sivaraj R, Rajendran V (2013) Bio-Fabrication of zinc oxide nanoparticles using leaf extract of Parthenium hysterophorus L. and its size-dependent antifungal activity against plant fungal pathogens." Spectrochimica Acta Part A: Molecular and Biomolecular Spectroscopy 112 : 384-387. https://doi.org/10.1016/j.saa.2013.04.072

[25] Padalia H, Moteriya P, Chanda S (2015) Green synthesis of silver nanoparticles from marigold flower and its synergistic antimicrobial potential. Arabian Journal of Chemistry, 8(5):732-741. https://doi.org/10.1016/j.arabjc.2014.11.015 https://doi.org/10.1016/j.arabjc.2014.11.015

[26] K. Radad, M. Al-Shraim, R. Moldzio, and W. D. Rausch, "Recent advances in benefits and hazards of engineered nanoparticles," Environmental Toxicology and Pharmacology, vol. 34, no. 3, pp. 661-672, 2012. https://doi.org/10.1016/j.etap.2012.07.011

[27] A. Roy and N. Bharadvaja, "Qualitative analysis of phytocompounds and synthesis of silver nanoparticles from Centella asiatica," Innovative Techniques in Agriculture, vol. 1, no. 2, pp. 88-95, 2017.

[28] P. Nagore, S. Ghotekar, K. Mane, A. Ghoti, M. Bilal, and A. Roy, "Structural properties and antimicrobial activities of Polyalthia longifolia leaf extract-mediated CuO nanoparticles," BioNanoScience, vol. 11, no. 2, pp. 579-589, 2021. https://doi.org/10.1007/s12668-021-00851-4

[29] B. Dréno, A. Alexis, B. Chuberre, M. Marinovich, Safety of titanium dioxide nanoparticles in cosmetics, J. Eur. Acad. Dermatology Venereol. 33 (2019) 34-46. https://doi.org/10.1111/jdv.15943

[30] S. Arango-Santander, A. Pelaez-Vargas, S.C. Freitas, C. García, A novel approach to create an antibacterial surface using titanium dioxide and a combination of dip-pen nanolithography and soft lithography, Sci. Rep. 8 (2018) 15818. https://doi.org/10.1038/s41598-018-34198-w

[31] T. Bruna, F. Maldonado-Bravo, P. Jara, N. Caro, Silver Nanoparticles and Their Antibacterial Applications, Int. J. Mol. Sci. 22 (2021) 7202. https://doi.org/10.3390/ijms22137202

[32] T. Maruyama, Carbon nanotubes, in: S. Thomas, C. Sarathchandran, S.A. Ilangovan, J.C.B.T.-H. of C.-B.N. Moreno-Piraján (Eds.), Micro Nano Technol., Elsevier, 2021: pp. 299-319. https://doi.org/10.1016/B978-0-12-821996-6.00009-9

[33] N. Gupta, A. Dixit, Carbon Nanotube Field-Effect Transistors (CNFETs): Structure, Fabrication, Modeling, and Performance, in: A. Hazra, R. Goswami (Eds.), Carbon Nanomater. Electron. Devices Appl., Springer Singapore, Singapore, 2021: pp. 199-214. https://doi.org/10.1007/978-981-16-1052-3_9

[34] M. Ćibo, A. Šator, A. Kazlagić, E. Omanović-Mikličanin, Application and Impact of Nanotechnology in Sport, in: M. Brka, E. Omanović-Mikličanin, L. Karić, V. Falan, A. Toroman (Eds.), 30th Sci. Conf. Agric. Food Ind., Springer International Publishing, Cham, 2020: pp. 349-362. https://doi.org/10.1007/978-3-030-40049-1_44

[35] A.K. Yetisen, H. Qu, A. Manbachi, H. Butt, M.R. Dokmeci, J.P. Hinestroza, M. Skorobogatiy, A. Khademhosseini, S.H. Yun, Nanotechnology in Textiles, ACS Nano. 10 (2016) 3042-3068. https://doi.org/10.1021/acsnano.5b08176

[36] V. Amenta, K. Aschberger, M. Arena et al., "Regulatory aspects of nanotechnology in the agri/feed/food sector in EU and non-EU countries," Regulatory Toxicology and Pharmacology, vol. 73, no. 1, pp. 463-476, 2015. https://doi.org/10.1016/j.yrtph.2015.06.016

[37] N. Prajitha, S. S. Athira, and P. V. Mohanan, "Bio-interactions and risks of engineered nanoparticles," Environmental Research, vol. 172, pp. 98-108, 2019. https://doi.org/10.1016/j.envres.2019.02.003

[38] Hiwa M. Ahmed, Arpita Roy,2Muhammad Wahab, Mohammed Ahmed, Gashaw Othman-Qadir,Basem H. Elesawy, Mayeen Uddin Khandaker, Mohammad Nazmul Islam,and Talha Bin Emran, Applications of Nanomaterials in Agrifood and Pharmaceutical Industry, Journal of Nanomaterials Volume 2021, Article ID 1472096, 10 pages. https://doi.org/10.1155/2021/1472096

Emerging Applications of Nanomaterials

Materials Research Foundations 141 (2023) 14-41

Materials Research Forum LLC

https://doi.org/10.21741/9781644902295-2

Chapter 2

Metallic Nanomaterials as Catalysts

Md Jafar Sharif[1]*, M. Jakir Hossain[2]

[1]Division of Chemistry, Department of Science and Humanities, Military Institute of Science & Technology (MIST), Mirpur, Dhaka-1216, Bangladesh

[2]Forest Chemistry Division, Bangladesh Forest Research Institute, Chattogram-4211, Bangladesh

*jafarbdu@gmail.com

Abstract

With the advancement of nanoscience and nanotechnology, metallic nanomaterials emerged as a bridge between homogeneous and heterogeneous catalysis due to their higher catalytic activity, selectivity, and easy separation while maintaining high stability. The novel catalytic properties of metallic nanomaterials originate from their unique geometric structures and electronic structures that are remarkably different from the corresponding bulk materials. Metallic nanomaterials have a high population of atoms with low coordination number and quantized energy levels which depends on the structural parameters such as the size and shape of the nanomaterials. For bimetallic nanomaterial, the composition is another key parameter to regulate the catalytic properties. In this chapter, factors affecting the catalytic properties of metallic nanomaterial are discussed in detail with several important applications.

Keywords

Nanoparticle, Nanocluster, Nanocatalyst, Heterogeneous Catalyst, Nanoalloy, Size and Shape Effect, Oxygen Reduction Reaction (ORR), CO Oxidation, CO_2 Hydrogenation

Contents

1.   Introduction

Catalysts transform raw chemicals into value-added chemicals in the most efficient way in terms of activity and selectivity. About 30 % of global GDP comes from catalytic processes and approximately 90 % of chemicals are produced industrially which involves the process of catalysis [1]. Catalysis plays a vital role in modern-day life in many ways, such as in the production of fuels, fertilizers, drugs, and fine chemicals [2]. Fundamental research on catalysis has also contributed immensely to chemistry knowledge, in so doing, research findings related to catalysis have received twenty Nobel Prizes [3].

Classically catalysts are of two different types; homogeneous catalysts (usually soluble molecular complex) and heterogeneous catalysts (usually solid particles, sometimes anchored on another solid particle). As homogeneous catalysts are applied in the same phase, access to the substrate is enhanced and as a result, showed high catalytic activity and selectivity. However, a homogeneous catalyst suffers from the disadvantage of separation of the catalyst from the products [4]. On the other hand, in heterogeneous catalysis, catalytic moieties are anchored on solid support materials which ease up the separation of the catalysts. In heterogeneous catalysis, the substrate and the catalysts are in different phases which limits the access of catalysts to substrates, consequently lowering the catalytic activity compared to homogeneous catalysts. Therefore, a desirable catalyst type would be the one that can offer the advantages of both types of catalysts. In this backdrop, with the advancement of nanoscience and nanotechnology, metallic nanomaterials i.e ligand/polymer-stabilized nanoparticles (NPs) or supported NPs have appeared as a bridge between homogeneous and heterogeneous catalysis, making available higher catalytic activity, selectivity, and easy separation while maintaining high stability [5].

In catalysis, one of the groundbreaking findings in 1987 by Haruta was the CO oxidation into CO_2 using molecular oxygen where Au NPs smaller than 5 nm were found to catalyze the oxidation whereas larger Au particles were almost inactive [6]. Surprisingly, after the discovery of Haruta, research works conducted on catalysis by nanomaterials have increased rapidly in the last three decades which are discussed here.

1.1 Why nanomaterials are the bridge between homogeneous and heterogeneous catalysis?

NPs have unique geometric and electronic structures compared to the corresponding bulk materials, as shown in Fig. 1. The energy levels of NPs are discrete compared to their counterpart in the bulk state due to the confinement of electrons in a small area. The electronic energy of the NPs can be tuned as a function of the size of the NPs. These essential differences in their electronic structure can trigger the metal to nonmetal transition because of their higher HOMO−LUMO gap for smaller (< 2 nm) NPs, which are usually termed, nanoclusters (NCs). The electronic properties of these NCs are very sensitive to atomicity (i.e. number of atoms). The conventional size effect is no longer applicable for NCs, a change in a single atom can drastically change their properties. Therefore, NCs less than 2 nm are referred to as the non-scalable region.

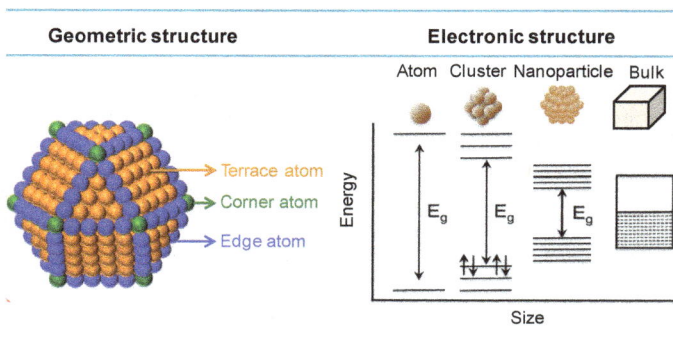

Figure 1. Illustration of geometric and electronic structure of metal NPs.

The geometric structures of metal NPs are also markedly different from the corresponding bulk materials. NPs have a large population of atoms with low coordination numbers as the number of atoms located on the surface is very high as shown in Table 1. The catalytic activity of the surface atoms depends on their location; the surface atoms on the terrace and low coordination (LC) sites such as edges and corners, may exhibit different catalytic activity. The turnover frequency (TOF) changes with the size because the relative populations of the terrace and LC sites change with the size. As shown in Table 1, the ratio of the LC sites increases with a decrease in size. If the catalytic reaction proceeds more efficiently at the LC sites than at the terrace, the TOF increases with a decrease in size. Conversely, if the terrace provides more active sites, the TOF increases with the increase in size. Hence, we can expect that the catalytic properties depend on the structure of facets because of the interactions with the substrates.

Table 1. Distribution of surface atoms in cuboctahedral NPs considering geometric shell model.

		Shell (n)	2	3	4	5	6
		Number of total atoms, N_{Tot}	13	55	147	309	561
Surface atoms		Total, N_{Sur}	12	42	92	162	252
		Low Coordination sites, N_{LC}	12	36	60	84	108
		terrace, N_{Ter}	0	6	32	78	144
		Dispersion (%)	92.3	76.4	62.6	52.4	44.9
		N_{LC} / N_{Sur} (%)	100	85.7	65.2	51.9	42.9
		N_{Ter} / N_{Sur} (%)	0	14.39	34.8	48.1	57.1

Therefore, nanomaterials can provide maximum exposure of surface atoms and low coordinated atoms, which are considered catalytically active sites, analogues to homogeneous catalysts and easy separation and reusability like heterogeneous catalysis. Thus, nanomaterials are considered the bridge between homogeneous and heterogeneous catalysis.

As the number of surface atoms in NPs is responsible for the catalytic performances, it can be regulated by tuning the size and shape of the NP. In addition, NCs often possess a non-close pack icosahedral structure. Because of these exceptionalities in structures, NPs show novel catalytic properties, which are not achievable in bulk materials. When the NPs are consisting of a single metal, their properties can be controlled by size and shape [7, 8]. In contrast, if the NPs consist of two or more metals, the composition of the metallic nanoalloy can be used as a key parameter to regulate different properties including catalytic properties besides the size and shape. In this chapter, factors affecting the catalytic properties of metallic nanomaterial are discussed in detail with several important applications of metallic nanomaterial.

2. Effect of geometric structure

The catalytic process is the surface functionality of the catalyst which takes place due to the interaction of the substrate molecule with the catalytically active sites which in turn depends upon the atomic arrangements of that specific surface of the catalysts. The atomic arrangements of the catalytic surfaces influence the substrate molecule, intermediates, and their adsorption energies, leading to a certain reaction pathway during the transformation of molecules [9]. Thus, the number of the surface atoms (terrace, edge, corner atoms) and their different atomic arrangements i.e the facets can dictate the catalytic process [10–12]. Therefore, the geometric structure of the NPs such as size and shape are the important structural parameters to regulate the catalytic performance of the NPs [13–17]. Atomic arrangements of the surfaces of the metal NPs with different shapes are shown in Fig. 2.

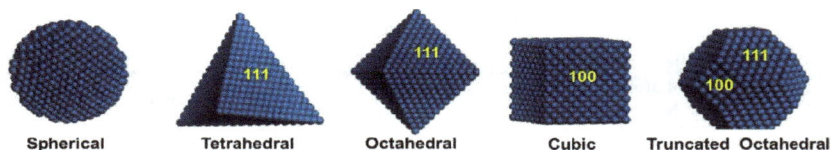

| Spherical | Tetrahedral | Octahedral | Cubic | Truncated Octahedral |

Figure 2. Schematic representation of different shapes with (100) and (111) facets. Reproduced from ref 17. Copyright 2010, American Chemical Society.

2.1 Controlling the size and shape

With the development of nanotechnology, wet-chemical colloidal synthesis has become the most effective tool to fabricate NPs of different shapes and sizes. Metal precursor, the stabilizer/capping agent (usually organic molecules, surfactants, and polymers), solvent, and the reducing agent are the four essential components of the colloidal synthesis technique. In general, metal precursors (metal salts) are dissolved in a solvent in presence of a suitable stabilizer and then the metal precursors are reduced to the zero-valent NPs. The role of the stabilizer is to protect the surface of the NPs and avoid aggregation into larger particles. The formation of colloidal NPs in the solution can be explained by the widely accepted LaMer diagram [18]. According to the LaMer model, the formation of the NPs involves three steps; generation of atoms, self-nucleation, and growth. At first, the population of generated atoms gradually increases with time due to the reduction of precursors. When the population of atoms reaches a point of super-saturation, self-nucleation started and small particles are formed. Subsequently, the population of atoms

Materials Research Forum LLC
https://doi.org/10.21741/9781644902295-2

started to decrease due to the formation of nuclei and when fell below a minimum saturation, the formation of nuclei stops. However, in this situation, the additional atoms will be deposited on the nuclei for the growth process of the small particles. Therefore, in order to have a narrow size distribution of the NPs, a quick nucleation phase and the slow growth phase is required. The concentration of precursor, the ratio of precursor to the stabilizer, and the temperature of the reaction are the parameter to control the nucleation and the growth phase. Therefore, the size and shape of the NPs can be regulated by the appropriate selection of the four essential components.

In order to obtain a more complicated shape, surface-specific capping agents (surfactants, ions, CO molecule etc.) are added to the nanocrystal seed to lower the surface energy of a particular facet of the nanocrystal. For example, CO molecules are used as surface specific capping agents for Pd (111) surface to lower the surface energy. Therefore, the final shape will be the one where (111) facets are expected to dominate [19].

2.2   Effect of size

The origin of the size effect in catalysis can be attributable to the unique geometric and electronic structure as discussed in the previous section. The coordination environment of the surface atoms changes drastically with a decrease in particle size. Tsukuda *et al.* studied the size effect on aerobic oxidation of alcohol Au NCs [20, 21]. Poly(N-vinyl-2-pyrrolidone) (PVP) stabilized Au NCs catalyzed *p*-hydroxybenzyl alcohol into *p*-hydroxybenzaldehyde where the catalytic activity increases with the decrease in the size of NCs [21]. It was found that the 1.3 nm Au NCs exhibit remarkably higher catalytic activity than the 9.5 nm Au NPs as shown in Fig. 3 (a). The kinetic isotope effect using α-deuterated *p*-hydroxy benzyl alcohol revealed that the C-H bond cleavage is the rate-determining step. The apparent activation energy obtained from the Arrhenius plot for this C–H bond cleavage for 1.3 nm Au NCs is lower than that of 9.5 nm Au NPs. It was proposed that on the surface of 1.3 nm NCs, molecular oxygen adsorbed as a superoxo-like species which abstracts a hydrogen atom from the alkoxide species.

Besides the size-dependent catalytic activity, the selectivity of a reaction is affected by the change in the size of the catalysts. One of the notable works on size-dependent selectivity is carried out by the Somorjai *et al.* on the Pt NPs catalyzed hydrogenation pyrrole [22]. Mesoporous SBA-15 silica supported Pt NPs in the size range of 0.8 nm to 5.0 nm were prepared using PAMAM dendrimer and PVP as a stabilizer. Selective formation of pyrrolidine increased with the decrease of the size of the Pt NPs below the 2 nm range as shown in Fig. 3 (b). This trend in selectivity for pyrrolidine is invariant with the size over 2 nm, where n-butylamine is the dominant product. It was predicted that there is a strong interaction between n-butylamine and the small Pt surface as the nitrogen of n-butylamine is more electron-rich than that of pyrrolidine and pyrrole. This interaction becomes stronger as the size of Pt NPs decreases, leading to n-butylamine product poisoning. Consequently, the selectivity for n-butylamine decreases with decreasing size. Similar size-dependent selectivity was observed using Pt NPs for the hydrogenation of furan and crotonaldehyde [23, 24].

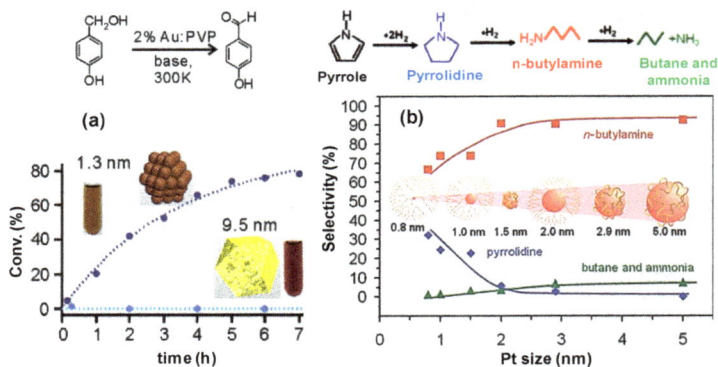

Figure 3. (a) Conversion of p-hydroxybenzyl alcohol over PVP stabilized Au NPs. Reproduced with permission from ref 20. Copyright 2005, American Chemical Society. (b) Size depended selective hydrogenation of pyrrole. Reproduced with permission from ref 21. Copyright 2008, American Chemical Society.

2.3 Effect of shape

NPs have different atomic arrangements on different crystallographic surfaces which offers surface-specific catalytic activity and selectivity. So, a nanocatalyst of particular material having different shapes shows catalytic performance in several ways depending on the dominant surface present on the nanocatalyst. One of the representative works on the shape-dependent catalytic performance is demonstrated by the El-Sayed group [11, 12]. PVP stabilized Pt NPs having tetrahedral (4.8± 0.1 nm), near-spherical (4.1± 0.1 nm), and polyacrylate stabilized cubic (7.1± 0.1 nm) stabilized Pt NPs were used to catalyze the electron-transfer reaction between hexacyanoferrate (III) ions and thiosulfate ions to form hexacyanoferrate (II) ions and tetrathionate ions. The tetrahedral NPs exhibit the highest activity and lowest activation energy; cubic NPs are the least active and have the highest activation energy whereas the near-spherical Pt NPs show an in-between activity and have intermediate activation energy as shown in Fig. 4. This shape-dependent catalytic activity was explained in terms of the fraction of the atoms located on the corners and edges of the different surfaces. The tetrahedral NPs have only (111) facets with ~35% edge and corner atoms whereas cubic NPs have only (100) facets with ~4% edge and corner atoms. The near-spherical NPs have both (100) and (111) facets with ~13 % edge and corner atoms. As the edge and corner atoms are considered as dominantly active sites, the percentage of the active sites reflected in the catalytic activity of the reaction and thereby tetrahedral NPs showed the highest catalytic activity and near-spherical NPs showed the in-between catalytic performance.

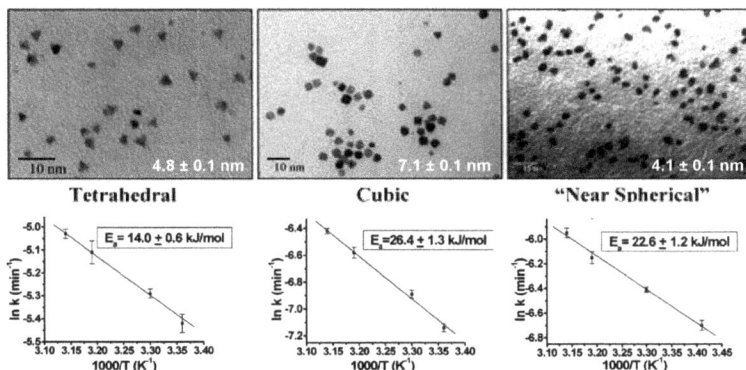

Figure 4. TEM images and the Arrhenius plots with activation energy of tetrahedral, cubic, and near spherical Pt NPs. Reproduced with permission from ref 11. Copyright 2004, American Chemical Society.

Somorjai *et al.* studied benzene hydrogenation by a surface monolayer consisting of tetradecyltrimethylammonium bromide (TTAB) stabilized Pt NPs of cubic and cuboctahedral shapes [25]. The product selectivity between cyclohexane and cyclohexene was found to be dependent on the shape of the NPs; both the products were formed on cuboctahedral NPs, where cyclohexane is the selective product formed on cubic NPs as shown in Fig. 5. This high selectivity of cyclohexane on cubic NPs which have only (100) surface, is due to the high adsorption energy of cyclohexene on Pt(100) make it available for further hydrogenation to cyclohexane. On the other hand, cuboctahedral NPs have (100) and (111) surfaces. As the adsorption energy of cyclohexene on the Pt(111) surface is low, cyclohexene can leave the surface without further hydrogenation. Similar catalytic selectivity were obtained on Pt(111) and Pt(100) single crystal surface [26, 27].

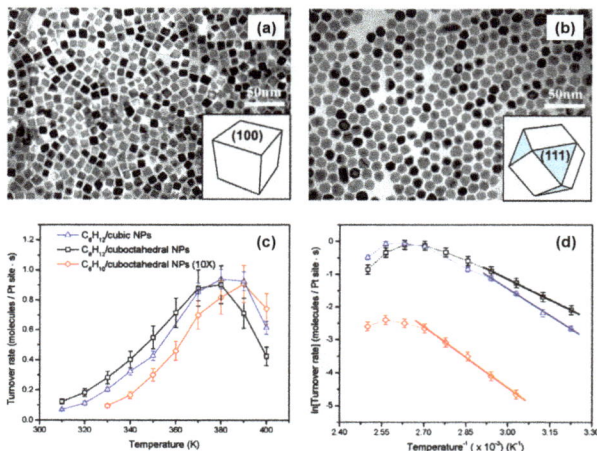

Figure 5: (a-b) TEM images and (c-d) Turnover rates of cyclohexane (C_6H_{12}) and cyclo- hexene (C_6H_{10}) formation on cubic and cuboctahedron Pt NPs. Reproduced with permission from ref 25. Copyright 2007, American Chemical Society.

3.   Effect of composition

In addition to the geometric structural parameters, the catalytic activity and selectivity can be controlled by the composition when the nanocatalysts are composed of more than one metal [28, 29]. As the electronic and geometric structures of bimetallic NPs are different compared to their constituent elements, a compositional variation in alloy NPs can change the geometry of active sites, adsorption energies, and as a consequence the catalytic performance [30, 31].

Still, it is challenging to predict how the chemical and electronic properties of an alloy surface will be changed relative to its constituent metals. However, this can be explained in terms of two factors: ligand effect and strain effect [32–34]. The ligand effect is the changes in the electronic environment of the alloy surface due to heteroatom bonding. The geometry of the alloy surface is modified due to the change in the metal-metal average distance which is referred to as the strain effect which also modulates the electronic structure of the metals by orbital overlap. When there is a strain in the surface atoms due to the presence of a heteroatom the overlapping among the d-orbitals is decreased, thereby energy of the d-band center is increased and the d-band becomes narrow [32, 35].

Therefore, the fabrication of alloy NPs by the combination of two or more metals with varying composition is an effective way to modify the electronic structure of metal NPs to optimize the activity and the selectivity of a reaction. Such change in catalytic properties by varying the compositions of an ensemble of atoms on the surface is termed as ensemble effect [33]. However, the improvement of the catalytic functions is frequently termed as synergism/synergistic effect although sometimes the reason for improvement remained unclear.

Bimetallic NPs can be categorized based on the mixing mode as random, ordered, core-shell, and phase separated alloys. Different mixing modes of two metals have different effects on the electronic interactions between them. Nevertheless, the modulation of geometric and electronic structures of random alloy NPs is quite complex, caused by the random distributions of the metal elements [36]. However, the modulation of electronic and geometric structures of ordered NPs can be well understood on the basis of their well-defined atomic distribution, and in the case of core shell NPs, the electronic properties of the metal shells in the NPs depend on the thickness of the shell [37–40].

$AgPd_x$ (where, $x = 2, 4, 6, 9, 19$) bimetallic NPs used to catalyze the dechlorination of 4-chlorophenol and the catalytic activities were found to be dependent on the composition [41]. AgPd NPs with well-regulated compositions and narrow size distributions were prepared by co-reduction of the corresponding Ag and Pd precursors in octadecylamine. Under mild conditions, the as-prepared NPs can catalyze the hydrodechlorination of 4-chlorophenol as shown in Fig. 6 (a). The catalytic activity was found to be dependent on the composition: $AgPd_x$ NPs catalysts showed a typical volcano curve in terms of the TOF value as shown in Fig. 6 (b). The TOF value increases with the addition of Pd ($x < 9$) until it reached a maximum value at the composition of AgPd9, and declines with further addition of Pd. It was proposed that 4-chlorophenol absorbs on the Ag sites and breaks the C−Cl bond whereas Pd sites facilitate the dissociation of molecular H_2 into H atom and regenerate metallic Ag from AgCl.

Atomically precise bimetallic Pd_1Au_{24} NCs were used to study the aerobic oxidation of benzyl alcohol [42]. Pd_1Au_{24} and Au_{25} NCs were prepared by immobilizing the dodecanethiolates protected $Pd_1Au_{24}(SR)_{18}$ and $Au_{25}(SR)_{18}$ on multiwall carbon nanotubes (CNTs) followed by removal of the ligands by heat treatment as shown in Fig. 6(c). The structure of the $Pd_1Au_{24}(SR)_{18}$ NCs was similar to the well-defined structure of $Au_{25}(SR)_{18}$ magic cluster with an exception of the replacement of the central Au atom by Pd atom. The prepared single Pd atom doped Pd_1Au_{24} NCs enhanced the conversion of aerobic oxidation of benzyl alcohol from 22 % to 74 %. This remarkable enhancement of the catalytic activity was explained in terms of the ligand effect and ensemble effect where Pd atom modulates the electronic environment of the Au atoms. Similar electron transfer for Pd atoms to Au atoms was demonstrated as the crown-jewel concept for aerobic oxidation of glucose [43].

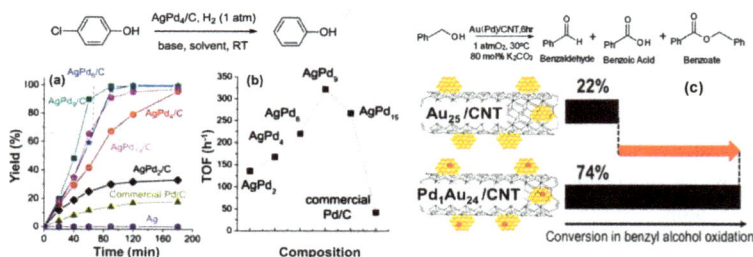

Figure 6. Time course of hydrodechlorination reaction catalyzed by different AgPdx, (b) composition dependent TOFs at 60 mins for hydrodechlorination reaction. Reproduced with permission from ref 41. Copyright 2013, American Chemical Society. (c) Schematic representation of benzyl alcohol oxidation by Pd_1Au_{24} NCs. Reproduced with permission from ref 42. Copyright 2012, American Chemical Society.

Another mentionable advantage of the nanosize effect is the formation of solid solution homogeneous alloy NPs of two metals which are not miscible in bulk state. These nanosize induced alloy NPs have unique atomic arrangement on their surface that does not exist in bulk state. Therefore, the unique distribution of atoms on the surface can show novel catalytic properties [44–45].

Pd_xRu_{1-x} bimetallic alloy NPs with atomically mixed structures were prepared over a wide range of compositions, by a chemical reduction method, although Ru and Pd are not miscible in the bulk state [46]. STEM−EDX elemental mapping confirms that the prepared alloy are NPs are solid solutions as shown in Fig. 7 (a-f). From the powder X-ray diffraction patterns both fcc and hcp structures were found to coexist in a single particle of Pd_xRu_{1-x}, in the composition range of 30−70% with respect to Pd. The prepared Pd_xRu_{1-x} NPs were used as the catalyst for the CO oxidation and it was found that the $Pd_{0.5}Ru_{0.5}$ NPs showed the highest catalytic activity, and this activity is also higher than that of Ru and Rh as shown in Fig. 7 (g). According to Langmuir−Hinshelwood mechanism, CO oxidation reaction takes place between chemisorbed O and CO; the adsorption of these species depends on the d band center [47, 48]. It was speculated that due to the atomically mixed alloying, the d band center of $Pd_{0.5}Ru_{0.5}$ was shifted to a favourable condition for CO and O coverage. Another possible reason is the coexistence of both fcc and hcp phases produces defects structures with vacancies, resulting high catalytic activity.

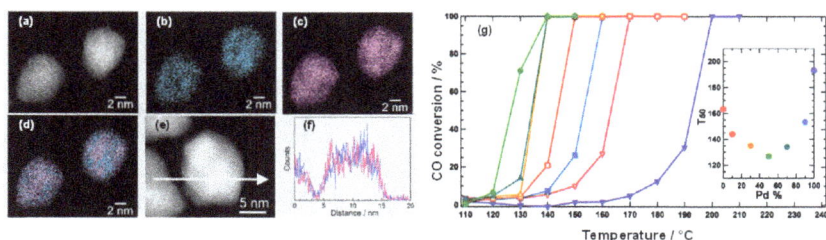

Figure 7. (a) HAADF-STEM image, (b) Pd-L STEM−EDX map, (c) Ru-L STEM−EDX map of $Pd_{0.5}Ru_{0.5}$ NPs. (d) The overlapping image of the (b) and (c). (f) Compositional line profiles of Pd (blue) and Ru (red) for the $Pd_{0.5}Ru_{0.5}$ NPs along the direction of the arrow shown in fugure (e). (g) CO conversion with temperature for Pd_xRu_{1-x} NPs supported on γ-Al_2O_3. Reproduced with permission from ref 46. Copyright 2017, American Chemical Society.

4. Role of support

The NPs are (catalytically active phase) usually anchored on high-surface-area solid supports for easy separation from the solution after the completion of the reaction. The commonly used solid supports are TiO_2, Al_2O_3, SiO_2, CeO_2, activated carbon, carbon nanotube, graphene oxide, zeolite and metal oxide framework etc. [19, 49]. Besides, stabilizing the NPs with high catalyst loading and maintaining the dispersion in reaction conditions, these solid supports affect the catalytic performance through metal-support interaction and charge transfer between metal-support. The catalytic performance of any supported nanocatalysts depends on the selection of support material and the amount of catalyst to be loaded on the support. Therefore, the realization of the interactions between NPs and support materials is of great importance to design highly active catalysts.

4.1 Stabilization of nanoparticles

One of the major concerns of nanocatalysts is the growth of the NPs i.e. sintering during the synthesis or the reaction at harsh conditions. The primary role of the support is to stabilize the metal NPs from sintering. The sintering of NPs can be understood by the Ostwald ripening or particle migration and coalescence mechanism. According to Ostwald ripening, different size of the NPs have different surface energy which is the driving force of the inter-particle movement of mobile species, consequently, the larger particles grow at the loss of smaller particles [50]. Therefore, the sintering process can be significantly suppressed if the size of the NPs has a narrow distribution. On the other hand, the particle migration and coalescence mechanism explain that the coalescence of NPs happens when NPs come in close to each other, the particle migration occurs due to the Brownian motion. When the particles are small, it follows the particle migration and coalescence mechanism and may shift to Ostwald ripening when the particle size becomes larger and immobile [51, 52].

Strong bonding of the NPs with the support can reduce the sintering rate by increasing the diffusion barrier. It is believed that the defects in the surface or vacancies in the surface of the support materials are the bonding sites which can prevent the NPs from sintering at high temperatures [53]. However, it is challenging to predict the sintering process in a quantitative measure.

Campbell *et al.* have studied the size depended stability of supported metal NPs and concluded that the thermal stability of metal NPs drastically increases with the increase of particle size below ~6 nm, and NPs having size 6 nm and above are relatively stable against sintering [54–56]. Campbell *et al.* also measured the adsorption enthalpies of different types of support by anchoring Ag NPs on different solid supports [55, 57]. The measured adsorption enthalpies for different oxide support were plotted as a function of the diameter of Ag NPs as shown in Fig. 8. The thermal stability increases with the size of the Ag particles and reaches a plateau at a larger size. Conversely, the Ag NPs of less than 4 nm shows higher thermal stability on the $Fe_3O_4(111)$ and reduced $CeO_2(111)$ surfaces compared to other supports. This higher stability was ascribed to the presence of the oxygen vacancies of $Fe_3O_4(111)$ and reduced $CeO_2(111)$ surface. Therefore, the stability of metal NPs on different support can be controlled by tuning the oxygen vacancies by introducing dopants.

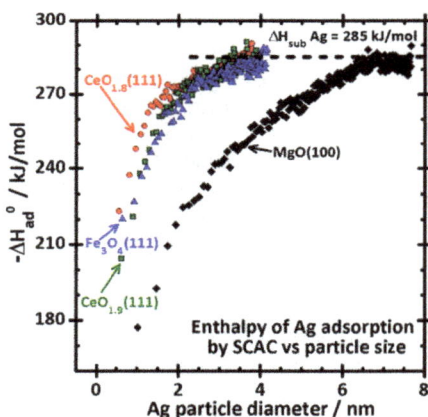

Figure 8. Heat of adsorption of Ag atom onto different metal oxide surfaces as a function of Ag particle diameter. Reproduced with permission from ref 57. Copyright 2013, American Chemical Society.

Materials Research Forum LLC
https://doi.org/10.21741/9781644902295-2

4.2 Electronic interaction between metal and the support

Electron transfer between metal and the support is a common phenomenon in supported metal NPs in order to balance the Fermi energy. Electrons can transfer from metal to support or support to metal which can perturb the charge density of the catalytically active site and thereby the adsorption and desorption energy of the reacting molecules [58]. Such electronic perturbations of the supported NPs are due to electron transfer are termed as electronic metal−support interaction (EMSI) [59]. Therefore, electron transfer between metal and support can affect the catalytic performance.

Electride is a low work function material, in which electrons serve as anions. Hosono *et al.* used a stable $[Ca_{24}Al_{28}O_{64}]^{4+}(e^-)_4$ (C12A7:e$^-$) electride consisting of positively charged subnanometer-sized cages where anionic electrons are trapped inside cages with small openings, as support for Ru catalyzed ammonia synthesis. The Ru loaded C12A7:e$^-$ catalyst shows very high catalytic activity at low temperatures due to the efficient donation from the support to Ru [60, 61]. Generally, CO oxidation takes place over Au NPs supported on a reducible oxide, such as TiO_2 and Fe_2O_3, in which the oxygen is activated on the support or support-metal interface (Mars−van Krevelen (MvK) type mechanism). However, the CO oxidation by C12A7:e$^-$ supported Ru NPs shows the highest catalytic activity and lowest activation energy compared to other support such as TiO_2 and Al_2O_3. [62, 63]. The high catalytic performance of Ru/C12A7:e$^-$ was due to the efficient electron transfer from C12A7:e$^-$ to the species adsorbed on the Ru surface but not due to the to the MvK type mechanism as shown in Fig. 9.

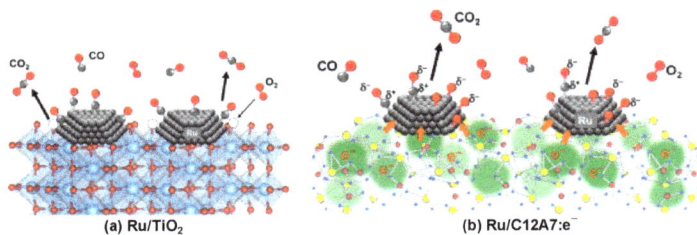

Figure 9. CO oxidation mechanism over (a) Ru/TiO2, (b) Ru/C12A7:e$^-$. Reproduced with permission from ref 62. Copyright 2015, American Chemical Society.

5. Single atom catalysts

Though significant progress has been attained in the last three decades in the field of catalysts with the evolution of nanocatalysts, precious metals such as Au, Pt, Pd, Rh, Ru, Ir should be utilized with maximum atoms efficiency as these metals are the least abundant on this earth. These metals are essential in fuel cells, the production of fine chemicals, reducing the vehicle's emission etc. [64]. As we discussed earlier, the catalytically active sites drastically increase with the decrease in the size of the particles which will reach a limiting value when the size is reduced to a single atom. To this end, an approach is to make use of all the metal atoms by preparing an atomically dispersed supported catalyst where every single atom can act as a catalytically active site as illustrated in Fig. 10. This type of catalyst is termed as single-atom catalyst (SAC) which combines the advantages of both homogeneous and heterogeneous catalysts in a most efficient way. The term single-atom catalyst (SAC) was introduced by Zhang *et al.* in 2011 for the first time in his work on the CO oxidation by Pt dispersed on FeO_x [65–67].

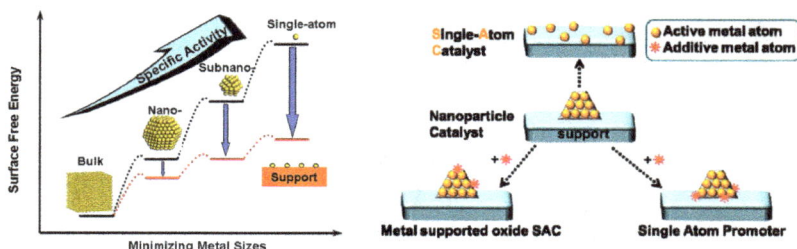

Figure 10. Schematic illustration showing variation of surface free energy and specific activity per metal atom with the particle size (left) and different types of single-atom catalysts (right). Reproduced from Ref 66, 67 Copyright 2013 & 2020, American Chemical Society.

The synthesis of SACs is challenging as the zero-valent single metal atoms are highly unstable and aggregate easily. The mass-selected soft-landing method is a previously used technique to prepare SACs on solid support materials. Pd_n clusters (n = 1 for Pd) were prepared by Heiz *et al.* by laser evaporation and then guided through a quadrupole mass analyzer and selectively deposited onto MgO surface [68, 69]. However, this method is not suitable for large scale synthesis for practical use because of the complicated instrumentation [70]. An alternative approach is a wet-chemical synthesis where the metal precursors are the single-atom metal species which need to be anchored on the solid supports through an interaction between metal and support to avoid their aggregation. To synthesize the SACs by wet-chemical synthesis by precipitation and impregnation method, an effective way is to reduce the metal loading on the support. Minimal metal precursors allow the spatial separation on the support and prevent aggregation during the synthesis and practical application in catalysis. In the SACs catalysts, the active site consists of a partially charged metal center coordinated with the atoms (O, N, S, P, and so on) on the solid-state supports such as FeO_x, CeO_2, TiO_2 etc.

These SACs catalysts exhibited improved catalytic performance in a variety of catalytic reactions such as CO oxidation, alcohol oxidation, hydrogenation, Fischer-Topsch synthesis, hydrogen production, and so on. The catalytic activity increases with the decrease of metal loadings or sizes and reaches a fixed value until the atomic dispersion is achieved [66].

Pt/FeO_x catalysts were prepared in the size range from a few nanometers to single atoms by changing the loading of Pt and the prepared catalysts were used for selective hydrogenation of 3-nitro-styrene [71]. All the catalysts show chemo-selective hydrogenation, however, the TOF values are nearly constant when the Pt loading is more than 0.08 %. Interestingly for a 0.08 % Pt loading, the Pt SACs show a notable higher TOF value which is 20 fold higher than that of Pt/TiO_2.

6. Important applications

6.1 CO_2 hydrogenations

The continuous rising of atmospheric concentration of CO_2 is a major concern for global warming and ocean acidification, results in a threat to our future life on this planet. As a result, recently much attention has been paid to reduce atmospheric CO_2 concentration which needs carbon sequestration and utilization [72-74]. CO_2 conversion to synthesize gas, alcohols, or hydrocarbons, can provide a renewable, carbon-neutral way for the production of fuels, and a convenient way for the portability of energy [75, 76]. The formation of CO through reverse water gas shift (RWGS) is an intermediate step for higher carbon products [77]. The RWGS is typically catalyzed by metal NPs or nanoalloys supported on different metal-oxide support. Cu, Pt, Rh NPs, and nanoalloys of these metals, supported on different metal-oxide such as TiO_2, CeO_2, Al_2O_3 were used as catalysts, based on a cooperative effect between metal and support material [78]. Investigation of RWGS reaction mechanism on Pt on CeO_2, concluded that dissociative chemisorption of the CO_2 molecule occurs at oxygen vacancies of CeO_2, adjacent to a Pt on CeO_2 boundary [79]. The role of the metallic Pt phase is to dissociate molecular hydrogen to atomic hydrogen, which eventually hydrogenates chemisorbed CO_2 molecule to CO.

Direct conversion of CO_2 into liquid fuels using a bi-functional catalyst consisting of indium oxides (In_2O_3) supported on zeolites gives high selective C_5+ hydrocarbons (78.6%) [80]. The In_2O_3 activates the molecular CO_2 and H_2 at the oxygen vacancies to methanol, and subsequently, C−C coupling occurs inside zeolite pore to produce C_5+ hydrocarbons [80, 81].

6.2 Selective hydrogenations

When the reactant contains more than one reducible groups, selective hydrogenation becomes important. Therefore, in order to transform one group while keeping other groups intact a selective catalyst is required [82, 83]. Noble metal (Au, Ni, Ag, Ru, Pt, and Ni etc.) NPs are often used for such chemo-selective hydrogenations of nitroaromatics [84]. Partially oxidized Ir NPs were used for the selective hydrogenation of nitroaromatics under mild conditions. It was proposed that molecular H_2 activated on the metallic part and nitroaromatic molecules adsorbed on the oxide phase through the NO_2 group [85].

Therefore, selective hydrogenation was achieved due to a cooperative effect between the oxide phase and the metallic phase.

FeO_x supported Pt single-atom and Pt NPs catalysts were prepared by varying the Pt loading [71]. The prepared catalysts show high chemo-selectivity and activity for the hydrogenation of 3-nitro-styene. The excellent catalytic performance was ascribed to efficient electron transfer from the Pt atoms to the FeO_x support. Positively charged Pt centers were formed due to the electron transfer. Properly reduced metal oxide surfaces, positively charged Pt centers, and the absence of metallic bonds in Pt, accelerate the preferential adsorption of nitro groups, results in superior catalytic performance.

6.3 Oxygen reduction reaction (ORR) at fuel cell

Polymer electrolyte fuel cells (PEFCs) convert the chemical energy of hydrogen, in the presence of oxygen to electrical energy, and only water as a byproduct providing a greener and cleaner environment without any greenhouse emission [86, 87]. Oxygen reduction reaction (ORR) is the key reaction in fuel cells which occurs at the cathodes. A high over-potential is necessary to continue the ORR due to the slow reaction rate which can be resolved by using catalytic coating over the electrode. To date, platinum NPs and the nanoalloys of Pt are the most practical catalysts for the ORR [88]. As the oxygenated species adsorbed strongly on the low coordinated site, it is expected that ORR activity will decrease with the size of the NPs since the smaller NPs have a higher number of low coordination sites [87], [89–92]. So, weakening the adsorption of oxygenated species can increase the ORR activity of Pt NPs, which can be achieved by modulating the electronic structure of Pt by alloying with nonprecious transition metals. The ORR activity has been studied by Pt_3M alloys; Pt_3Co and Pt_3Ni showed the most favorable effect on the adsorption energy for the oxygenated species on Pt_3M surface [93]. Monodisperse Pt_3Co NPs of size range from 3 to 9 nm were applied as the catalyst for ORR. The specific ORR activity was found to depend on the size and increases with increasing the size of the catalysts. After balancing the specific surface area and activity, it was found that 4.5 nm catalyst shows the maximum mass activity. Contrary to the linear increase of specific activity with the size of the pure Pt and Pt-rich alloy NPs, $PtNi_3$ NPs show an unfamiliar volcano-shaped size dependency on the specific activity of ORR [94]. The highest specific activity was shown by 6.7 and 7.8 nm NPs because of the highest Ni retention during the long catalyst cycles.

6.4 CO oxidation

CO oxidation is an important catalytic reaction to reduce emission from automobiles and improve indoor air quality as well as a fundamental understanding of catalytic mechanisms [95]. The pioneering work by Haruta *et al.* on Au NPs catalyzed CO oxidation reaction, encouraged the scientific community to study CO oxidation using the NPs of other noble metals supported on various types of oxide materials. The catalytic activity for CO oxidation strongly depended on the size of the NPs and the type of the support, the interfacial area between metal-support, and the electronic interaction between metal-support [96, 97]. Goodman *et al.* demonstrated the size depended CO oxidation catalyzed by Au NPs within the size range of 1-5 nm supported on TiO_2. They found size dependence on CO oxidation, and Au NPs with an average size of 3 nm exhibited the highest catalytic activity [98]. It was explained that the size dependency is due to quantum size effects related to small sized Au NPs.

Haruta *et al.* have studied the role of the support effect for CO oxidation by Au NPs supported on Co_3O_4, NiO, $Mg(OH)_2$, $Be(OH)_2$, TiO_2, and α-Fe_2O_3. All the catalysts are very active at low temperature. Among different types of support, the TOF values are independent on the type of support in the case of TiO_2, α-Fe_2O_3, and Co_3O_4, and increase with a decrease in size below 4 nm. It was proposed that an interface between Au and metal oxide is essential for the high catalytic activity, where CO adsorbed on Au NPs migrates toward the perimeter adjacent to the oxide support to react with the adsorbed oxygen.

An unusual size effect was observed for CO oxidation catalyzed by Ru NPs stabilized by PVP in the size range of 2 to 6 nm [99]. It was found that the catalytic activity increases with the size of the NPs and the NPs having the size of 6 nm shows 8 times higher catalytic activity than the corresponding 2 nm NPs. Under the reaction conditions, the stability of the Ru NPs changes with the formation of the RuO_2 surface oxide layer. The smaller Ru NPs oxidized to a greater extent than the larger NPs, thereby smaller NPs become less active compared to larger NPs. Therefore, the unusual size dependency on CO oxidation by Ru NPs can be correlated with the stability of the catalytically active metallic core-oxide shell structure of Ru NPs.

6.5   Oxidation of hydrocarbons

The oxidation of C−H bond in hydrocarbons is an important reaction for obtaining valuable functionalized chemicals. As an example, large scale production of CH_3OH by oxidation of CH_4 is an alternative way for the valorization of natural gas. A Cu based catalyst: Cu_3 clusters in zeolite can catalyze the oxidation of CH_4 to CH_3OH selectively [100].

The catalytic oxidation of toluene is another important step in the industrial process to remove volatile organic compounds. Pt NCs encapsulating within carbon nanotube were used to catalyze the oxidation of toluene [101]. The prepared Pt NCs exhibit 3 times higher catalytic activity than the corresponding Pt NCs supported on the exterior walls of carbon nanotube and carbon black support. Such higher catalytic activity was attributed to the higher stability of encapsulated Pt NCs in the presence of oxygen under reaction conditions [102].

Aerobic oxidation of cyclohexane is an essential step in nylon-6 and nylon-6,6 industrial processes. Au clusters Au_n (n = 10, 18, 25, 39) with tunable size on hydroxyapatite (HAP) were synthesized to study cyclohexane oxidation [103]. A volcano shape size dependency on the oxidation of cyclohexane was obtained by Au_n/HAP catalysts. The TOF values increase with an increase of atomicity from 10 up to 39, then started to decrease from 39 to ~85. It is speculated that this volcano-shaped size dependency may originate from the electronic state transition from the molecular like state to the metallic state.

6.6 Fischer–Tropsch (FT) synthesis

The Fischer–Tropsch (FT) synthesis is an alternative way to produce long-chain hydrocarbons from synthesis gas obtained from natural, coal, and biomass conversion [104, 105]. FT process is also an effective way to convert the synthesis gas into light olefins (C_2–C_4). Though it was developed hundred years ago, still controlling the selectivity of the FT reaction is very important for producing the transportation fuels and value-added chemicals [19]. The catalytic performance of FT reaction is highly sensitive to the structural parameters such as particle size and shape of the metal NPs [105–107]. The effect of the size of the NPs studied extensively for cobalt, iron and ruthenium based catalysts to control the selectivity of the reaction.

Jong et al. investigated the size effects on FT synthesis by Co NPs supported on carbon nanofibers in the range of 2.6–27 nm [106]. It was found that the catalytic performance was independent of the size of catalysts larger than 6–8 nm, although both activity and selectivity were strongly dependent on the size of the catalysts for the smaller size. When the particle size decreased from 17 to 2.6 nm, the TOF value decreased significantly and the C_5+ selectivity decreased from 85 to 51 wt %. It was explained that for smaller Co NPs, the surface residence times of reversibly bonded CH_x and OH_x species increased, whereas it remains constant for larger particles, resulting in lowering of TOF value for smaller particles. Jong et al. also investigated the size effect on FT synthesis by iron carbide NPs. In contrast to the size of by Co or Fe catalyst, the catalytic activity increases when the size of the catalyst decreases from 7 to 2 nm while the selectivity for lower olefins remains unchanged [105].

References

[1] G. Hutchings, M. Davidson, R. Catlow, C. Hardacre, N. Turner, P. Collier, Modern Developments in Catalysis. World Scientific, Europe, 2017. https://doi.org/10.1142/q0035

[2] M. Khalil, G. T. M. Kadja, M. M. Ilmi, Advanced nanomaterials for catalysis: Current progress in fine chemical synthesis, hydrocarbon processing, and renewable energy, J. Ind. Eng. Chem. 93, (2021) 78-100. https://doi.org/10.1016/j.jiec.2020.09.028

[3] C. W Jones, Another Nobel Prize for Catalysis: Frances Arnold in 2. ACS Catal. 8 (2018) 10913–10913. https://doi.org/10.1021/acscatal.8b04266

[4] D. J. Cole-Hamilton, Homogeneous Catalysis--New Approaches to Catalyst Separation, Recovery, and Recycling, Science 299 (2003) 1702-1706. https://doi.org/10.1126/science.1081881

[5] V. Polshettiwar, T. Asefa, Introduction to Nanocatalysis in: Nanocatalysis Synthesis and Applications, John Wiley & Sons, Ltd, New Jersey, 2013, pp. 1-9. https://doi.org/10.1002/9781118609811.ch1

[6] M. Haruta, T. Kobayashi, H. Sano, N. Yamada, Novel gold catalysts for the oxidation of carbon monoxide at a temperature far below 0°C, Chem. Lett. 16 (1987) 405−408. https://doi.org/10.1246/cl.1987.405

[7] C. N. R. Rao, G. U. Kulkarni, P. J. Thomas, P. P. Edwards, Size-Dependent Chemistry: Properties of Nanocrystals, Advances in Chemistry, 12 (2002) 227-233. https://doi.org/10.1142/9789812835734_0021

[8] M. A. El-Sayed, Small Is Different: Shape-, Size-, and Composition-Dependent Properties of Some Colloidal Semiconductor Nanocrystals, Acc. Chem. Res., 37 (2004) 326-333. https://doi.org/10.1021/ar020204f

[9] S. Cao, F. F. Tao, Y. Tang, Y. Li, J. Yu, Size- and shape-dependent catalytic performances of oxidation and reduction reactions on nanocatalysts, Chem. Soc. 45 (2016) 4747-4765. https://doi.org/10.1039/C6CS00094K

[10] K. An, G. A. Somorjai, Size and Shape Control of Metal Nanoparticles for Reaction Selectivity in Catalysis, Chem Cat Chem, 4 (2012) 1512-1524. https://doi.org/10.1002/cctc.201200229

[11] R. Narayanan, M. A. El-Sayed, Shape-dependent catalytic activity of platinum nanoparticles in colloidal solution, Nano Lett., 4 (2004) 1343-1348. https://doi.org/10.1021/nl0495256

[12] R. Narayanan, M. A. El-Sayed, Catalysis with transition metal nanoparticles in colloidal solution: Nanoparticle shape dependence and stability, J. Phys. Chem. B, 109 (2005) 12663-12676. https://doi.org/10.1021/jp051066p

[13] F. J. Perez-Alonso, D. N. McCarthy, A. Nierhoff, P. Hernandez-Fernandez, C. Strebel, I. E. Stephens, J. H. Nielsen, I. Chorkendorff, The effect of size on the oxygen electroreduction activity of mass-selected platinum nanoparticles. Angew. Chem. 124 (2012), 4719−4721. https://doi.org/10.1002/ange.201200586

[14] K. J. J. Mayrhofer, B. B. Blizanac, M. Arenz, V. R. Stamenkovic, P. N. Ross, N. M. Markovic, The Impact of Geometric and Surface Electronic Properties of Pt-Catalysts on the Particle Size Effect in Electrocatalysis, J. Phys. Chem. B, 109 (2005)14433-14440. https://doi.org/10.1021/jp051735z

[15] Y. Wu, S. Cai, D. Wang, W. He, Y. Li, Syntheses of Water-Soluble Octahedral, Truncated Octahedral, and Cubic Pt–Ni Nanocrystals and Their Structure–Activity Study in Model Hydrogenation Reactions, J. Am. Chem. Soc., 134 (2012) 8975–8981. https://doi.org/10.1021/ja302606d

[16] H. Xiaoqing et al., High-performance transition metal–doped Pt3Ni octahedra for oxygen reduction reaction, Science 348 (2015) 1230–1234. https://doi.org/10.1126/science.aaa8765

[17] H. M. Lu, X. K. Meng, Theoretical model to calculate catalytic activation energies of platinum nanoparticles of different sizes and shapes, J. Phys. Chem. C, 114 (2010) 1534-1538. https://doi.org/10.1021/jp9106475

[18] V. K. LaMer, R. H. Dinegar, Theory, Production and Mechanism of Formation of Monodispersed Hydrosols, J. Am. Chem. Soc.,72 (1950) 4847-4854. https://doi.org/10.1021/ja01167a001

[19] Z. Li et al., Well-Defined Materials for Heterogeneous Catalysis: From Nanoparticles to Isolated Single-Atom Sites, Chem. Rev.,120 (2020) 623-682. https://doi.org/10.1021/acs.chemrev.9b00311

[20] H. Tsunoyama, H. Sakurai, Y. Negishi, T. Tsukuda, Size-specific catalytic activity of polymer-stabilized gold nanoclusters for aerobic alcohol oxidation in water, J. Am. Chem. Soc., 127 (2005) 9374-9375. https://doi.org/10.1021/ja052161e

[21] H. Tsunoyama, H. Sakurai, T. Tsukuda, Size effect on the catalysis of gold clusters dispersed in water for aerobic oxidation of alcohol, Chem. Phys. Lett., 429 (2006) 528-532. https://doi.org/10.1016/j.cplett.2006.08.066

[22] J. N. Kuhn, W. Huang, C. K. Tsung, Y. Zhang, G. A. Somorjai, Structure sensitivity of carbon-nitrogen ring opening: Impact of platinum particle size from below 1 to 5 nm upon pyrrole hydrogenation product selectivity over monodisperse platinum nanoparticles loaded onto mesoporous silica, J. Am. Chem. Soc., 130 (2008) 14026-14027. https://doi.org/10.1021/ja805050c

[23] C. J. Kliewer et al., Furan Hydrogenation over Pt(111) and Pt(100) Single-Crystal Surfaces, Pt Nanoparticles from 1 to 7 nm: A Kinetic and Sum Frequency Generation Vibrational Spectroscopy Study, J. Am. Chem. Soc.,132 (2010) 13088-13095. https://doi.org/10.1021/ja105800z

[24] M. E. Grass, R. M. Rioux, G. A. Somorjai, Dependence of Gas-Phase Crotonaldehyde Hydrogenation Selectivity and Activity on the Size of Pt Nanoparticles (1.7-7.1 nm) Supported on SBA-15, Catal. Letters, 128 (2009) 1-8. https://doi.org/10.1007/s10562-008-9754-4

[25] K. M. Bratlie, H. Lee, K. Komvopoulos, P. Yang, G. A. Somorjai, Platinum nanoparticle shape effects on benzene hydrogenation selectivity, Nano Lett., 7 (2007) 3097-3101. https://doi.org/10.1021/nl0716000

[26] K. M. Bratlie, C. J. Kliewer, G. A. Somorjai, Structure Effects of Benzene Hydrogenation Studied with Sum Frequency Generation Vibrational Spectroscopy and Kinetics on Pt(111) and Pt(100) Single-Crystal Surfaces, J. Phys. Chem. B, 110 (2006) 17925-17930. https://doi.org/10.1021/jp062623q

[27] K. M. Bratlie, L. D. Flores, G. A. Somorjai, In Situ Sum Frequency Generation Vibrational Spectroscopy Observation of a Reactive Surface Intermediate during High-Pressure Benzene Hydrogenation, J. Phys. Chem. B, 110 (2006) 10051-10057. https://doi.org/10.1021/jp0612735

[28] T. S. Rodrigues, A. G. M. Da Silva, P. H. C. Camargo, Nanocatalysis by noble metal nanoparticles: Controlled synthesis for the optimization and understanding of activities, J. Mater. Chem. A, 7 (2019) 5857-5874. https://doi.org/10.1039/C9TA00074G

[29] J. Gu, Y.-W. Zhang, F. (Feng) Tao, Shape control of bimetallic nanocatalysts through well-designed colloidal chemistry approaches, Chem. Soc. Rev., 41(2012) 8050-8065. https://doi.org/10.1039/c2cs35184f

[30] W. Yu, M. D. Porosoff, J. G. Chen, Review of Pt-Based Bimetallic Catalysis: From Model Surfaces to Supported Catalysts, Chem. Rev., 112 (2012) 5780-5817. https://doi.org/10.1021/cr300096b

[31] N. Takehiro, P. Liu, A. Bergbreiter, J. K. Nørskov, R. J. Behm, Hydrogen adsorption on bimetallic PdAu(111) surface alloys: Minimum adsorption ensemble, ligand and ensemble effects, and ensemble confinement, Phys. Chem. Chem. Phys., 16 (2014) 23930-23943. https://doi.org/10.1039/C4CP02589J

[32] J. R. Kitchin, J. K. Nørskov, M. A. Barteau, J. G. Chen, Modification of the surface electronic and chemical properties of Pt(111) by subsurface 3d transition metals, J. Chem. Phys., 120 (2004) 10240-10246. https://doi.org/10.1063/1.1737365

[33] P. Liu, J. K. Nørskov, Ligand and ensemble effects in adsorption on alloy surfaces, Phys. Chem. Chem. Phys., 3 (2001) 3814-3818. https://doi.org/10.1039/b103525h

[34] M. Cui et al., AgPd nanoparticles for electrocatalytic CO2reduction: Bimetallic composition-dependent ligand and ensemble effects, Nanoscale, 12 (2020) 14068-14075. https://doi.org/10.1039/D0NR03203D

[35] J. K. Nørskov, T. Bligaard, J. Rossmeisl, C. H. Christensen, Towards the computational design of solid catalysts, Nat. Chem., 1 (2009) 37-46. https://doi.org/10.1038/nchem.121

[36] D. Kim, J. Resasco, Y. Yu, A. M. Asiri, P. Yang, Synergistic geometric and electronic effects for electrochemical reduction of carbon dioxide using gold-copper bimetallic nanoparticles, Nat. Commun., 5 (2014) 4948. https://doi.org/10.1038/ncomms5948

[37] S. Furukawa, T. Komatsu, Intermetallic Compounds: Promising Inorganic Materials for Well-Structured and Electronically Modified Reaction Environments for Efficient Catalysis, ACS Catal., 7 (2017) 735-765. https://doi.org/10.1021/acscatal.6b02603

[38] S. Penner, M. Armbrüster, Formation of Intermetallic Compounds by Reactive Metal-Support Interaction: A Frequently Encountered Phenomenon in Catalysis, ChemCatChem, 7 (2015) 374-392. https://doi.org/10.1002/cctc.201402635

[39] Y. Yan, J. S. Du, K. D. Gilroy, D. Yang, Y. Xia, H. Zhang, Intermetallic Nanocrystals: Syntheses and Catalytic Applications, Adv. Mater., 29 (2017). 1605997. https://doi.org/10.1002/adma.201605997

[40] L. Liu, A. Corma, Metal Catalysts for Heterogeneous Catalysis: From Single Atoms to Nanoclusters and Nanoparticles, Chem. Rev., 118 (2018) 4981-5079. https://doi.org/10.1021/acs.chemrev.7b00776

[41] H. Rong, S. Cai, Z. Niu, Y. Li, Composition-Dependent Catalytic Activity of Bimetallic Nanocrystals: AgPd-Catalyzed Hydrodechlorination of 4-Chlorophenol, ACS Catal., 3 (2013) 1560-1563. https://doi.org/10.1021/cs400282a

[42] S. Xie, H. Tsunoyama, W. Kurashige, Y. Negishi, T. Tsukuda, Enhancement in Aerobic Alcohol Oxidation Catalysis of Au25 Clusters by Single Pd Atom Doping, ACS Catal., 2 (2012) 1519-1523. https://doi.org/10.1021/cs300252g

[43] H. Zhang, T. Watanabe, M. Okumura, M. Haruta, N. Toshima, Catalytically highly active top gold atom on palladium nanocluster, Nat. Mater., 11 (2012) 49-52. https://doi.org/10.1038/nmat3143

[44] M. J. Sharif, S. Yamazoe, T. Tsukuda, Selective hydrogenation of 4-nitrobenzaldehyde to 4-aminobenzaldehyde by colloidal RhCu bimetallic nanoparticles, Top. Catal., 57 (2014) 10-13. https://doi.org/10.1007/s11244-014-0269-5

[45] K. Kusada, H. Kobayashi, R. Ikeda, Y. Kubota, M. Takata, S. Toh, T. Yamamoto, S. Matsumura, N. Sumi, K. Sato, K. Nagaoka, H. Kitagawa, Solid Solution Alloy Nanoparticles of Immiscible Pd and Ru Elements Neighboring on Rh: Changeover of the Thermodynamic Behavior for Hydrogen Storage and Enhanced CO-Oxidizing Ability, J. Am. Chem. Soc., 136 (2014) 1864-1871. https://doi.org/10.1021/ja409464g

[46] B. Huang, H. Kobayashi, T. Yamamoto, S. Matsumura, Y. Nishida, K. Sato, K. Nagaoka, S. Kawaguchi, Y. Kubota, H. Kitagawa, Solid-Solution Alloying of Immiscible Ru and Cu with Enhanced CO Oxidation Activity, J. Am. Chem. Soc.,139, no. 13, pp. 4643-4646, Apr. 2017. https://doi.org/10.1021/jacs.7b01186

[47] K. Liu, A. Wang, T. Zhang, Recent Advances in Preferential Oxidation of CO Reaction over Platinum Group Metal Catalysts, ACS Catal., 2 (2012) 1165-1178. https://doi.org/10.1021/cs200418w

[48] B. Hammer, J. K. Nørskov, Electronic factors determining the reactivity of metal surfaces, Surf. Sci., 343 (1995) 211-220. doi: https://doi.org/10.1016/0039-6028(96)80007-0. https://doi.org/10.1016/0039-6028(96)80007-0

[49] M. J. Ndolomingo, N. Bingwa, R. Meijboom, Review of supported metal nanoparticles: synthesis methodologies, advantages and application as catalysts, J. Mater. Sci., 55 (2020) 6195-6241. https://doi.org/10.1007/s10853-020-04415-x

[50] T. W. Hansen, A. T. DeLaRiva, S. R. Challa, A. K. Datye, Sintering of Catalytic Nanoparticles: Particle Migration or Ostwald Ripening?, Acc. Chem. Res., 46 (2013) 1720-1730. https://doi.org/10.1021/ar3002427

[51] P. Wynblatt, N. A. Gjostein, Supported metal crystallites, Prog. Solid State Chem., 9 (1975) 21-58. doi: https://doi.org/10.1016/0079-6786(75)90013-8. https://doi.org/10.1016/0079-6786(75)90013-8

[52] D. D. Beck, C. J. Carr, A study of thermal aging of PtAl2O3 using temperature-programmed desorption spectroscopy, J. Catal.,110 (1998) 285-297. https://doi.org/10.1016/0021-9517(88)90320-X

[53] N. Lopez, On the origin of the catalytic activity of gold nanoparticles for low-temperature CO oxidation, J. Catal., 223 (2004) 232-235. https://doi.org/10.1016/j.jcat.2004.01.001

[54] J. A. Farmer, C. T. Campbell, Ceria maintains smaller metal catalyst particles by strong metal-support bonding, Science, 329 (2010) 933-936. https://doi.org/10.1126/science.1191778

[55] C. T. Campbell, J. C. Sharp, Y. X. Yao, E. M. Karp, T. L. Silbaugh, Insights into catalysis by gold nanoparticles and their support effects through surface science studies of model catalysts, Faraday Discuss., 152 (2011) 227-239. https://doi.org/10.1039/c1fd00033k

[56] C. T. Campbell, J. R. V. Sellers, Enthalpies and Entropies of Adsorption on Well-Defined Oxide Surfaces: Experimental Measurements, Chem. Rev., 113 (2013) 4106-4135. https://doi.org/10.1021/cr300329s

[57] C. T. Campbell, The Energetics of Supported Metal Nanoparticles: Relationships to Sintering Rates and Catalytic Activity, Acc. Chem. Res., 46 (2013) 1712-1719. https://doi.org/10.1021/ar3003514

[58] X. Lin et al., Charge-Mediated Adsorption Behavior of CO on MgO-Supported Au Clusters, J. Am. Chem. Soc., 132 (2010) 7745-7749. https://doi.org/10.1021/ja101188x

[59] C. T. Campbell, Catalyst-support interactions: Electronic perturbations., Nat. Chem., 4 (2012) 597-8. https://doi.org/10.1038/nchem.1412

[60] M. Kitano et al., Ammonia synthesis using a stable electride as an electron donor and reversible hydrogen store., Nat. Chem., 4 (2012) 934-40. https://doi.org/10.1038/nchem.1476

[61] M. Kitano et al., Electride support boosts nitrogen dissociation over ruthenium catalyst and shifts the bottleneck in ammonia synthesis, Nat. Commun., 6 (2015) 6731. https://doi.org/10.1038/ncomms7731

[62] M. J. Sharif et al., Electron Donation Enhanced CO Oxidation over Ru-Loaded 12CaO•7Al2O3 Electride Catalyst, J. Phys. Chem. C, 119 (2015) 11725-11731. https://doi.org/10.1021/acs.jpcc.5b02342

[63] H. Hosono, M. Kitano, Advances in Materials and Applications of Inorganic Electrides, Chem. Rev., 121 (2021) 3121-3185. https://doi.org/10.1021/acs.chemrev.0c01071

[64] A. Wang, J. Li, T. Zhang, Heterogeneous single-atom catalysis, Nat. Rev. Chem., 2 (2018) 65-81. https://doi.org/10.1038/s41570-018-0010-1

[65] B. Qiao et al., Single-atom catalysis of CO oxidation using Pt1/FeOx, Nat. Chem., 3 (2011) 634-641. https://doi.org/10.1038/nchem.1095

[66] X.-F. Yang, A. Wang, B. Qiao, J. Li, J. Liu, T. Zhang, Single-Atom Catalysts: A New Frontier in Heterogeneous Catalysis, Acc. Chem. Res., 46 (2013) 1740-1748. https://doi.org/10.1021/ar300361m

[67] L. Zhang, M. Zhou, A. Wang, T. Zhang, Selective Hydrogenation over Supported Metal Catalysts: From Nanoparticles to Single Atoms, Chem. Rev.,120 (2020) 683-733. https://doi.org/10.1021/acs.chemrev.9b00230

[68] S. Abbet, U. Heiz, H. Häkkinen, U. Landman, CO Oxidation on a Single Pd Atom Supported on Magnesia, Phys. Rev. Lett., 86 (2001) 5950-5953. https://doi.org/10.1103/PhysRevLett.86.5950

[69] S. Abbet, A. Sanchez, U. Heiz, W.-D. Schneider, Tuning the Selectivity of Acetylene Polymerization Atom by Atom, J. Catal., 198 (2001) 122-127. https://doi.org/10.1006/jcat.2000.3105

[70] D. K. Böhme and H. Schwarz, Gas-Phase Catalysis by Atomic and Cluster Metal Ions: The Ultimate Single-Site Catalysts, Angew. Chemie Int. Ed., 44 (2005) 2336-2354. https://doi.org/10.1002/anie.200461698

[71] H. Wei et al., FeOx-supported platinum single-atom and pseudo-single-atom catalysts for chemoselective hydrogenation of functionalized nitroarenes, Nat. Commun., 5 (2014) 5634. https://doi.org/10.1038/ncomms6634

[72] D. S. J., C. Ken, M. H. Damon, Future CO2 Emissions and Climate Change from Existing Energy Infrastructure, Science, 329 (2010) 1330-1333. https://doi.org/10.1126/science.1188566

[73] J. Rogelj et al., Paris Agreement climate proposals need a boost to keep warming well below 2 °C, Nature, 534 (2016) 631-639. https://doi.org/10.1038/nature18307

[74] Q. Lai et al., Catalyst-TiO(OH)2 could drastically reduce the energy consumption of CO2 capture, Nat. Commun., 9 (2018) 2672. https://doi.org/10.1038/s41467-018-05145-0

[75] R. P. Ye et al., CO2 hydrogenation to high-value products via heterogeneous catalysis, Nat. Commun., 10 (2019) 5698. https://doi.org/10.1038/s41467-019-13638-9

[76] X. Jiang, X. Nie, X. Guo, C. Song, J. G. Chen, Recent Advances in Carbon Dioxide Hydrogenation to Methanol via Heterogeneous Catalysis, Chem. Rev., 120 (2020) 7984-8034. https://doi.org/10.1021/acs.chemrev.9b00723

[77] Y. A. Daza, J. N. Kuhn, RSC Advances comparison of catalysts, mechanisms and their consequences for CO2 conversion to liquid fuels, RSC Adv., 6 (2016) 49675-49691. https://doi.org/10.1039/C6RA05414E

[78] W. Wang, S. P. Wang, X. B. Ma, J. L. Gong, Recent advances in catalytic hydrogenation of carbon dioxide, Chem. Soc. Rev., 40 (2011) 3703-3727. https://doi.org/10.1039/c1cs15008a

[79] A. Goguet, F. C. Meunier, D. Tibiletti, J. P. Breen, R. Burch, Spectrokinetic Investigation of Reverse Water-Gas-Shift Reaction Intermediates over a Pt/CeO2 Catalyst, J. Phys. Chem. B, 108 (2004) 20240-20246. https://doi.org/10.1021/jp047242w

[80] P. Gao et al., Direct conversion of CO2 into liquid fuels with high selectivity over a bifunctional catalyst, Nat. Chem., 9 (2017) 1019-1024. https://doi.org/10.1038/nchem.2794

[81] O. Martin et al., Indium Oxide as a Superior Catalyst for Methanol Synthesis by CO2 Hydrogenation, Angew. Chemie Int. Ed., 55 (2016) 6261-6265. https://doi.org/10.1002/anie.201600943

[82] A. Corma, P. Serna, P. Concepción, J. J. Calvino, Transforming Nonselective into Chemoselective Metal Catalysts for the Hydrogenation of Substituted Nitroaromatics, J. Am. Chem. Soc., 130 (2008) 8748-8753. https://doi.org/10.1021/ja800959g

[83] M. Boronat, P. Concepción, A. Corma, S. González, F. Illas, P. Serna, A Molecular Mechanism for the Chemoselective Hydrogenation of Substituted Nitroaromatics with Nanoparticles of Gold on TiO2 Catalysts: A Cooperative Effect between Gold and the Support, J. Am. Chem. Soc., 129 (2007) 16230-16237. https://doi.org/10.1021/ja076721g

[84] C. Avelino, S. Pedro, Chemoselective Hydrogenation of Nitro Compounds with Supported Gold Catalysts, Science, 313 (2006) 332-334. https://doi.org/10.1126/science.1128383

[85] M. J. Sharif, P. Maity, S. Yamazoe, T. Tsukuda, Selective hydrogenation of nitroaromatics by colloidal iridium nanoparticles, Chem. Lett., 42 (2013), 1023-1025. https://doi.org/10.1246/cl.130333

[86] K. Kodama, T. Nagai, A. Kuwaki, R. Jinnouchi, Y. Morimoto, Challenges in applying highly active Pt-based nanostructured catalysts for oxygen reduction reactions to fuel cell vehicles, Nat. Nanotechnol., 16 (2021) 140-147. https://doi.org/10.1038/s41565-020-00824-w

[87] R. L. Borup et al., Recent developments in catalyst-related PEM fuel cell durability, Curr. Opin. Electrochem., 21 (2020) 192-200. https://doi.org/10.1016/j.coelec.2020.02.007

[88] C. Xie, Z. Niu, D. Kim, M. Li, P. Yang, Surface and Interface Control in Nanoparticle Catalysis, Chem. Rev., 120 (2020) 1184-1249. https://doi.org/10.1021/acs.chemrev.9b00220

[89] F. J. Perez-Alonso *et al.*, The Effect of Size on the Oxygen Electroreduction Activity of Mass-Selected Platinum Nanoparticles, Angew. Chemie Int. Ed., 51, (2012) 4641-4643. https://doi.org/10.1002/anie.201200586

[90] M. Shao, A. Peles, K. Shoemaker, Electrocatalysis on Platinum Nanoparticles: Particle Size Effect on Oxygen Reduction Reaction Activity, Nano Lett., 11 (2011) 3714-3719. https://doi.org/10.1021/nl2017459

[91] K. Qadir *et al.*, Intrinsic relation between catalytic activity of CO oxidation on Ru nanoparticles and Ru oxides uncovered with ambient pressure XPS., Nano Lett., 12 (2012) 5761-8. https://doi.org/10.1021/nl303072d

[92] D. Li *et al.*, Functional links between Pt single crystal morphology and nanoparticles with different size and shape: the oxygen reduction reaction case, Energy Environ. Sci., 7 (2014) 4061-4069. https://doi.org/10.1039/C4EE01564A

[93] C. Wang *et al.*, Monodisperse Pt3Co Nanoparticles as a Catalyst for the Oxygen Reduction Reaction: Size-Dependent Activity, J. Phys. Chem. C, 113 (2009) 19365-19368. https://doi.org/10.1021/jp908203p

[94] L. Gan, S. Rudi, C. Cui, M. Heggen, P. Strasser, Size-Controlled Synthesis of Sub-10 nm PtNi3 Alloy Nanoparticles and their Unusual Volcano-Shaped Size Effect on ORR Electrocatalysis, Small, 12 (2016) 3189-3196. https://doi.org/10.1002/smll.201600027

[95] B. K. Min, C. M. Friend, Heterogeneous gold-based catalysis for green chemistry: low-temperature CO oxidation and propene oxidation, Chem. Rev., (2007) 2709-24. https://doi.org/10.1021/cr050954d

[96] M. Valden, X. Lai, D. W. Goodman, Onset of catalytic activity of gold clusters on titania with the appearance of nonmetallic properties, Science, 281 (1998) 1647-1650. https://doi.org/10.1126/science.281.5383.1647

[97] M. Comotti, W.-C. Li, B. Spliethoff, F. Schüth, Support effect in high activity gold catalysts for CO oxidation., J. Am. Chem. Soc., 128 (2006) 917-24. https://doi.org/10.1021/ja0561441

[98] V. M., L. X., G. D. W., Onset of Catalytic Activity of Gold Clusters on Titania with the Appearance of Nonmetallic Properties, Science, 281 (1998) 1647-1650. https://doi.org/10.1126/science.281.5383.1647

[99] S. H. Joo, J. Y. Park, J. R. Renzas, D. R. Butcher, W. Huang, G. A Somorjai, Size effect of ruthenium nanoparticles in catalytic carbon monoxide oxidation., Nano Lett., 10 (2010) 2709-2713. https://doi.org/10.1021/nl101700j

[100] S. Grundner *et al.*, Single-site trinuclear copper oxygen clusters in mordenite for selective conversion of methane to methanol, Nat. Commun., 6 (2015) 7546. https://doi.org/10.1038/ncomms8546

[101] F. Zhang *et al.*, Tailoring the Oxidation Activity of Pt Nanoclusters via Encapsulation, ACS Catal., 5(2015) 1381-1385. doi: 10.1021/cs501763k. https://doi.org/10.1021/cs501763k

[102] X. Pan, X. Bao, Confinement Effects in Nanosupports in: Nanomaterials in: Catalysis, First Ed. Wiley-VCH Verlag GmbH & Co. KGaA, Germany, 2012, pp. 415-441. https://doi.org/10.1002/9783527656875.ch11

[103] Y. Liu, H. Tsunoyama, T. Akita, S. Xie, T. Tsukuda, Aerobic Oxidation of Cyclohexane Catalyzed by Size-Controlled Au, ACS Catal., 1 (2011) 2-6. doi: 10.1021/cs100043j https://doi.org/10.1021/cs100043j

[104] F. Jiao *et al.*, Selective conversion of syngas to light olefins, Science, 351 (2016) 1065-1068. https://doi.org/10.1126/science.aaf1835

[105] H. M. Torres Galvis, J. H. Bitter, T. Davidian, M. Ruitenbeek, A. I. Dugulan, and K. P. De Jong, Iron particle size effects for direct production of lower olefins from synthesis gas, J. Am. Chem. Soc., 134 (2012) 16207-16215. https://doi.org/10.1021/ja304958u

[106] G. L. Bezemer *et al.*, Cobalt particle size effects in the Fischer-Tropsch reaction studied with carbon nanofiber supported catalysts, J. Am. Chem. Soc., 128 (2006) 3956-3964. https://doi.org/10.1021/ja058282w

Emerging Applications of Nanomaterials
Materials Research Foundations 141 (2023) 42-74

Materials Research Forum LLC
https://doi.org/10.21741/9781644902295-3

Chapter 3

Green Chemical Synthesis in the Presence of Nanoparticles as Catalysts

Abeda Sultana Touchy[1]*, S.M.A. Hakim Siddiki[1]*

[1]Department of Chemistry, Tokyo Metropolitan University, 1-1 Minami Osawa, Hachioji, Tokyo 192-0397, Japan

abeda@tmu.ac.jp, hakim@tmu.ac.jp

Abstract

As part of the heterogeneous catalysis concept, nanoparticle catalysis is the advanced key technology to connect the divergence between classical chemical synthesis methods and environmentally benign sustainable synthesis processes. The demonstration of nanoparticle catalysts for the sustainable and mild synthesis of chemicals is a fast-growing area in catalysis. This chapter will focus on a series of catalytic systems for hydrogen transfer-type or so-called borrowing hydrogen reactions using supported metal catalysts to synthesize chemicals directly and will highlight a group of coupling reactions that generates C-O, C-C, C-N, and C-S bonds. These catalytic coupling reactions possess a general mechanistic aspect: dehydrogenation of poor electrophile alcohols/amines to activated electrophiles, condensation (self or cross or with different nucleophiles), and hydrogenation of condensate. The feasibility of these catalytic coupling reactions is abided with the proper catalyst design where supports are enriched with acidic, basic, or amphoteric properties that promote the condensation reactions, and metal nanoparticle sites are responsible for the hydrogen transfer from the alcohols or amines followed by the re-hydrogenation of the condensation product accordingly. Compared to the state-of-the-art homogeneous catalytic systems, these heterogeneous metal nanoparticle-catalyzed reactions possess catalytic reusability, catalytic efficiency, and sustainability advantages.

Keywords

Nanoparticles Catalysis, Coupling Reactions, Green Synthesis, Hydrogen borrowing/Dehydrogenative Reactions

Contents

1.   Introduction

Nanocatalysis has already been established as an essential part of nanoscience [1]. Nanoparticles can be a replacement for conventional materials and exhibit functional and stable heterogeneous catalytic features or as co-catalyst materials for various catalytic groups [2]. The catalytic-active nanoparticles have a higher surface area and increased exposed active sites and, thereby, functional contact areas with reactants comparable to those of homogeneous catalytic systems. Nanoparticles with nanostructured catalytic properties can act as heterogeneous catalysts, primarily reusable and easily separable from the catalyst-product mixtures. The catalytic activity and selectivity of nanoparticle catalysts are tunable and nearly similar to the actions of state-of-the-art homogeneous catalysis. The chemical and physical modification employed by various synthetic methods can improve the activity and stability of nanoparticle catalysts. Nanocatalysts synthesized under controlled size, shape and morphologies possess better catalytic activity, selectivity, and peace for significant reactions [3,4].

All catalytic reactions and interactions are supposed to be under the nanoscale, but not all catalysts are nanocatalysts: nanoparticles, nanomaterial catalysis, by default, proceeds through a nanoscale range interaction. Nanoparticle catalysts are exceptionally sensitive to the nano-structure. The catalytic activity and selectivity depend on the nanoparticles' size, shape, composition, and support materials. The unexpected results show that gold nanoparticles smaller than 5 nm are active catalysts even at sub-ambient temperatures. In contrast, bulk and more considerable size gold are inactive [3,4]. The benefit of the increased surface-to-volume ratio and accessibility to specific sites (steps, edges, corners, or planes) of nanoparticles to catalysis are now well-known [5,6]. This also encompasses the synthesis of metallic and metal oxide nanomaterials with different unique structures and shapes, and their applications are essential either as catalysts or as support materials for active catalytic species. The continuous development of nanocatalysis is being decorated and enriched with nanocrystalline and nano-bimetallic, nano-intermetallic catalytic systems. Further research in this area or the developments is also currently ongoing. The nano-catalytic reaction mechanisms are, to some extent, complex and not yet entirely predictable [7], which makes nanocatalysis a fascinating science with lots of room for further investigation. Nanocatalysis can stand as a feature of "green" science because nanocatalysis primarily allows conducting chemical transformations in an environmentally benign manner [8]. The mechanical separation of the catalysts (magnetic nanocatalyst recovery) [9]. Mixing ionic liquids with metal nanoparticles [10], creating electrodes with nanocatalysts with improved redox properties for energy-relevant applications, encapsulating nanoparticles in nanoporous cavities of zeolites and organic molecular frameworks, and controlling the shapes of metal and metal oxide nanoparticles to enhance their catalytic activities all these directly assist the catalytic reactions with minimum use of solvents and mostly no use of additives [11].

This chapter aims to give readers a broad perspective and current information about nanomaterial systems developed over the past several years involving nanomaterials/nanoparticles as catalysts for the sustainable green synthesis of fine chemicals. Considerable development of catalytic reactions has been explored with the demonstration of enormous coupling reactions under the borrowing hydrogen/hydrogen-transfer type and acceptorless dehydrogenative type methodology employing state-of-the-art homogeneous catalysts and sustainable heterogeneous catalysts. This chapter focuses on the recent development of coupling reactions under borrowing hydrogen/transfer hydrogen methodology using heterogeneous nanoparticle catalysts. It also includes the typical synthesis and characterization methods of the catalysts and their applications in different chemical reactions with possible reaction mechanisms.

2. Chemical synthesis under borrowing hydrogen transfer hydrogen type coupling reactions with alcohols

To minimization of environmental pollution, a sustainable society is desired, and the green chemistry for chemical synthesis is rapidly growing. The developments of the catalytic process for green chemistry are essential. In contrast to the conventional catalytic methods, one-pot synthesis of chemicals with heterogeneous catalysis is prospective as it ensures low energy consumption and reuse of the catalysts. Ideal organic synthesis requires catalytic activity, selectivity, atom economy, and step efficiency, which encourages the development of new heterogeneous catalytic chemical synthesis methodologies. The result of nanoparticle catalysis followed by heterogeneous catalysts for sustainable synthesis of fine chemicals is highly desirable to the benign environment and green synthesis of chemicals. We will focus on the recent demonstrations borrowing hydrogen-type coupling reactions employing metal/support nanoparticle catalysts and application to the one-pot synthesis of fine (**Scheme 1**) [12–26]. The selected reactions include (1) transfer hydrogenation of carbonyl compounds and transfer dehydrogenation of alcohols, (2) C-C bond formation reactions, (3) C-N bond formation reactions, and (4) C-S bond formation reactions.

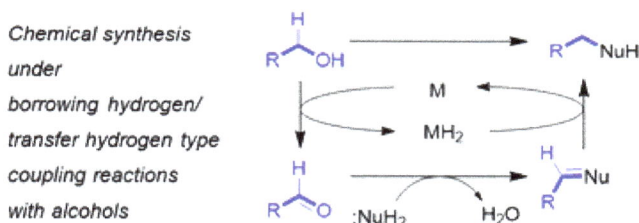

Scheme 1. General reaction path for metal-catalyzed borrowing hydrogen type or transfer hydrogen type coupling reactions of nucleophiles with alcohols.

Emerging Applications of Nanomaterials   Materials Research Forum LLC
Materials Research Foundations 141 (2023) 42-74   https://doi.org/10.21741/9781644902295-3

2.1   Hydrogenation of carbonyl compounds and dehydrogenation of alcohols under the transfer-hydrogenation method

The borrowing hydrogen or hydrogen transfer type reaction methodology consists of (1) hydrogenation of carbonyl compounds by hydrogen donor alcohols (2) dehydrogenation of alcohols by hydrogen acceptor carbonyl compounds. Here we have focused on the nanoparticle catalysts for the above two reactions and concentrated on the possible catalyst design concept. The transfer-type hydrogenation of aldehydes or ketones by 2-propanol [27–29] to the corresponding alcohols provides an efficient and selective hydrogenation method of carbonyl groups in organic synthesis. Compared to conventional methods, hydrogen donor alcohol is low-cost and straightforward to handle under catalytic reaction conditions. A group of noble metal catalysts (homogeneous), including Ru/PbO [30–32] and Ir [27, 36] complexes, are influential in this reaction. Still, they struggle to recover and reuse expensive catalysts and need basic co-catalysts and costly ligands.

On the other hand, supported noble metal nanoparticles (Pt, Pd, Ru, Ir, Au) [34–42] catalysts are efficient and reusable for this catalytic conversion. Additionally, non-noble metal nanoparticles (Ni, Co, Cu) catalysts[43–47] were applied by several groups. However, these catalytic methods have difficulties with the need for basic additives for liquid phase reaction, tolerance to the limited number of substrates, high temperature, and catalyst loading.

To some extent, non-noble-metal nanocatalysts were adequate for reacting benzylic and α,β-unsaturated carbonyl compounds without any additives. However, these systems lack tolerances for less reactive aliphatic aldehydes or ketones compounds.[43,48–53] A non-noble metal nanocatalyst, Ni/CeO_2 [54], was found to be a suitable and reusable catalyst for the transfer type hydrogenation of activated ketones and less reactive aliphatic ketones under additive-free conditions (**Scheme 2**). Only 1 mol% of Ni/CeO_2 catalyst resulted in a high yield (>92%) of 2-octanol with five successive runs during the hydrogenation of 2-octanone. The efficiency of the catalysts' performances varies with the acidic or basic surface sites of support material in the following order: $Ni^\circ/CeO_2 > Ni^\circ/Al_2O_3 > Ni^\circ/SiO_2$, and the catalytic efficiency increases with the number of basic surface sites, which is estimated by CO_2 desorption experiment. The high activity of Ni/CeO_2 is associated with the basic property of CeO_2. The initial step of the reaction is proposed considering the *in situ* infrared (IR) studies of the adsorption of 2-propanol on CeO_2:

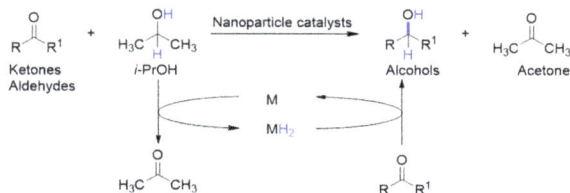

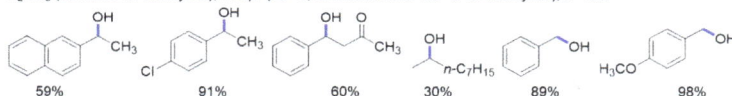

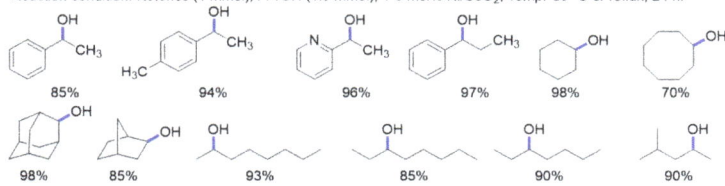

Scheme 2. Transfer type hydrogenation of ketones with hydrogen source 2-propanol using (A) TiO₂ supported Au nanoparticle catalysts, (B) CeO₂ supported Ni - nanoparticle catalysts.

The basic surface sites of CeO_2 catalyst, i.e., surface oxygen abstracts a proton from 2-propanol, and the generated isopropoxide on Lewis acidic Ce^{4+} site may undergo metallic Ni-promoted dehydrogenation to acetone and transfer hydrogenation of the ketones to alcohols.

Transfer dehydrogenation of alcohols and hydrogenation of alkenes

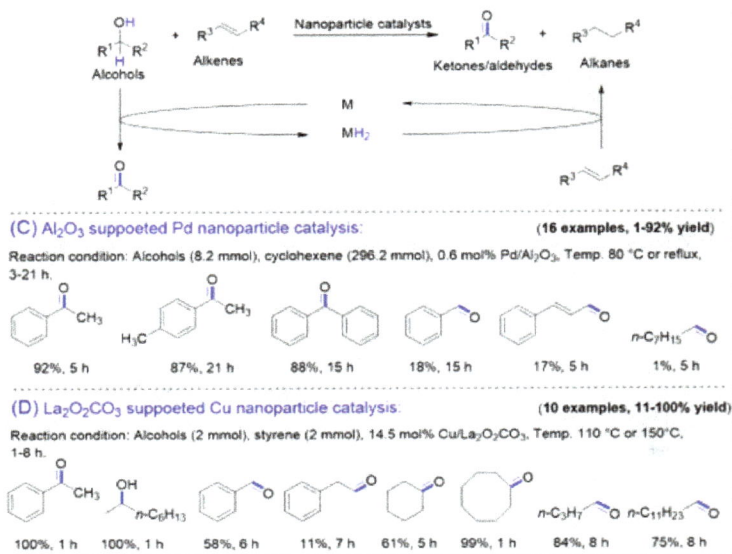

Scheme 3. Transfer dehydrogenation of alcohols with alkenes using (C) Al_2O_3 supported Pd nanoparticle catalysts, (D) $La_2O_2CO_3$ supported Cu nanoparticle catalysts.

Cu/Al_2O_3 catalyzed transfer hydrogenation mechanism explored a cooperative reaction pathway combining metallic Cu nanoparticles amphoteric sites on Al_2O_3.[47] The transfer dehydrogenation of alcohols by acceptor carbonyl compounds (Oppenauer-type oxidation) is a sustainable green method of alcohol oxidation under anaerobic conditions. Sn- or Zr-zeolite catalysts [52] and hydrous zirconia catalysts [55]) are excellent heterogeneous nanocatalysts catalysts for this reaction.

Pd/Al_2O_3 [68, 69] and $Cu/La_2O_2CO_3$ [58–60] nanoparticle catalytic systems are competent for the transfer dehydrogenation type reaction for a series of alcohols in the presence of an alkenes acceptor. These reactions afforded a high yield of ketones (**Scheme 3C**) from secondary alcohols. In contrast, to the activated alcohols, a low yield of aldehydes is observed for primary alcohols using Pd/Al_2O_3 catalyst. Exceptionally high yields of carbonyl compounds are achieved for secondary and primary, including activated and unactivated primary alcohols (**Scheme 3D**) with $Cu/La_2O_2CO_3$ catalyst [60] system. This $Cu/La_2O_2CO_3$ catalyst showed reusability for at least five cycles without any activity loss. The mechanistic studies suggest that Cu nanoparticle and Lewis acid-base pair site on the

Materials Research Forum LLC

https://doi.org/10.21741/9781644902295-3

metal-support interface are indispensable for this dehydrogenative reaction with hydrogen generation [60]. At the same time, the Cu nanoparticle reduces the styrene double bonds using the released hydrogen from alcohols. It can be concluded that metal oxides having acid-base sites loaded with non-noble metal nanoparticles, such as Ni/CeO_2 [54] and $Cu/La_2O_2CO_3$ [60], are active catalysts to govern the transfer hydrogenation of carbonyl compounds and dehydrogenation of alcohols.

2.2 Selective c-alkylation reactions with alcohols under borrowing hydrogen methodology

2.2.1 A-alkylation of ketones with alcohols

The conventional C–C bond formation reaction through Alkylation at α-carbon of enolates is one of the classical ways of organic synthesis. However, these classical methods suffer from some disadvantages of using strong bases and low atom efficiency due to the leaving group of the alkylating agent and the generation of substantial stoichiometric waste. Extensive attention has been directed to this α-alkylation reaction of ketones using alcohol as an alkylating agent under the well-known environmentally benign borrowing hydrogen methods.

Different homogeneous and heterogeneous nanoparticle catalysts are developed to perform the α-alkylation reaction of ketones with alcohols. (homogeneous Ir [61,62], Ru [63–65] and heterogeneous nanoparticle (Ru [66], Pd[79, 80], Cu [69], Au [70], Ni [21]).

The developed nanocatalysts systems indeed require a large amount of basic additives. It can be seen in Scheme 4A that AlO(OH) supported Pd catalyst tolerates broad substrate scope with a high turnover number and reusability for α-alkylation of methyl ketones with primary alcohols in the presence of 3 equivalent of basic additive K_3PO_4 as a co-catalysts [68]. A simple and general hybrid nanoparticle catalyst Ag/Mo ($Ag_6Mo_{10}O_{33}$) Scheme 4B was successfully employed for the Alkylation of aromatic ketones with activated benzylic alcohols using 20 mol% basic co-catalysts [71]. For the first time, nickel is used to activate primary alcohols in the α-alkylation of ketones. The nickel-metal used in stoichiometric amount (100 mol%) in the form of nanoparticles has shown to be a potential alternative in these reactions to noble-metal catalysts such as those derived from palladium, ruthenium, or iridium, as shown in Scheme 4C [72].

Selective C-alkylation reactions: *α-Alkylation of ketones with alcohols*

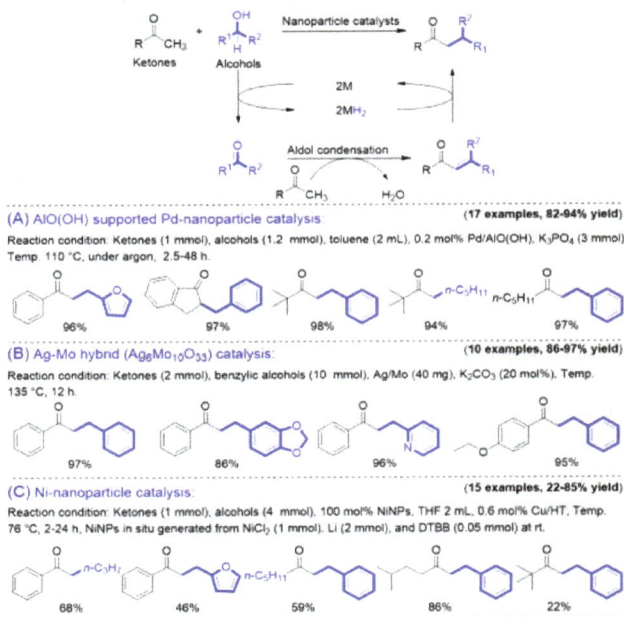

Scheme 4. *α-Alkylation of ketones with alcohols (A) AlO(OH) supported Pd nanoparticle catalysis, (B) Ag-Mo nanoparticle hybrid catalysis, (C) Ni nanoparticle catalysis.*

Kaneda and his coworkers reported the first additive-free α-alkylation of acetophenone with benzyl alcohol employing hydrotalcite-supported grafted Ru (Ru/HT) [66]. An additive-free CeO_2-supported Pt nanoparticle catalyst system was highly active for α-alkylation of different methyl ketones with aliphatic secondary alcohols Scheme 5A [73]. The acid-base sites on the surface accelerate the catalytic activity, and cerium oxide gave higher activity than other metal oxides (alumina, titania, silica). Even with a very low mol% (0.2) of Pt/CeO_2 catalyst, a wide range of aliphatic ketones successfully alkylated to the corresponding α-alkylated ketones in moderate to high yields (65–99%). A 99% yield of the α-alkylated product afforded a TON of 495 in a reaction of 2-heptanone with 2-heptanol. The reaction proceeds under a sequential mechanism, i.e., dehydrogenation of alcohols to ketones followed by the CeO_2 catalyzed aldol condensation with another ketone to generate an α,β-unsaturated ketone, which is finally undergoes hydrogenated by the borrowed hydrogen on the catalyst species as Pt-H. TiO_2 [74], a non-noble metal-based additive-free nano catalytic system Cu/HT was also found to be effective for the α-alkylation of acetophenones, aliphatic ketones with linear aliphatic and benzyl alcohols (Scheme 5B) [75] in 86-97% yield.

(A) CeO$_2$ suppoeted Pt nanoparticle catalysis: (7 examples, 65-99% yield)

Reaction condition: Ketones (1 mmol), alcohols (1 mmol), 0.2 mol% Pt/CeO$_2$, o-xylene 2.0 mL, Temp. 130 °C, 12-48 h.

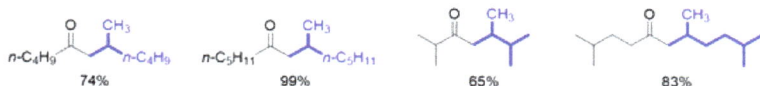

74%   99%   65%   83%

(B) Hydrotalcite supported Cu Catalysis: (9 examples, 86-97% yield)

Reaction condition: Ketones (10 mmol), alcohols (10 mmol), 0.6 mol% Cu/HT, Temp. 185 °C, 15 h.

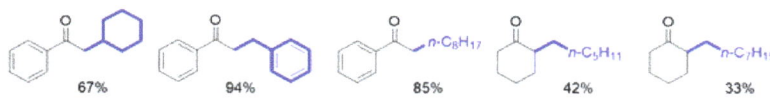

67%   94%   85%   42%   33%

Scheme 5. α-Alkylation of ketones with alcohols (A) CeO$_2$ supported Pt nanoparticle catalysis, (B) hydrotalcite supported Cu nanoparticle.

Catalytic α-alkylation of nitriles with alcohols is another application of C-C bond formation reaction; a few homogeneous (Ru, Ir) [15, 88–91] and heterogeneous (Ru, Pd) [78, 92, 93] catalytic systems were found to be effective for this selective coupling reaction. Hydrotalcite-supported Ru nanoparticle catalyst [78, 92] is one of the pioneer demonstrations of solid catalysts in this aspect. A series of nitriles and primary alcohols successfully produce α-alkylated nitrile compounds (Scheme 6). The Ru metal and base sites on the hydrotalcite promote the catalytic cycle. Ru metal nanoparticles dehydrogenate the alcohols to aldehydes, and base sites catalyze the condensation reaction of phenyl acetonitrile and aldehydes to form α,β-unsaturated nitriles, finally, re-hydrogenation of the nitrile intermediates give the desired product. Magnesium oxide-supported Pd catalysts (Pd/MgO), a reusable catalytic system, were also found to be effective for the selective α-alkylation reaction of active methylene compounds with activated alcohols (benzyl alcohols) [81].

Selective C-alkylation reactions: *α-Alkylation of nitriles with alcohols*

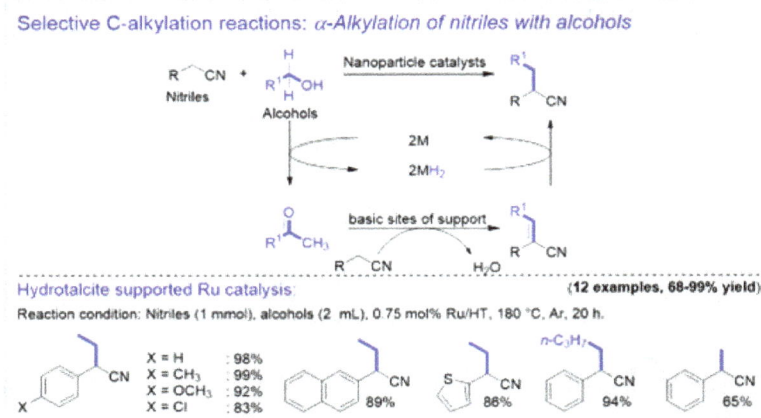

Scheme 6. α-Alkylation of nitriles with alcohols using hydrotalcite-supported Ru nanoparticle catalyst.

2.2.2 Selective C3 alkylation of oxindoles with primary alcohols

Oxindole derivatives, including C3 alkylated oxindoles, are ubiquitous functional intermediates, vastly using biological activity in medicinal compounds. The classical C3 alkylation methods of oxindoles use alkyl halides as an alkylating agent, reflecting low selectivity, generation of dialkylated compounds as a byproduct, and quantitative amount of inorganic salt wastes, and use of hazardous reagents. The catalytic C3 alkylation reaction of oxindoles with alcohols is an atom-efficient one-pot synthesis method for C3-substituted-oxindoles. In the presence of basic co-catalysts KOH or NaOH (10–20 mol%), Ir and Ru complex catalysts [94, 95] catalyze this reaction. Still, the major limitations are difficulties in catalyst reusability, catalyst/product separation, the need for expensive ligands, and basic additives. A supported Ir complex catalyst using basic co-catalysts KOH as an additive was demonstrated as a reusable catalyst for this reaction [84]. An improved catalytic method for this reaction is the first additive-free heterogeneous metal nanoparticle catalytic system (Pt/CeO$_2$) afforded the highest yield of C3 alkylated product [85]. The catalytic activity of Pt-impregnated catalysts varies with the support materials' nature. Basic supports cerium oxide and magnesium oxide produce a higher yield of the C3 alkylated oxindole than amphoteric aluminum oxide and acidic silica-alumina supports. The reusable catalyst was adequate for the selective Alkylation of oxindoles with various alcohols without any additives, as mentioned in Scheme 7A). The catalytic cycle was driven by the so-called borrowing-hydrogen pathway, where the metallic Pt^0 nanoparticle site dehydrogenates the alcohols to aldehydes, which undergoes cerium oxide-catalyzed aldol condensation with oxindoles to give alkenyl-oxindoles. The borrowed hydrogen on the catalysts species (Pt-H) reduces these condensed intermediates to the desired alkylated compounds.

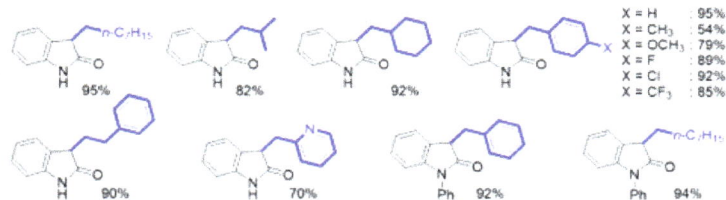

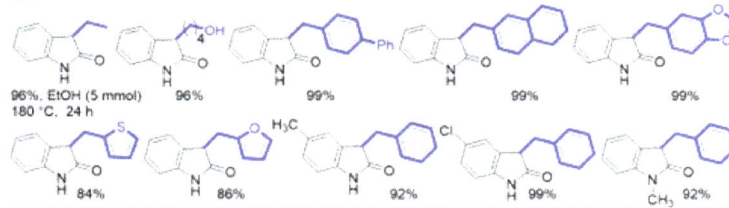

Scheme 7. C3-Alkylation of oxindoles with alcohol using (A) CeO$_2$-supported Pt nanoparticle catalysis, (B) CeO$_2$-supported Ni phosphide nanoalloy catalysis.

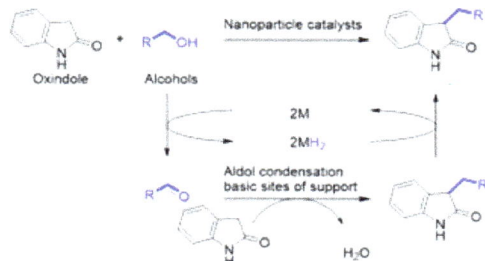

Mitsudome et al. are the pioneer to report a highly active non-precious metal-based heterogeneous nano-Ni$_2$P/CeO$_2$ catalyst for this C3 alkylation reaction of oxindoles with alcohols (Scheme 6B) [86]. A CeO$_2$-supported   Ni$_2$P nanoalloy catalyst efficiently promoted this C3 Alkylation of oxindoles with alcohols. The nano-Ni$_2$P/CeO$_2$ catalyst was

Emerging Applications of Nanomaterials   Materials Research Forum LLC
Materials Research Foundations 141 (2023) 42-74   https://doi.org/10.21741/9781644902295-3

separable and reusable without any remarkable loss in activity. Compared to state-of-the-art catalytic systems, this catalyst possesses high activity and selectivity. The catalytic cycle is directed by metal alloy (nano-Ni_2P) and amphoteric support (CeO_2), where nano-Ni_2P dehydrogenates the alcohol to aldehyde and nano-Ni_2P or CeO_2 promotes the condensation reaction of oxindoles with the generated aldehyde to produce an alkenyl oxindole. CeO_2 then transfers the hydrogen from nano-Ni_2P and reduces the C3 alkenyl oxindole to form the desired product [86].

2.2.3 Selective C3 alkylation of indoles with primary alcohols

Indoles with C3 substituents are essential building blocks for synthesizing various pharmaceutical and biologically important compounds. The classical synthetic methods for forming C3 substituted indoles suffer from needing a stoichiometric amount of Lewis acids, which is poor in regio-control of alkylation, discharge of huge salt wastes, and need for hazardous reagents. A non-catalytic method for C3 alkylation indoles with activated benzylic alcohols using a stoichiometric base (KOH) is exhibited by Ramónm and his group [87]. Transition-metal catalyzed processes for the regio-selective C3 Alkylation of indoles have been rarely exploited. Grigg et al. reported an early example of employing Ir complex catalysts and activated-benzylic alcohols as alkylating agents [88]. Beller and coworkers showed Ru catalyzed reaction system that afforded the Alkylation of indoles with benzylic and aliphatic primary amines as alkylating agents [89]. Grigg's catalytic system was the most atom economic method for C3 Alkylation of indoles with the generation of only water as a byproduct. However, both ways suffer from drawbacks, including limited substrate scope, low TON, catalysts reusability, and need for additives e.g., a strong base (KOH), and a large excess molar amount of alkylating alcohols. Siddiki and his coworker demonstrated the first additive-free reusable general catalytic method for regioselective C3 Alkylation of indoles with alcohols using an alumina-supported Pt nanoparticle catalyst (Scheme 8) [90]. Compared to the other noble and non-noble metals (Ir, Re, Rh, Pd, Ru, and Ni) loaded alumina catalysts, Pt/Al_2O_3 showed the highest yield of the C3-alkylated product. This catalytic reaction method has broad substrate scope for various alcohols and indoles, including N-substituted indoles, higher TON, simple catalyst/product separation, reusability of the catalyst, and additive-free. The kinetic isotopic effect study for the model reaction of indole and α-deutero benzyl alcohol results in a moderate k_H/k_D value of 1.7, suggesting the cleavage of the α-C–H bond of benzyl alcohol is the slowest step. The kinetic studies indicated that the catalytic cycle is driven by borrowing hydrogen mechanism, not by a Friedel-Crafts-type reaction [90].

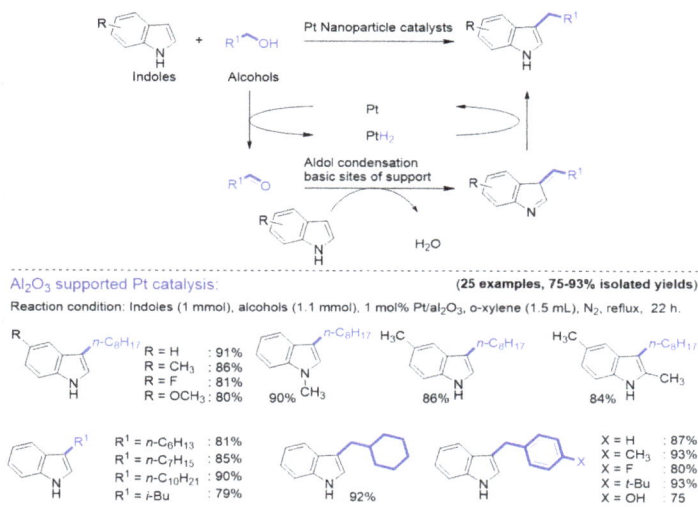

Scheme 8. C3-Alkylation of indoles with primary alcohol using Al_2O_3 supported Pt nanoparticle catalyst.

2.2.4 Selective self and cross-coupling reactions of secondary and primary alcohols

The self-coupling reaction of primary alcohols into β-alkylated alcohols is a proper synthetic method to form higher carbon-containing alcohols. A group of researchers exploited a series of alcohol self-coupling catalytic processes. For example, coupling of ethanol in a flow reaction system using solid catalysts at elevated temperatures of 300–500 °C [91] resulted in poor selectivity, whereas the metal complex catalyzed homogeneous system Ru [92] and Ir [93] is highly selective to the conversion of ethanol into *n*-butanol.

The particular self-coupling reaction of different aliphatic alcohols using an Ir [94] complex catalyst has also been reported. Non-noble-based Cu-nanoparticle catalyst was found to be effective for self-coupling 2-propanol into isobutyl ketone with moderately good selectivity at elevated temperatures (>200 °C) [106, 107]. Improved homogeneous catalytic method demonstrated by using Pd [97], Ir [98], and Ru[99] complexes in liquid phase reaction system to transform1-phenyl ethanol to β-alkylated alcohol or β-alkylated ketone at low temperatures (<130 °C), for a limited number of alcohols using an excess amount of basic co-catalyst. Shimizu et al. demonstrated the non-noble-based ceria-supported Ni nanoparticle catalyst that showed high regio-selectivity, catalytic activity, and good reusability for self-coupling various aliphatic alcohols to β-alkylated ketones at 130–

144 °C in the absence of any additives [100]. Noble metal catalysts Pt/CeO$_2$ was superior in catalytic activity than Ni/CeO$_2$ (54 times higher TON) catalyst for this self-coupling reaction (Scheme 9) [73].

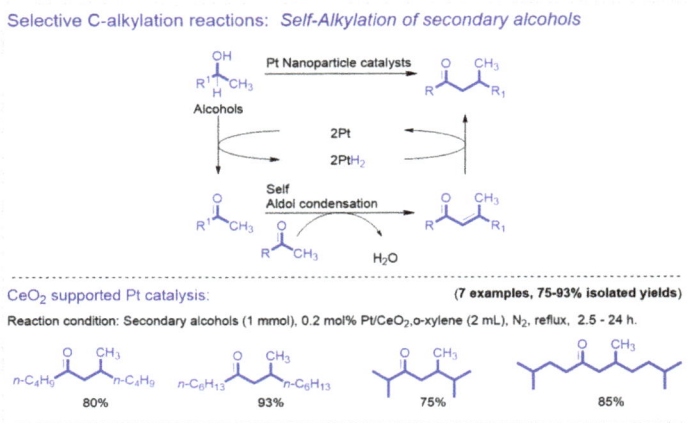

Scheme 9. Self-coupling of secondary alcohols using CeO$_2$-supported Pt nanoparticle catalysts.

β-Alkylated alcohols and ketones can be synthesized through cross-alkylation reaction. Secondary alcohols and primary alcohols, a group of homogeneous catalytic methods, are already reported using Ru [101–104] and Ir [105–108] complex catalysts without any hydrogen acceptor or donor. A sustainable nanoparticle catalytic method using alumina-supported silver in the presence of Cs$_2$CO$_3$, a basic co-catalyst, produced β-alkylated ketones (Scheme 10) with high efficiency and selectivity [109]. This catalytic system is applicable for a series of secondary and primary alcohols, hetero-atom-containing alcohols in good to high yield of corresponding alkylated ketones. Compared to other noble metal catalysts (Pt, Ru, Rh) supported on alumina, this Ag/Al$_2$O$_3$ catalyst showed higher activity. For a model alkylation reaction, only 0.092 mol% of Ag metal was effective in synthesizing a 74% yield of 1,3-diphenyl propan-1-one in the reaction of 1-phenyl ethanol with benzyl alcohols, which corresponds to a TON of 800. This coupling reaction also follows the similar borrowing hydrogen mechanistic method, where alkylating partner alcohols are dehydrogenated to produce carbonyl compounds aldehydes or ketones and Ag-H$_2$ species; the carbonyl compounds are then coupled with ketones under Cs$_2$CO$_3$ catalyzed aldol condensation to produce the corresponding α,β-unsaturated ketones, which is hydrogenated by the borrowed hydrogen on the Ag species that is Ag–H to give β-alkylated ketones. Studies on the structure-activity relationship exhibited that the reaction proceeds by the cooperation of coordinatively unsaturated Ag site and acid-base sites of the Al$_2$O$_3$ support.

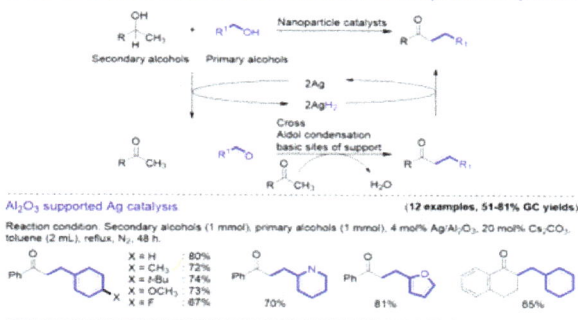

Scheme 10. Cross-coupling of secondary alcohols with primary alcohols using Al_2O_3-supported Ag nanoparticle catalysts.

A relevant catalytic system hydrotalcite-supported Au nanoparticle was demonstrated by Cao et al. for this reaction [110]. The support nature affects the condensation reaction of intermediate aldehydes or ketones. Here in this hydrotalcite-supported bimetallic catalysts Au–Pd/HT (Au: Pd = 13: 1), hydrotalcite co-catalyzed the synthesis of β-alkylated ketone from secondary and primary benzylic alcohols in the absence of any additives. Another mechanical separable catalytic system, Fe_3O_4, supported IrO_2, demonstrated by Ramón et al.. This catalytic system produced a high yield of β-alkylated alcohols in cross-reaction of primary alcohols (Scheme 11) [111]. The catalyst was removed from the reaction mixture by a magnet and could be reused up to ten times without losing its activity.

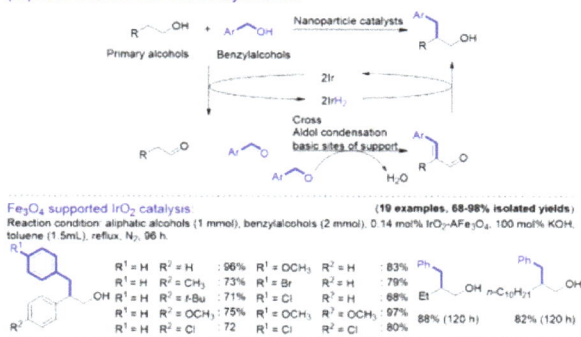

Scheme 11. Cross-coupling of secondary alcohols with primary alcohols using Al_2O_3-supported Ag nanoparticle catalysts.

2.2.5 Selective alkylation at CH₃ group of 2-methyl quinoline with primary alcohols

Selective C-alkylation reactions with alcohols under the borrowing hydrogen methodology are also applicable for the more challenging alkylation reaction through CH3 activation. An Ir complex catalyzed homogeneous catalytic method demonstrated by Obora et al. for the alkylation of 2-methyl quinolines with alcohols using phosphine ligand and base as additives [112] to afford a high yield of alkyl quinolines. The system has difficulties in reusability, low TON for the catalyst, a need for a stoichiometric amount of strong base, and catalysts/product separation. The first additive-free CH₃ alkylation method for 2-methyl quinoline with alcohols using γ-Al$_2$O$_3$ supported Pt nanoparticle catalysts (Scheme 12) exhibited by Chaudhari et al.[113]. Pt/γ-Al$_2$O$_3$ was the highest active catalyst of other noble and non-noble metal Ir, Re, Rh, Pd, Ag, and Ni-loaded/γ-Al$_2$O$_3$ catalysts and Pt-loaded on different supports (CeO$_2$, MgO, HBEA zeolite, carbon). This method has advantages over the previous homogeneous system in higher TON, catalyst reusability, and greener (additive-free) conditions. The borrowing-hydrogen methodology also drives the reaction, followed by the formation of aldehyde by dehydrogenation of alcohol undergoes Al$_2$O$_3$ catalyzed aldol condensation with 2-methyl quinoline to give the alkene intermediate, which is finally hydrogenated by Pt-H species.

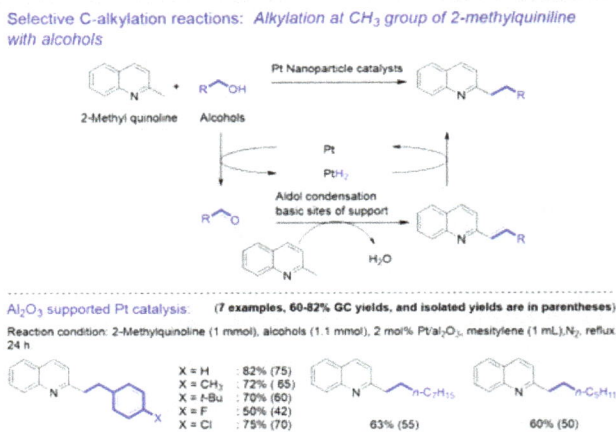

Scheme 12. Selective CH₃ Alkylation of 2-methyl quinoline with alcohols using Al₂O₃ supported Pt nanoparticle catalysts.

Materials Research Forum LLC
https://doi.org/10.21741/9781644902295-3

2.3 Selective *N*-alkylation reactions with alcohols under borrowing hydrogen methodology (C-N and C-S bond formation reactions)

2.3.1 *N*-alkylation of amines with alcohols

N-alkylated amines are being synthesized via alkylation of primary and secondary amines as very useful functional moieties in synthesizing a wide range of molecules used in pharmaceuticals, agrochemicals, surfactants, and biologically active compounds [2,3,35]. Conventionally, *N*-alkylamines are synthesized by the reaction of amines with alkyl, but the methods suffer from the use of hazardous halides and the generation of a stoichiometric amount of salt waste. The different research groups demonstrated an improved sustainable catalytic method under borrowing hydrogen methodology using low-cost and available starting material alcohols as an alkylating agent. Alcohols transform into activated carbonyl compounds via catalytic dehydrogenation, and imine formation is followed by the condensation of amines and carbonyl compounds. The imine is finally reduced to an amine [12,13,114,115]. Noble metal-based reusable nanoparticle catalysts (Pd [116], Ru [117,118], Ag [71,119], Au [120], Pt [121], and Ir [122]) are also effective for N-alkylation reactions. Still, some of them suffer from low selectivity to mono-alkylation, low turnover number (TON), and the necessity of high reaction temperature or co-catalysts as additives.

Apart from the noble metals, non-noble-metal catalysts, such as Fe [123], Cu [124,125], Ni [19], Mn [126] and zeolite [127] are also demonstrated for this same reaction. These catalysts generally suffer from low catalytic activity, limited substrate scope, or harsh reaction conditions, i.e., the need for elevated temperatures under an H_2 atmosphere or a large amount of basic co-catalyst as additives.

Non-noble metal-based other sustainable catalysts Ni/θ-Al$_2$O$_3$ [128] and Cu–Ag/γ-Al$_2$O$_3$ [129] (Cu: Ag = 95: 5) are also effective for N-alkylation reaction even in the absence of any additives. Ni/θ-Al$_2$O$_3$ (1 mol%) effectively catalyzed the mono-alkylation of various amines and anilines with aliphatic and benzyl alcohols (Scheme 13) [128]. More challenging reactions of primary and secondary amines with primary aliphatic alcohols were also successful under this catalytic system. The noble metal-free heterogeneous metal nanoparticle catalysts are effective for this reaction. Ni nanoparticle catalysts' activity depended on the support materials' nature. Compared to the basic, acidic, or neutral metal oxide supports, the amphoteric Al$_2$O$_3$ support was found to be the highest active, suggesting that the acid-base pair sites on the supports are essential for this catalytic system. For a series of Ni/θ-Al$_2$O$_3$ catalysts with different particle sizes, the turnover frequency (TOF) per surface Ni increased with decreasing the Ni mean particle size. These results suggest that the metal–support interface is the active site.

Selective *N*-alkylation reactions of amines with alcohols

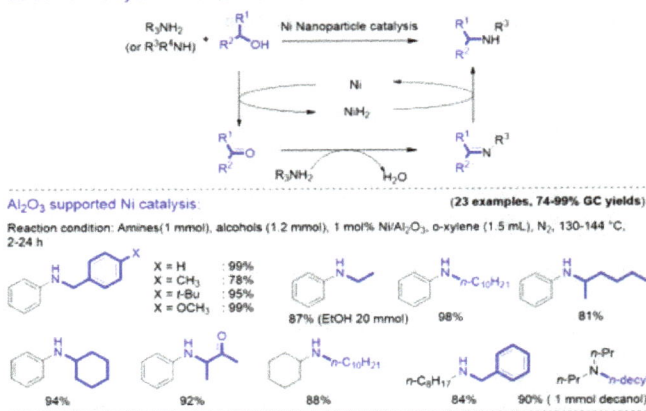

Scheme 13. Selective N-alkylation of primary and secondary amines with primary alcohols using Al$_2$O$_3$-supported Ni nanoparticle catalysts

The bimetallic reusable Pt–Sn/γ-Al$_2$O$_3$ catalyst directly synthesized diamines or cyclic amines from amines and diols under additive-free conditions (Scheme 14) [130]. A series of diols were selectively transformed into the corresponding cyclic amines or diamines under this catalytic system.

Selective *N*-alkylation reactions of amines with diols

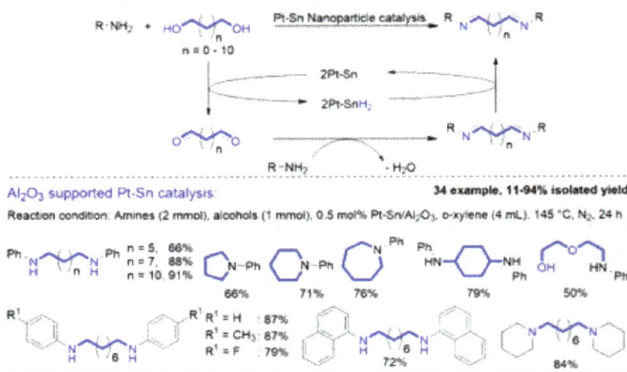

Scheme 14. Selective N-alkylation of primary amines with diols using Al$_2$O$_3$ supported Pt-Sn nanoparticle catalysts.

Emerging Applications of Nanomaterials   Materials Research Forum LLC
Materials Research Foundations 141 (2023) 42-74   https://doi.org/10.21741/9781644902295-3

2.3.2 Selective *N*-alkylation of indoles with alcohols

The selective N-alkylation of indoles compared to the C3-alkylation is relatively difficult due to the higher nucleophilicity at the C3 position. A base-catalyzed *N*-alkylation reaction of indoles with alkyl halide is a classical approach that suffers from generating a stoichiometric amount of salt waste. An environmentally benign alkylation method using alcohol as an alkylating agent overcomes the difficulties of the classical approach efficiently. Improved *N*-alkylation methods are demonstrated by different research groups employing the excess amount of RANEY® nickel [131]; Brønsted acidic assisted Ru-based Shvo catalyst [89] are better atom-efficient methods compared with the classical ones. However, these homogeneous and heterogeneous methods are still in difficulties with low TON, catalysts/product separation, and the need for additives. The first reusable and additive-free N-alkylation of indoles with alcohols was explored by Siddiki et al. using an HBEA zeolite-supported Pt nanoparticle catalyst (Scheme 15) [132]. A series of alcohols, including aliphatic and benzylic, produced the corresponding N-alkylated indoles with good to high isolated yields.

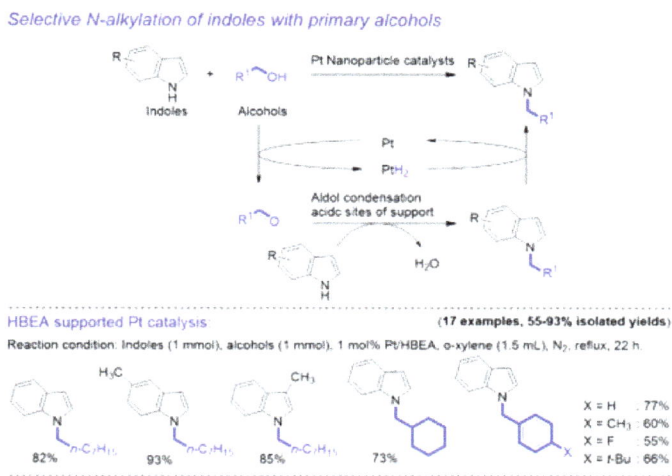

Scheme 15. Regioselective N-alkylation of indoles with alcohols using HBEA-supported Pt nanoparticle catalysts.

Interestingly, this work and C3-alkylation of indoles [102] (Scheme 7) are significant examples of the regioselective alkylation of indoles using Pt nanoparticle catalysts, where support nature directs the regioselectivity. Zeolite HBEA support Pt catalyst produces N-alkylated indoles (Scheme 15), whereas Al_2O_3-supported Pt gives C3-alkylated indoles

Materials Research Forum LLC
https://doi.org/10.21741/9781644902295-3

(Scheme 7) in reaction with alcohols. The controlled experiments also supported the mechanistic insight of this selective alkylation reaction. A bench model alkylation reaction of indole with n-octanol and reaction of indole with *n*-octanal (in situ generated from n-octanol through metal nanoparticle catalyzed dehydrogenation) was studied using Pt-free support oxides (HBEA or Al$_2$O$_3$), the results show that HBEA gave enamine, whereas Al$_2$O$_3$ gave C3-adduct. Interaction between the Brønsted acid site of HBEA and the C=C bond of indole may reduce the nucleophilicity of the C3 carbon and, instead, the nucleophilic attack of the *N* atom of indole to aldehyde results in the formation of the enamine intermediate [132].

2.3.3 Quinolines synthesis from nitroarenes and alcohols (selective N and C-alkylation)

Quinolines and their derivatives are an important class of bioactive compounds prescribed as antimalarial, antibacterial, antihypertensive, and anti-inflammatory drugs [133]. The classical routes to prepare the functionalized quinolines include several named reactions, such as Skraup, Doebner–Von Miller, Conrad–Limpach, Friedl_nder, and Pfitzinger syntheses based on the reaction of substituted anilines with carbonyl compounds [133]. The quinoline derivatives are synthesized from nitroarenes and primary aliphatic alcohols using homogeneous Ru or Rh [134] catalysts. Cao et al. [133] demonstrated a new heterogeneous iridium-based catalyst (Ir/TiO$_2$) that facilitates the one-pot synthesis of quinolines from nitroarenes and alcohols under mild conditions through a facile sequential transfer reduction–condensation–dehydrogenation pathway (Scheme 16). The combination of sub-nanosized iridium clusters with oxide support TiO$_2$ possessing a suitable surface acidity is assigned as an optimal catalyst.

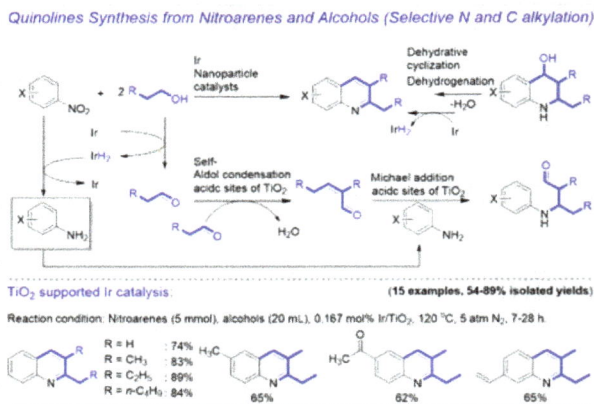

Scheme 16. Quinoline synthesis from nitroarenes and alcohols through selective N- and C- alkylation reaction using TiO$_2$ supported Ir nanoparticle catalysts.

Emerging Applications of Nanomaterials   Materials Research Forum LLC
Materials Research Foundations 141 (2023) 42-74   https://doi.org/10.21741/9781644902295-3

Conclusions

Heterogeneous metal nanoparticle catalysis was exhaustively studied for the environmentally benign synthesis of fine chemicals through coupling reactions, including carbon-carbon, carbon-nitrogen, and carbon-sulfur bond formation under hydrogen transfer/borrowing hydrogen methodology. The method is highly efficient in synthesizing chemicals from alcohols, amines, and ammonia without needing any oxidizing or reducing agents, and the discharge byproduct is only water. A group of noble and non-noble metal nanoparticle catalyst systems is easy to synthesize and superior in catalytic activity compared to state-of-the-art homogeneous catalysis. Conventional homogeneous catalysis systems need additives and expensive ligands in liquid-phase reactions. The catalytic activity decreases with-atom economy and efficiency. The catalysts are designed based on the structure-activity relationship of the metal nanoparticles and support different surface functionality, where metal dehydrogenates the alcohols or amines and support promotes the condensation of reactions of aldehydes/ketones/imines with nucleophiles. However, the general knowledge of multifunctional catalysis at the metal-support interface sites still needs more information. Mechanistic insight based on theoretical studies and new synthesis methods for structurally well-functionalized catalysts will accelerate the rational development of new heterogeneous metal nanoparticle catalysts for this multistep reaction.

References

[1]   K. Philippot, P. Serp, Concepts in Nanocatalysis, Nanomater. Catal. First Ed. (2012) 1–54. https//doi.org/10.1002/9783527656875.ch1

[2]   D. Astruc, ed., Nanoparticles and Catalysis, wiley, 2008. https://www.wiley.com/en-ie/Nanoparticles+and+Catalysis-p-9783527315727

[3]   T. Ishida, M. Haruta, Gold catalysts: Towards sustainable chemistry, Angew. Chemie - Int. Ed. 46 (2007) 7154–7156. https//doi.org/10.1002/anie.200701622

[4]   A.S.K. Hashmi, G.J. Hutchings, Gold Catalysis, Angew. Chemie - Int. Ed. 45 (2006) 7896–7936. https//doi.org/10.1002/anie.200602454

[5]   G.A. Somorjai, H. Frei, J.Y. Park, Advancing the frontiers in nanocatalysis, biointerfaces, and renewable energy conversion by innovations of surface techniques, J. Am. Chem. Soc. 131 (2009) 16589–16605. https//doi.org/10.1021/ja9061954

[6]   C. Burda, X. Chen, R. Narayanan, M.A. El-Sayed, Chemistry and properties of nanocrystals of different shapes, 2005. https//doi.org/10.1021/cr030063a

[7]   E. de Smit, I. Swart, J.F. Creemer, C. Karunakaran, D. Bertwistle, H.W. Zandbergen, F.M.F. de Groot, B.M. Weckhuysen, Nanoscale Chemical Imaging of the Reduction Behavior of a Single Catalyst Particle, Angew. Chemie. 121 (2009) 3686–3690. https//doi.org/10.1002/ange.200806003

[8]   V. Polshettiwar, R.S. Varma, Green chemistry by nano-catalysis, Green Chem. 12 (2010) 743–75. https//doi.org/10.1039/b921171c

[9]   V. Polshettiwar, R. Luque, A. Fihri, H. Zhu, M. Bouhrara, J.M. Basset,
Magnetically recoverable nanocatalysts, Chem. Rev. 111 (2011) 3036–3075.
https//doi.org/10.1021/cr100230z

[10]   A. Fihri, M. Bouhrara, B. Nekoueishahraki, J.M. Basset, V. Polshettiwar,
Nanocatalysts for Suzuki cross-coupling reactions, Chem. Soc. Rev. 40 (2011) 5181–
5203. https//doi.org/10.1039/c1cs15079k

[11]   X. Xie, Y. Li, Z.Q. Liu, M. Haruta, W. Shen, Low-temperature oxidation of CO
catalysed by Co 3 O 4 nanorods, Nature. 458 (2009) 746–749.
https//doi.org/10.1038/nature07877

[12]   T.D. Nixon, M.K. Whittlesey, J.M.J. Williams, Transition metal catalysed
reactions of alcohols using borrowing hydrogen methodology, Dalt. Trans. (2009)
753–762. https//doi.org/10.1039/b813383b

[13]   M.H.S.A. Hamid, P.A. Slatford, J.M.J. Williams, Borrowing hydrogen in the
activation of alcohols, Adv. Synth. Catal. 349 (2007) 1555–1575.
https//doi.org/10.1002/adsc.200600638

[14]   R. Grigg, T.R.B. Mitchell, S. Sutthivaiyakit, N. Tongpenyai, Transition metal-
catalysed N-alkylation of amines by alcohols, J. Chem. Soc. Chem. Commun. (1981)
611. https//doi.org/10.1039/c39810000611

[15]   A. Balker, J. Kijenski, Catalytic Synthesis of Higher Aliphatic Amines from the
Corresponding Alcohols, 1985. https//doi.org/10.1080/01614948508064235

[16]   L. Liu, A. Corma, Metal Catalysts for Heterogeneous Catalysis: From Single
Atoms to Nanoclusters and Nanoparticles, Chem. Rev. 118 (2018) 4981–5079.
https//doi.org/10.1021/acs.chemrev.7b00776

[17]   S.M.A. Hakim Siddiki, T. Toyao, K.I. Shimizu, Acceptorless dehydrogenative
coupling reactions with alcohols over heterogeneous catalysts, Green Chem. 20 (2018)
2933–2952. https//doi.org/10.1039/c8gc00451j

[18]   K. Shimizu, Heterogeneous catalysis for the direct synthesis of chemicals by
borrowing hydrogen methodology, Catal. Sci. Technol. 5 (2015) 1412–1427.
https//doi.org/10.1039/C4CY01170H

[19]   G. Guillena, D.J. Ramón, M. Yus, Hydrogen autotransfer in the N-alkylation of
amines and related compounds using alcohols and amines as electrophiles, Chem. Rev.
110 (2010) 1611–1641. https//doi.org/10.1021/cr9002159

[20]   G. Guillena, D.J. Ramón, M. Yus, Alcohols as electrophiles in C-C bond-forming
reactions: The hydrogen autotransfer process, Angew. Chemie - Int. Ed. 46 (2007)
2358–2364. https//doi.org/10.1002/anie.200603794

[21]   F. Alonso, P. Riente, M. Yus, Nickel Nanoparticles in Hydrogen Transfer
Reactions, Acc. Chem. Res. 44 (2011) 379–391. https//doi.org/10.1021/ar1001582

[22]   K.I. Fujita, R. Yamaguchi, Cp*Ir complex-catalyzed hydrogen transfer reactions directed toward environmentally benign organic synthesis, Synlett. (2005) 560–571. https//doi.org/10.1055/s-2005-862381

[23]   Y. Obora, Y. Ishii, Iridium-catalyzed reactions involving transfer hydrogenation, addition, N-heterocyclization, and alkylation using alcohols and diols as key substrates, Synlett. (2011) 30–51. https//doi.org/10.1055/s-0030-1259094

[24]   Y. Obora, Recent advances in α-alkylation reactions using alcohols with hydrogen borrowing methodologies, ACS Catal. 4 (2014) 3972–3981. https//doi.org/10.1021/cs501269d

[25]   J. Choi, A.H.R. MacArthur, M. Brookhart, A.S. Goldman, Dehydrogenation and related reactions catalyzed by iridium pincer complexes, Chem. Rev. 111 (2011) 1761–1779. https//doi.org/10.1021/cr1003503

[26]   C. Gunanathan, D. Milstein, Metal-ligand cooperation by aromatization-dearomatization: A new paradigm in bond activation and "green" catalysis, Acc. Chem. Res. 44 (2011) 588–602. https//doi.org/10.1021/ar2000265

[27]   R.A.W. Johnstone, A.H. Wilby, I.D. Entwistle, Heterogeneous Catalytic Transfer Hydrogenation and Its Relation to Other Methods for Reduction of Organic Compounds, Chem. Rev. 85 (1985) 129–170. https//doi.org/10.1021/cr00066a003

[28]   G. Zassinovich, G. Mestroni, S. Giadiali, Asymmetric Hydrogen Transfer Reactions Promoted by Homogeneous Transition Metal Catalysts, Chem. Rev. 92 (1992) 1051–1069. https//doi.org/10.1021/cr00013a015

[29]   J.S. Cha, Recent developments in Meerwein-Ponndorf-Verley and related reactions for the reduction of organic functional groups using aluminum, boron, and other metal reagents: A review, Org. Process Res. Dev. 10 (2006) 1032–1053. https//doi.org/10.1021/op068002c

[30]   R. Noyori, S. Hashiguchi, Asymmetric Transfer Hydrogenation Catalyzed by Chiral Ruthenium Complexes, Acc. Chem. Res. 30 (1997) 97–102. https//doi.org/10.1021/ar9502341

[31]   T.L. Lambat, R.G. Chaudhary, A.A. Abdal, R.K. Mishra S.H. Mahmood, S. Banerjee, Mesoporous PbO nanoparticle-catalyzed synthesis of arylbenzodioxy xanthenedione scaffolds under solvent-free conditions in a ball mill, RSC Advances 9, 31683-31690

[32]   T. Ikariya, A.J. Blacker, Asymmetric transfer hydrogenation of ketones with bifunctional transition metal-based molecular catalysts, Acc. Chem. Res. 40 (2007) 1300–1308. https//doi.org/10.1021/ar700134q

[33]   R. Malacea, R. Poli, E. Manoury, Asymmetric hydrosilylation, transfer hydrogenation and hydrogenation of ketones catalyzed by iridium complexes, Coord. Chem. Rev. 254 (2010) 729–752. https//doi.org/10.1016/j.ccr.2009.09.033

[34] F. Alonso, P. Riente, F. Rodríguez-Reinoso, J. Ruiz-Martínez, A. Sepúlveda-Escribano, M. Yus, Platinum nanoparticles supported on titania as an efficient hydrogen-transfer catalyst, J. Catal. 260 (2008) 113–118. https//doi.org/10.1016/j.jcat.2008.09.009

[35] M.J. Gracia, J.M. Campelo, E. Losada, R. Luque, J.M. Marinas, A.A. Romero, Microwave-assisted versatile hydrogenation of carbonyl compounds using supported metal nanoparticles, Org. Biomol. Chem. 7 (2009) 4821–4824. https//doi.org/10.1039/b913695a

[36] J.Q. Yu, H.C. Wu, C. Ramarao, J.B. Spencer, S. V. Ley, Transfer hydrogenation using recyclable polyurea-encapsulated palladium: Efficient and chemoselective reduction of aryl ketones, Chem. Commun. 3 (2003) 678–679. https//doi.org/10.1039/b300074p

[37] K. Yamaguchi, T. Koike, M. Kotani, M. Matsushita, S. Shinachi, N. Mizuno, Synthetic scope and mechanistic studies of Ru(OH)x/Al 2O3-catalyzed heterogeneous hydrogen-transfer reactions, Chem. - A Eur. J. 11 (2005) 6574–6582. https//doi.org/10.1002/chem.200500539

[38] M.L. Kantam, B.P.C. Rao, B.M. Choudary, B. Sreedhar, Selective transfer hydrogenation of carbonyl compounds by ruthenium nanoclusters supported on alkali-exchanged zeolite beta, Adv. Synth. Catal. 348 (2006) 1970–1976. https//doi.org/10.1002/adsc.200505497

[39] B. Baruwati, V. Polshettiwar, R.S. Varma, Magnetically recoverable supported ruthenium catalyst for hydrogenation of alkynes and transfer hydrogenation of carbonyl compounds, Tetrahedron Lett. 50 (2009) 1215–1218. https//doi.org/10.1016/j.tetlet.2009.01.014

[40] C. Hammond, M.T. Schümperli, S. Conrad, I. Hermans, Hydrogen transfer processes mediated by supported iridium oxide nanoparticles, ChemCatChem. 5 (2013) 2983–2990. https//doi.org/10.1002/cctc.201300253

[41] F.Z. Su, L. He, J. Ni, Y. Cao, H.Y. He, K.N. Fan, Efficient and chemoselective reduction of carbonyl compounds with supported gold catalysts under transfer hydrogenation conditions, Chem. Commun. (2008) 3531–3533. https//doi.org/10.1039/b807608a

[42] L. He, J. Ni, L.C. Wang, F.J. Yu, Y. Cao, H.Y. He, K.N. Fan, Aqueous room-temperature gold-catalyzed chemoselective transfer hydrogenation of aldehydes, Chem. - A Eur. J. 15 (2009) 11833–11836. https//doi.org/10.1002/chem.200901261

[43] F. Alonso, P. Riente, M. Yus, Hydrogen-transfer reduction of carbonyl compounds promoted by nickel nanoparticles, Tetrahedron. 64 (2008) 1847–1852. https//doi.org/10.1016/j.tet.2007.11.093

[44] J. Tanna, R.G. Chaudhary, N.V. Gandhare, A.R. Rai, S. Yerpude, H.D. Juneja, Copper nanoparticles catalysed an efficient one-pot multicomponents synthesis of

chromenes derivatives and its antibacterial activity, J. Expt. Nanosci. 11 (2016) 884–900

[45]   Chaudhary RG, A Tanna J, V Gandhare N, R Rai A, D Juneja H. Synthesis of nickel NPs: Microscopic investigation, an efficient catalyst and effective antibacterial activity. Adv Mater Lett. 2015;1;6(11):990-8

[46]   T. Subramanian, K. Pitchumani, Transfer hydrogenation of carbonyl compounds and carbon-carbon multiple bonds by zeolite supported Cu nanoparticles, Catal. Sci. Technol. 2 (2012) 296–300. https//doi.org/10.1039/c1cy00383f

[47]   L. Huang, Y. Zhu, C. Huo, H. Zheng, G. Feng, C. Zhang, Y. Li, Mechanistic insight into the heterogeneous catalytic transfer hydrogenation over Cu/Al2O3: Direct evidence for the assistant role of support, J. Mol. Catal. A Chem. 288 (2008) 109–115. https//doi.org/10.1016/j.molcata.2008.03.026

[48]   M. Kidwai, V. Bansal, A. Saxena, R. Shankar, S. Mozumdar, Ni-nanoparticles: an efficient green catalyst for chemoselective reduction of aldehydes, Tetrahedron Lett. 47 (2006) 4161–4165. https//doi.org/10.1016/j.tetlet.2006.04.048

[49]   J.R. Ruiz, C. Jiménez-Sanchidrián, J.M. Hidalgo, Meerwein-Ponndorf-Verley reaction of acetophenones with 2-propanol over MgAl mixed oxide: The substituent effect, Catal. Commun. 8 (2007) 1036–1040. https//doi.org/10.1016/j.catcom.2006.10.007

[50]   R. Radhakrishan, D.M. Do, S. Jaenicke, Y. Sasson, G.K. Chuah, Potassium phosphate as a solid base catalyst for the catalytic transfer hydrogenation of aldehydes and ketones, ACS Catal. 1 (2011) 1631–1636. https//doi.org/10.1021/cs200299v

[51]   Y. Zhu, G. Chuah, S. Jaenicke, Chemo- and regioselective Meerwein-Ponndorf-Verley and Oppenauer reactions catalyzed by Al-free Zr-zeolite beta, J. Catal. 227 (2004) 1–10. https//doi.org/10.1016/j.jcat.2004.05.037

[52]   A. Corma, M.E. Domine, S. Valencia, Water-resistant solid Lewis acid catalysts: Meerwein-Ponndorf-Verley and Oppenauer reactions catalyzed by tin-beta zeolite, J. Catal. 215 (2003) 294–304. https//doi.org/10.1016/S0021-9517(03)00014-9

[53]   A. Ramanathan, D. Klomp, J.A. Peters, U. Hanefeld, Zr-TUD-1: A novel heterogeneous catalyst for the Meerwein-Ponndorf-Verley reaction, J. Mol. Catal. A Chem. 260 (2006) 62–69. https//doi.org/10.1016/j.molcata.2006.06.057

[54]   K. Shimura, K. Shimizu, Transfer hydrogenation of ketones by ceria-supported Ni catalysts, Green Chem. 14 (2012) 2983–2985. https//doi.org/10.1039/c2gc35836k

[55]   R. Chorghade, C. Battilocchio, J.M. Hawkins, S. V. Ley, Sustainable flow oppenauer oxidation of secondary benzylic alcohols with a heterogeneous zirconia catalyst, Org. Lett. 15 (2013) 5698–5701. https//doi.org/10.1021/ol4027107

[56]   M. Hayashi, K. Yamada, S.Z. Nakayama, H. Hayashi, S. Yamazaki, Environmentally benign oxidation using a palladium catalyst system, Green Chem. 2 (2000) 257–260. https//doi.org/10.1039/b003887n

[57]   C. Keresszegi, T. Mallat, A. Baiker, Selective transfer dehydrogenation of aromatic alcohols on supported palladium, New J. Chem. 25 (2001) 1163–1167. https//doi.org/10.1039/b102463a

[58]   F. Zaccheria, N. Ravasio, R. Psaro, A. Fusi, Anaerobic oxidation of non-activated secondary alcohols over Cu/Al 2O3, Chem. Commun. (2005) 253–255. https//doi.org/10.1039/b413634a

[59]   R. Shi, F. Wang, Tana, Y. Li, X. Huang, W. Shen, A highly efficient Cu/La2O3 catalyst for transfer dehydrogenation of primary aliphatic alcohols, Green Chem. 12 (2010) 108–11. https//doi.org/10.1039/b919807p

[60]   F. Wang, R. Shi, Z.Q. Liu, P.J. Shang, X. Pang, S. Shen, Z. Feng, C. Li, W. Shen, Highly efficient dehydrogenation of primary aliphatic alcohols catalyzed by Cu nanoparticles dispersed on rod-shaped La2O2CO 3, ACS Catal. 3 (2013) 890–894. https//doi.org/10.1021/cs400255r

[61]   K. Taguchi, H. Nakagawa, T. Hirabayashi, S. Sakaguchi, Y. Ishii, An Efficient Direct α-Alkylation of Ketones with Primary Alcohols Catalyzed by [Ir(cod)CI]2/PPh3/KOH System without Solvent, J. Am. Chem. Soc. 126 (2004) 72–73. https//doi.org/10.1021/ja037552c

[62]   S. Ogawa, Y. Obora, Iridium-catalyzed selective α-methylation of ketones with methanol, Chem. Commun. 50 (2014) 2491–2493. https//doi.org/10.1039/c3cc49626k

[63]   C.S. Cho, B.T. Kim, T.J. Kim, S.C. Shim, An unusual type of ruthenium-catalyzed transfer hydrogenation of ketones with alcohols accompanied by C-C coupling, J. Org. Chem. 66 (2001) 9020–9022. https//doi.org/10.1021/jo0108459

[64]   R. Martínez, G.J. Brand, D.J. Ramón, M. Yus, [Ru(DMSO)4] Cl2 catalyzes the α-alkylation of ketones by alcohols, Tetrahedron Lett. 46 (2005) 3683–3686. https//doi.org/10.1016/j.tetlet.2005.03.158

[65]   T. Kuwahara, T. Fukuyama, I. Ryu, RuHCl(CO)(PPh 3) 3 -Catalyzed α-Alkylation of Ketones with Primary Alcohols, Org. Lett. 14 (2012) 4703–4705. https//doi.org/10.1021/ol302145a

[66]   K. Motokura, D. Nishimura, K. Mori, T. Mizugaki, K. Ebitani, K. Kaneda, A Ruthenium-Grafted Hydrotalcite as a Multifunctional Catalyst for Direct α-Alkylation of Nitriles with Primary Alcohols, J. Am. Chem. Soc. 126 (2004) 5662–5663. https//doi.org/10.1021/ja0491811

[67]   C.S. Cho, A palladium-catalyzed route for α-alkylation of ketones by primary alcohols, J. Mol. Catal. A Chem. 240 (2005) 55–60. https//doi.org/10.1016/j.molcata.2005.06.043

[68]   M.S. Kwon, N. Kim, S.H. Seo, I.S. Park, R.K. Cheedrala, J. Park, Recyclable palladium catalyst for highly selective α alkylation of ketones with alcohols, Angew. Chemie - Int. Ed. 44 (2005) 6913–6915. https//doi.org/10.1002/anie.200502422

[69]   Y.M.A. Yamada, Y. Uozumi, A solid-phase self-organized catalyst of nanopalladium with main-chain viologen polymers: α-alkylation of ketones with primary alcohols, Org. Lett. 8 (2006) 1375–1378. https//doi.org/10.1021/ol060166q

[70]   S. Kim, S.W. Bae, J.S. Lee, J. Park, Recyclable gold nanoparticle catalyst for the aerobic alcohol oxidation and C-C bond forming reaction between primary alcohols and ketones under ambient conditions, Tetrahedron. 65 (2009) 1461–1466. https//doi.org/10.1016/j.tet.2008.12.005

[71]   X. Cui, Y. Zhang, F. Shi, Y. Deng, Organic ligand-free alkylation of amines, carboxamides, sulfonamides, and ketones by using alcohols catalyzed by heterogeneous Ag/Mo oxides, Chem. - A Eur. J. 17 (2011) 1021–1028. https//doi.org/10.1002/chem.201001915

[72]   F. Alonso, P. Riente, M. Yus, Alcohols for the α-alkylation of methyl ketones and indirect aza-wittig reaction promoted by nickel nanoparticles, European J. Org. Chem. (2008) 4908–4914. https//doi.org/10.1002/ejoc.200800729

[73]   C. Chaudhari, S.M.A.H. Siddiki, K.-I. Shimizu, Self-coupling of secondary alcohols and α-alkylation of methyl ketones with secondary alcohols by Pt/CeO$_2$ catalyst, Top. Catal. 57 (2014). https//doi.org/10.1007/s11244-014-0268-6

[74]   A. Fischer, P. Makowski, J.O. Müller, M. Antonietti, A. Thomas, F. Goettmann, High-surface-area TiO2 and TiN as catalysts for the C-C coupling of alcohols and ketones., ChemSusChem. 1 (2008) 444–449. https//doi.org/10.1002/cssc.200800019

[75]   M. Dixit, M. Mishra, P.A. Joshi, D.O. Shah, Clean borrowing hydrogen methodology using hydrotalcite supported copper catalyst, Catal. Commun. 33 (2013) 80–83. https//doi.org/10.1016/j.catcom.2012.12.027

[76]   C. Löfberg, R. Grigg, M.A. Whittaker, A. Keep, A. Derrick, Efficient solvent-free selective monoalkylation of arylacetonitriles with mono-, bis-, and tris-primary alcohols catalyzed by a Cp*Ir complex, J. Org. Chem. 71 (2006) 8023–8027. https//doi.org/10.1021/jo061113p

[77]   T. Naota, H. Taki, M. Mizuno, S. Murahashi, Ruthenium-catalyzed aldol and Michael reactions of activated nitriles, J. Am. Chem. Soc. 111 (1989) 5954–5955. https//doi.org/10.1021/ja00197a073

[78]   M. Morita, Y. Obora, Y. Ishii, Alkylation of active methylene compounds with alcohols catalyzed by an iridium complex, Chem. Commun. (2007) 2850–2852. https//doi.org/10.1039/b702293j

Materials Research Forum LLC
https://doi.org/10.21741/9781644902295-3

[79]   T. Sawaguchi, Y. Obora, Iridium-catalyzed α-alkylation of acetonitrile with primary and secondary alcohols, Chem. Lett. 40 (2011) 1055–1057. https//doi.org/10.1246/cl.2011.1055

[80]   K. Motokura, N. Fujita, K. Mori, T. Mizugaki, K. Ebitani, K. Jitsukawa, K. Kaneda, Environmentally friendly one-pot synthesis of α-alkylated nitriles using hydrotalcite-supported metal species as multifunctional solid catalysts, Chem. - A Eur. J. 12 (2006) 8228–8239. https//doi.org/10.1002/chem.200600317

[81]   A. Corma, T. Ródenas, M.J. Sabater, Monoalkylations with alcohols by a cascade reaction on bifunctional solid catalysts: Reaction kinetics and mechanism, J. Catal. 279 (2011) 319–327. https//doi.org/10.1016/j.jcat.2011.01.029

[82]   T. Jensen, R. Madsen, Ruthenium-Catalyzed Alkylation of Oxindole with Alcohols, J. Org. Chem. 74 (2009) 3990–3992. https//doi.org/10.1021/jo900341w

[83]   R. Grigg, S. Whitney, V. Sridharan, A. Keep, A. Derrick, Iridium catalysed C-3 alkylation of oxindole with alcohols under solvent free thermal or microwave conditions, Tetrahedron. 65 (2009) 4375–4383. https//doi.org/10.1016/j.tet.2009.03.065

[84]   G. Liu, T. Huang, Y. Zhang, X. Liang, Y. Li, H. Li, C-3 alkylation of oxindole with alcohols catalyzed by an indene-functionalized mesoporous iridium catalyst, Catal. Commun. 12 (2011) 655–659. https//doi.org/10.1016/j.catcom.2010.12.021

[85]   C. Chaudhari, S.M.A.H. Siddiki, K. Kon, A. Tomita, Y. Tai, K.I. Shimizu, C-3 alkylation of oxindole with alcohols by Pt/CeO2 catalyst in additive-free conditions, Catal. Sci. Technol. 4 (2014) 1064–1069. https//doi.org/10.1039/c3cy00911d

[86]   S. Fujita, K. Imagawa, S. Yamaguchi, J. Yamasaki, S. Yamazoe, T. Mizugaki, T. Mitsudome, A nickel phosphide nanoalloy catalyst for the C-3 alkylation of oxindoles with alcohols, Sci. Rep. 11 (2021) 1–10. https//doi.org/10.1038/s41598-021-89561-1

[87]   R. Cano, M. Yus, D.J. Ramón, Environmentally friendly and regioselective C3-alkylation of indoles with alcohols through a hydrogen autotransfer strategy, Tetrahedron Lett. 54 (2013) 3394–3397. https//doi.org/10.1016/j.tetlet.2013.04.062

[88]   S. Whitneys, R. Grigg, A. Derrick, A. Keep, [Cp*IrCl2]2-catalyzed indirect functionalization of alcohols: Novel strategies for the synthesis of substituted indoles, Org. Lett. 9 (2007) 3299–3302. https//doi.org/10.1021/ol071274v

[89]   S. Bähn, S. Imm, K. Mevius, L. Neubert, A. Tillack, J.M.J. Williams, M. Beller, Selective ruthenium-catalyzed N-alkylation of indoles by using alcohols, Chem. - A Eur. J. 16 (2010) 3590–3593. https//doi.org/10.1002/chem.200903144

[90]   S.M.A.H. Siddiki, K. Kon, K.I. Shimizu, General and selective C-3 alkylation of indoles with primary alcohols by a reusable Pt nanocluster catalyst, Chem. - A Eur. J. 19 (2013) 14416–14419. https//doi.org/10.1002/chem.201302464

[91]   J.T. Kozlowski, R.J. Davis, Heterogeneous catalysts for the guerbet coupling of alcohols, ACS Catal. 3 (2013) 1588–1600. https//doi.org/10.1021/cs400292f

[92]   G.R.M. Dowson, M.F. Haddow, J. Lee, R.L. Wingad, D.F. Wass, Catalytic conversion of ethanol into an advanced biofuel: Unprecedented selectivity for n-butanol, Angew. Chemie - Int. Ed. 52 (2013) 9005–9008. https//doi.org/10.1002/anie.201303723

[93]   K. Koda, T. Matsu-ura, Y. Obora, Y. Ishii, Guerbet Reaction of Ethanol to n - Butanol Catalyzed by Iridium Complexes , Chem. Lett. 38 (2009) 838–839. https//doi.org/10.1246/cl.2009.838

[94]   T. Matsu-Ura, S. Sakaguchi, Y. Obora, Y. Ishii, Guerbet reaction of primary alcohols leading to β-alkylated dimer alcohols catalyzed by iridium complexes, J. Org. Chem. 71 (2006) 8306–8308. https//doi.org/10.1021/jo061400t

[95]   S.A. El-Molla, Dehydrogenation and condensation in catalytic conversion of iso-propanol over CuO/MgO system doped with Li2O and ZrO2, Appl. Catal. A Gen. 298 (2006) 103–108. https//doi.org/10.1016/j.apcata.2005.09.029

[96]   G. Torres, C.R. Apesteguía, J.I. Di Cosimo, One-step methyl isobutyl ketone (MIBK) synthesis from 2-propanol: Catalyst and reaction condition optimization, Appl. Catal. A Gen. 317 (2007) 161–170. https//doi.org/10.1016/j.apcata.2006.10.010

[97]   O. Kose, S. Saito, Cross-coupling reaction of alcohols for carbon-carbon bond formation using pincer-type NHC/palladium catalysts, Org. Biomol. Chem. 8 (2010) 896–900. https//doi.org/10.1039/b914618k

[98]   S. Musa, L. Ackermann, D. Gelman, Dehydrogenative cross-coupling of primary and secondary alcohols, Adv. Synth. Catal. 355 (2013) 3077–3080. https//doi.org/10.1002/adsc.201300656

[99]   I.S. Makarov, R. Madsen, Ruthenium-catalyzed self-coupling of primary and secondary alcohols with the liberation of dihydrogen, J. Org. Chem. 78 (2013) 6593–6598. https//doi.org/10.1021/jo4008699

[100] K. Shimura, K. Kon, S.M.A. Hakim Siddiki, K.I. Shimizu, Self-coupling of secondary alcohols by Ni/CeO2 catalyst, Appl. Catal. A Gen. 462–463 (2013) 137–142. https//doi.org/10.1016/j.apcata.2013.04.040

[101] C.S. Cho, B.T. Kim, H.-S. Kim, T.-J. Kim, S.C. Shim, Ruthenium-Catalyzed One-Pot β-Alkylation of Secondary Alcohols with Primary Alcohols, Organometallics. 22 (2003) 3608–3610. https//doi.org/10.1021/om030307h

[102] S.C. Chan, X.R. Wen, C.S. Sang, Pd/C-catalyzed oxidative alkylation of secondary alcohols with primary alcohols, Bull. Korean Chem. Soc. 26 (2005) 1611–1613. https//doi.org/10.5012/bkcs.2005.26.10.1611

[103] G.R.A. Adair, J.M.J. Williams, Oxidant-free oxidation: Ruthenium catalysed dehydrogenation of alcohols, Tetrahedron Lett. 46 (2005) 8233–8235. https//doi.org/10.1016/j.tetlet.2005.09.083

[104] R. Martínez, D.J. Ramón, M. Yus, RuCl2(DMSO)4 catalyzes the β-alkylation of secondary alcohols with primary alcohols through a hydrogen autotransfer process, Tetrahedron. 62 (2006) 8982–8987. https//doi.org/10.1016/j.tet.2006.07.012

[105] K.I. Fujita, C. Asai, T. Yamaguchi, F. Hanasaka, R. Yamaguchi, Direct β-alkylation of secondary alcohols with primary alcohols catalyzed by a Cp*Ir complex, Org. Lett. 7 (2005) 4017–4019. https//doi.org/10.1021/ol051517o

[106] A. Prades, M. Viciano, M. Sanaú, E. Peris, Preparation of a series of "Ru(p-cymene)" complexes with different N-heterocyclic carbene ligands for the catalytic β-alkylation of secondary alcohols and dimerization of phenylacetylene, Organometallics. 27 (2008) 4254–4259. https//doi.org/10.1021/om800377m

[107] M. Viciano, M. Sanaú, E. Peris, Ruthenium janus-head complexes with a triazolediylidene ligand. Structural features and catalytic applications, Organometallics. 26 (2007) 6050–6054. https//doi.org/10.1021/om7007919

[108] A.P. Da Costa, M. Viciano, M. Sanaú, S. Merino, J. Tejeda, E. Peris, B. Royo, First Cp*-functionalized N-heterocyclic carbene and its coordination to iridium. Study of the catalytic properties, Organometallics. 27 (2008) 1305–1309. https//doi.org/10.1021/om701186u

[109] K.I. Shimizu, R. Sato, A. Satsuma, Direct C-C cross-coupling of secondary and primary alcohols catalyzed by a γ-alumina-supported silver subnanocluster, Angew. Chemie - Int. Ed. 48 (2009) 3982–3986. https//doi.org/10.1002/anie.200901057

[110] X. Liu, R.S. Ding, L. He, Y.M. Liu, Y. Cao, H.Y. He, K.N. Fan, C-C cross-coupling of primary and secondary benzylic alcohols using supported gold-based bimetallic catalysts, ChemSusChem. 6 (2013) 604–608. https//doi.org/10.1002/cssc.201200804

[111] R. Cano, M. Yus, D.J. Ramón, First practical cross-alkylation of primary alcohols with a new and recyclable impregnated iridium on magnetite catalyst, Chem. Commun. 48 (2012) 7628–7630. https//doi.org/10.1039/c2cc33101b

[112] Y. Obora, S. Ogawa, N. Yamamoto, Iridium-catalyzed alkylation of methylquinolines with alcohols, J. Org. Chem. 77 (2012) 9429–9433. https//doi.org/10.1021/jo3019347

[113] C. Chaudhari, S.M.A. Hakim Siddiki, K.I. Shimizu, Alkylation of 2-methylquinoline with alcohols under additive-free conditions by Al2O3-supported Pt catalyst, Tetrahedron Lett. 54 (2013) 6490–6493. https//doi.org/10.1016/j.tetlet.2013.09.077

[114] S. Bähn, S. Imm, L. Neubert, M. Zhang, H. Neumann, M. Beller, The catalytic amination of alcohols, ChemCatChem. 3 (2011) 1853–1864. https//doi.org/10.1002/cctc.201100255

[115] A.C. Marr, Organometallic hydrogen transfer and dehydrogenation catalysts for the conversion of bio-renewable alcohols, Catal. Sci. Technol. 2 (2012) 279–287. https//doi.org/10.1039/c1cy00338k

[116] A. Corma, T. Ródenas, M.J. Sabater, A bifunctional PdVMgO solid catalyst for the one-pot selective N-monoalkylation of amines with alcohols, Chem. - A Eur. J. 16 (2010) 254–260. https//doi.org/10.1002/chem.200901501

[117] J.W. Kim, K. Yamaguchi, N. Mizuno, Heterogeneously catalyzed selective N-alkylation of aromatic and heteroaromatic amines with alcohols by a supported ruthenium hydroxide, J. Catal. 263 (2009) 205–208. https//doi.org/10.1016/j.jcat.2009.01.020

[118] K. Yamaguchi, J. He, T. Oishi, N. Mizuno, The "borrowing hydrogen strategy" by supported ruthenium hydroxide catalysts: Synthetic scope of symmetrically and unsymmetrically substituted amines, Chem. - A Eur. J. 16 (2010) 7199–7207. https//doi.org/10.1002/chem.201000149

[119] K. Shimizu, M. Nishimura, A. Satsuma, γ-alumina-supported silver cluster for N-benzylation of anilines with alcohols, ChemCatChem. 1 (2009) 497–503. https//doi.org/10.1002/cctc.200900209

[120] L. He, X.B. Lou, J. Ni, Y.M. Liu, Y. Cao, H.Y. He, K.N. Fan, Efficient and clean gold-catalyzed one-pot selective n-alkylation of amines with alcohols, Chem. - A Eur. J. 16 (2010) 13965–13969. https//doi.org/10.1002/chem.201001848

[121] W. He, L. Wang, C. Sun, K. Wu, S. He, J. Chen, P. Wu, Z. Yu, Pt-Sn/γ-Al2O3-catalyzed highly efficient direct synthesis of secondary and tertiary amines and imines, Chem. - A Eur. J. 17 (2011) 13308–13317. https//doi.org/10.1002/chem.201101725

[122] H. Ohta, Y. Yuyama, Y. Uozumi, Y.M.A. Yamada, In-Water Dehydrative Alkylation of Ammonia and Amines with Alcohols by a Polymeric Bimetallic Catalyst, Org. Lett. 13 (2011) 3892–3895. https//doi.org/10.1021/ol201422s

[123] R. Martínez, D.J. Ramón, M. Yus, Selective N-monoalkylation of aromatic amines with benzylic alcohols by a hydrogen autotransfer process catalyzed by unmodified magnetite, Org. Biomol. Chem. 7 (2009) 2176–2181. https//doi.org/10.1039/b901929d

[124] P.R. Likhar, R. Arundhathi, M.L. Kantam, P.S. Prathima, Amination of alcohols catalyzed by copper-aluminium Hydrotalcite: A green synthesis of amines, European J. Org. Chem. (2009) 5383–5389. https//doi.org/10.1002/ejoc.200900628

[125] T. Yamakawa, I. Tsuchiya, D. Mitsuzuka, T. Ogawa, Alkylation of ethylenediamine with alcohols by use of Cu-based catalysts in the liquid phase, Catal. Commun. 5 (2004) 291–295. https//doi.org/10.1016/j.catcom.2004.03.004.s

[126] X. Yu, C. Liu, L. Jiang, Q. Xu, Manganese dioxide catalyzed N-alkylation of sulfonamides and amines with alcohols under air, Org. Lett. 13 (2011) 6184–6187. https//doi.org/10.1021/ol202582c

[127] M.M. Reddy, M.A. Kumar, P. Swamy, M. Naresh, K. Srujana, L. Satyanarayana, A. Venugopal, N. Narender, N-Alkylation of amines with alcohols over nanosized zeolite beta, Green Chem. 15 (2013) 3474–3483. https//doi.org/10.1039/c3gc41345d

[128] K.I. Shimizu, N. Imaiida, K. Kon, S.M.A. Hakim Siddiki, A. Satsuma, Heterogeneous Ni catalysts for N-alkylation of amines with alcohols, ACS Catal. 3 (2013) 998–1005. https//doi.org/10.1021/cs4001267

[129] K.I. Shimizu, K. Shimura, M. Nishimura, A. Satsuma, Silver cluster-promoted heterogeneous copper catalyst for N-alkylation of amines with alcohols, RSC Adv. 1 (2011) 1310–1317. https//doi.org/10.1039/c1ra00560j

[130] L. Wang, W. He, K. Wu, S. He, C. Sun, Z. Yu, Heterogeneous bimetallic Pt-Sn/γ-Al2O3 catalyzed direct synthesis of diamines from N-alkylation of amines with diols through a borrowing hydrogen strategy, Tetrahedron Lett. 52 (2011) 7103–7107. https//doi.org/10.1016/j.tetlet.2011.10.100

[131] F. DE ANGELIS, M. GRASSO, R. NICOLETTI, N -Alkylation of Indole with Secondary Alcohols, Synthesis (Stuttg). 1977 (1977) 335–336. https//doi.org/10.1055/s-1977-24388

[132] S.M.A. Hakim Siddiki, K. Kon, K.I. Shimizu, Selective N-alkylation of indoles with primary alcohols using a pt/HBEA catalyst†, Green Chem. 17 (2015) 173–177. https//doi.org/10.1039/c4gc01419g

[133] L. He, J.Q. Wang, Y. Gong, Y.M. Liu, Y. Cao, H.Y. He, K.N. Fan, Titania-supported iridium subnanoclusters as an efficient heterogeneous catalyst for direct synthesis of quinolines from nitroarenes and aliphatic alcohols, Angew. Chemie - Int. Ed. 50 (2011) 10216–10220. https//doi.org/10.1002/anie.201104089

[134] W.J. Boyle, F. Mares, Rhodium and Molybdenum Complexes as Catalysts for Conversion of Nitrobenzene and Aliphatic Alcohols to Alkylquinolines, Organometallics. 1 (1982) 1003–1006. https//doi.org/10.1021/om00067a020

Emerging Applications of Nanomaterials
Materials Research Foundations 141 (2023) 75-100

Materials Research Forum LLC
https://doi.org/10.21741/9781644902295-4

Chapter 4

Emerging Nano-Enable Materials in the Sports Industry

Khandaker Tanzim Rahman[2], Tanvir Siddike Moin[2], Mohammed Farhad Mahmud Chowdhury[3], M. Nuruzzaman Khan[1,*]

[1]Department of Applied Chemistry and Chemical Engineering, Faculty of Engineering and Technology, University of Dhaka, Dhaka, 1000, Bangladesh

[2]Institute of Leather Engineering and Technology, University of Dhaka, Dhaka, 1000, Bangladesh

[3]Bangladesh University of Textiles, Dhaka, Bangladesh

*mnuruzzaman.khan@du.ac.bd

Abstract

Within the last decade, there have been numerous innovations in integrating nanotechnology for various sports applications. Nanotechnology is a technology discipline that deals with dimensions and tolerances of less than 100 nanometers, particularly manipulating single atoms and molecules. Nanotechnology has profoundly impacted sports competition like any other revolutionary innovation in materials science. The emergence of nanomaterials and nanotechnology has greatly improved athlete's performance. Sports equipment is becoming more humanized with the development of nanotechnology because it is more convenient, protective, and rational. Commercial nanotechnology-enable sports products, including ski goggle, ski wax, tennis racket, tennis ball, golf ball, bicycle, sportswear, shoe, and more, have offered several benefits to sports sectors in comparison to traditional sports equipment and clothes. These have enhanced the athletes' performance to multifunctional features of sportswear including water resistance, anti-microbial, anti-odor, anti-stain, anti-UV, heat, and cold resistance. This chapter has focused on minimizing the gap between understanding the scientific implications and applications of nanotechnology to sports equipment and clothing, and the characterization and impact of nanomaterials in the sports industry.

Keywords

Nanotechnology, Nanomaterials, Sports Equipment, Sportswear

Contents

Emerging Applications of Nanomaterials   Materials Research Forum LLC
Materials Research Foundations 141 (2023) 75-100   https://doi.org/10.21741/9781644902295-4

1.   Introduction

How much money is spent on sports searching for national pride of a country? Approximately a billion dollars are poured into sports each year in developed countries. Britain, which won 67 medals, spent 274.46 million pounds ($376 million) on British athletes in Rio de Janeiro [1], while China's General Administration of Sports had a budget of 10bn yuan ($1.5bn) in 2019 [2]. There was a strong positive correlation (0.80) between the absolute amount of money invested in elite sports, and the absolute number of medals won in Tokyo 2020 [3]. Countries typically splurge on their athletic teams to win medals. Nevertheless, how has this improved their athlete's performance? Does nanotechnology offer athletes more than a sporting chance?

Nanotechnology encompasses the technology of manipulation, production, and application of entities at scales ranging from individual atoms or molecules to around 100 nanometers. This "i" dot alone can encompass 1 million nanoparticles. They become more chemically reactive due to a relatively larger surface-area-to-mass ratio. Furthermore, below 50 nm, the laws of classical physics give way to quantum phenomena, resulting in different optical, electrical, and magnetic behaviors.

Nanomaterials have different material properties than bulk materials because of the high surface area to volume ratio, allowing more atoms to interact with other materials and potentially develop nanoscale quantum effects. Different nanomaterials such as carbon nanotubes (CNTs), carbon nanofibres (CNFs), carbon nanoparticles (CNPs), silica nanoparticles (SNPs), fullerenes, nano-clay, and nano-nickel, etc. are being incorporated into various sports equipment to improve the performance of athletes. In sporting equipment, these nanomaterials are responsible for added advantages such as high strength and stiffness, durability, reduced weight, abrasion resistance, etc. [4]. These also offer several properties such as waterproof, breathability, comfort, anti-bacterial, anti-fungal, UV-protection, photocatalytic, self-cleaning, flame retardant, UV shielding, protection from heat and cold, enhanced blood circulation, etc. So, nanomaterials incorporated in sports equipment have become much more durable, with the added benefit of tailoring the mass distribution of the individual piece of equipment.

In the case of sportswear, the nanomaterial is used as nanofibres, nanocomposite fibres, nano-finishing, etc. Under the nanocomposite fibres, metal matrix composites, ceramic matrix composites, and polymer matrix composites are prominent. Nanofinishing has sparked fresh ideas for various applications that would be difficult to complete using traditional methods. On the other hand, carbon nanotubes (CNTs) contribute a significant share of sports equipment making, along with nanosilicate, fullerenes, nanoclay, and nanotitanium. With expanding importance and a plethora of studies in this field, new methods and concepts for making nanotechnology sports products are in high demand [5]). Current research on nanomaterials for the sports industry is based on carbon nanotubes (CNTs), carbon nanofibres (CNFs), nanoclays, and nanocomposites. However, more research is needed in this field to understand these materials and their properties better.

This chapter discussed nanotechnology enable sportswear, sports equipment, and current advances in nanotechnology based sports sectors, which could provide a new dimension to engineering and innovation. This chapter also briefly illustrated the latest publications, scientific newsletters, and technical product brochures related to sports industries. The reader will meticulously go through the know-how, applications, and future impact of nanotechnology in this sector. The special properties developed by nanoengineering in sports materials are also discussed. Several nanomaterials and nanocomposites based sportswear, sports shoe, and sporting equipment are also discussed in this chapter. A short technical brief on the instruments, e.g., Scanning Electron Microscopy (SEM), X-ray diffraction analysis (XRD), UV-spectroscopy, and Transmission electron microscopy (TEM), that characterize nanomaterials used in sports industries and different methods of testing are discussed. Finally, the safety aspect of the nanomaterials used in sports industries and their associated challenges in the sports industry and future research directions for nanotechnology are also examined for sustainable growth of this emerging sector.

2.   Properties of nano-enable sports materials

Sporting goods containing different nanomaterials possess higher performance, increased flexibility and durability, and lightweight. The athlete's performance is also influenced by the equipment, protecting them from danger. The following section thoroughly discusses some of the most important properties of nano-enabled sports clothes and shoes. The properties of nanotechnology-enhanced sports textiles and equipment are illustrated in Figure 1.

2.1   Waterproof

The invention of water-resistant, breathable textiles that protected from wind, rain, and the loss of body heat propelled the expansion of sportswear. Water resistant fabrics keep water out of the breathable fibres while enabling water vapor to pass through. Breathable fabrics include tightly woven fabrics, microporous membranes, coatings, and smart breathable fabrics. The actual waterproof breathable fabric was made from densely woven cloth or polymeric and resin coating. On the other hand, nanotechnology has opened up new possibilities for developing water-resistant, breathable fabrics [6]. A waterproof polyester fabric coated with nanosilicate was able to stay dry after two months in the water due to the reduced resistance between water and fabric. Swimsuits are in high demand as a result. LSMZ™ are anti-bacterial, water-resistant, permeable sports shoes made of a nano-membrane of TiO_2 and ZnO nanostructured materials in fluorinates perfluoroalkyl substrate with Lotus effect and single-directed water vapor ability [7].

2.2   Anti-bacterial property

Sweating in sports activity creates an exceptional environment for microbes to thrive and emit an undesirable stench. Infectious disease-causing bacteria such as Staphylococcus aureus is commonly discovered in sports teams. Anti-bacterial athletic apparel can thus protect players from microbes and bad odors. This also keeps the apparel from decaying and getting damaged. Sportswear and socks were made with chitosan based fibres that have moisture-controlling, anti-bacterial, and anti-fungal properties [8]. The major anti-bacterial mechanism is to destroy microbe's lipids, proteins, and DNA. In most cases, the anti-bacterial effect is achieved using nanomaterials during the finishing stage or incorporating nanoparticles into fibres during the spinning process. Anti-bacterial finishes for textile fabrics have included organic metals, quaternary ammonium compounds, and organic silicones. However, due to the small particle size and high specific surface area, Ag-NPs are commonly used in anti-bacterial sports gear, sports shoes, and insoles [9]. For example, anti-bacterial sports clothes and shoes have been made using Silverclear™ as a disinfectant and anti-bacterial materials [7]. Anti-bacterial sportswear must have a quick onset of action against microorganisms. According to a study conducted by the Hohenstein Institute in Germany, efficacy should remain permanent during sports activity, excluding skin reactions (allergy, irritation) or harming skin microbiota should also be carefully considered [10].

2.3   UV protection

Nanoparticles with UV-protection properties in sportswear have been developed in response to the significant risk of UV-related skin lesions, particularly in outdoor activities. Among the organic and inorganic UV-protective materials, semiconductors such as TiO_2, ZnO, SiO_2, and Al_2O_3 have gained greater attention. They are chemically stable, inexpensive, readily available, and non-toxic. Rather than reflecting or scattering, UV protection is primarily concerned with the potential of UV radiation absorption [11]. The homogeneous dispersion of nanoparticles on fabric surfaces can improve the performance of UV-blocking compounds. Multilayered fabric systems with a very light coating of zinc oxide polyurethane nanocomposite fibres web could be used in outdoor sportswear to give UV protection and anti-bacterial properties [12].

2.4   Self-cleaning

The invention of self-cleaning sports clothes is one of the successful applications of nanotechnology in the sports textile sector. Photocatalytic nanoparticles like TiO_2 and ZnO are known for producing hydrophilic surfaces that clean themselves. The photocatalytic activity is responsible for the photo-excitation of semiconductor nanoparticles under light irradiation with energy greater than or equal to their bandgap. An electron-hole pair is formed between the valence and conduction bands [13]. The photoinduced electrons may be transported to oxygen. For this, superoxide, hydroperoxy (HOO-), and hydroxyl (OH⁻) radicals may be produced. The generated hydroxyl radicals provide self-cleaning and anti-stain qualities with an oxidizing role. Sportswear could benefit from photocatalytic self-cleaning textiles. There are various types of self-cleaning textiles available commercially. For instance, "Nanosphere™" is a self-cleaning fabric produced by Schoeller Textile AG for athletic activities and climbing tents [14].

2.5   Heat and cold protection

Sports activities like skiing, snowboarding, diving, mountaineering, cycling, etc., require sportswear with heat and cold insulation properties. The relationship between physical body heat, ambient factors, and physical activity should be considered. The incorporation of phase change materials (PCMs) in textiles has recently gained popularity due to the development of thermo-regulated smart textiles [15].

PCMs include organic chemicals such as alkyl hydrocarbons and non-paraffin materials such as fatty acids, glycolic acids, alcohols, and inorganic materials such as hydrated inorganic salts. PCM nanocapsules have smaller particle sizes, and faster heat transfer speed than microcapsules and are more suitable for such applications. Commonly used shells encapsulating PCMs are urea-formaldehyde, melamine-formaldehyde, diacid silicone, polymethacrylate, and polystyrene. The PCM nanocapsules are added at the fibre spinning stage or coated on textile fibres in a finishing process using a polyurethane binder [16]. Composite electrospinning and coaxial electrospinning are used to make PCM based nanofibres [17]. PCMs in sportswear can absorb and release surplus body heat during athletic exercise, reducing thermal heat. PCMs in sportswear could absorb heat as soon as the temperature rises and release it the instant it drops. The number of PCMs used, and the amount of thermal insulation provided by clothing is determined by the type and length of physical activity.

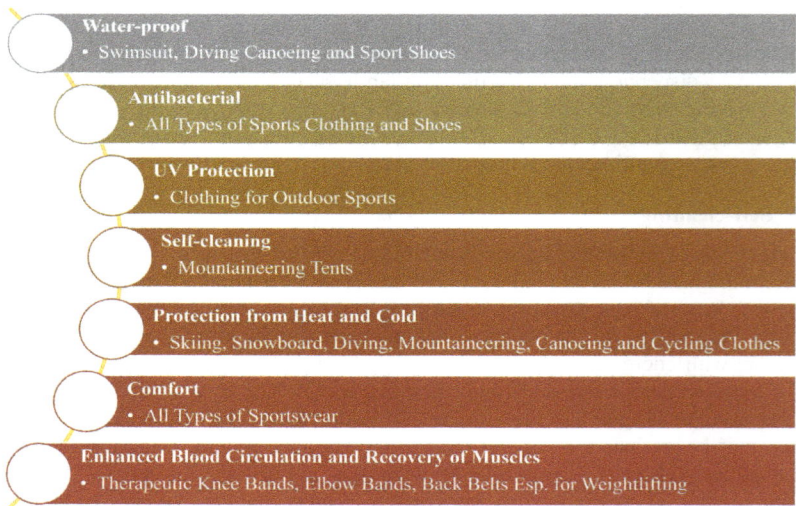

Figure 1: Properties of nanotechnology-enhanced sports textile and equipment.

2.6   Well-being

Fabrics that can transport moisture from the wearer's skin to the material surface and subsequently release it to the atmosphere are known as moisture transport fabrics [18]. These functional fabrics convey sweat and moisture from the wearer's skin to the material's outer surface, keeping the body dry and comfortable and preventing polyester cloth from adhering to the wearer's body. Improved moisture control was observed using environmentally friendly plasma technology [19]. It has been designed as functional sportswear with hydrophobic exterior and hydrophilic interior sides. This design was realized to produce useful, intelligent, and comfortable sportswear that absorbs and quickly releases perspiration from the body. Although weaving separate hydrophobic or hydrophilic threads on the inner and outer surfaces of the garment and fabric treatment with hydrophilic or hydrophobic chemicals are offered for manufacturing functional sportswear [20].

3.   Nanomaterials and nanocomposites in sportswear and sports shoe

There is no exact classification of nanomaterial used in the sports industry. These may be classified according to their end applications, properties, or source of nanomaterials. Nanomaterials used in sportswear can be divided into three categories: nanofibres, nanocomposite fibres, and sportswear nano-finishing, which are all described in brief in the next sections.

3.1   Nanofibres

When polymer fibre diameter shrinks from micrometres like 10–100 μm to nanometres, amazing characteristics like excellent instrumental features and the large area flexibility in surface functionalities, compared with the other known style of the fabric, have appeared. Nanofibres having less than 100 nm diameters and excellent instrumental features provide mass applications, including sports apparel and shoes like waterproof and windproof sportswear for canoeing, cycling, and mountaineering [21]. They are manufactured by electrostatic spinning methods, interfacial polymerization, and electrospinning. Carbon nanotubes (CNTs), like single-walled nanotubes (SWNTs), multi-walled nanotubes (MWNTs), and carbon nanofibres (CNFs), provide several unique mechanical and electronic features [7]. The incorporation of CNTs into electrospun nanofibres was reported [22, 23], where dielectrophoretic force is predicted to align CNTs or their bundles. Therefore, high shear forces are induced in the polymer solution during the spinning.

3.2   Nanocomposite fibres

Nanocomposite fibres are manufactured using nanofillers like carbon nanotubes, nano-clay, and metal oxides. Nanoparticles provide a large surface area and interact with polymer matrix. For improved mechanical and physical properties, the nanofillers are incorporated with polymer matrix. Furthermore, the addition of organic surfactants and compatibilizers improves it. As a result, nanocomposite materials having the desired physical and mechanical properties viz, toughness, and abrasion resistance are manufactured. The fibres are given multifunctional qualities such as conductivity, anti-microbial, and anti-static capabilities [23]. Carbon fibres are used to manufacture various sporting equipment, such as tennis and badminton racquets. Strong strings add to the impact strength of the racquets. They are 50% more resistant than traditional carbon racquets. They minimize vertical bending and twisting of the structure when the ball strikes it. CNT also produces baseball bats that are more durable, lighter, and play better [5]. Nanocomposites are divided into three types: a) metal matrix composites, b) ceramic matrix composites, and c) polymer matrix composites.

3.2.1 Metal matrix composites (MMCS)

Metal and ceramic particles are used to produce metal matrix composites. The application of MMCs can improve a wide range of properties over alloys. Ceramic particles and fibres made of boron, carbon, and other metals produce reinforcing phases in MMCs. Metal matrix composites are commonly employed in a range of applications. Composite materials in this category include graphite-coated cast iron, steel with a high carbide content, and tungsten carbides, which are made up of carbides and metallic binders. Metal matrix composites become appealing to use as constructional and functional materials when the property profile of traditional materials either fails to fulfill the higher criteria of specific demands or is the solution to the problem [24].

3.2.2 Ceramic matrix composites

The majority of the degree in this set of composites is taken up by a ceramic, which comprises oxides, bromides, nitrides, and silicides. Metal is usually incorporated as a second component. To get particular nanoscopic properties, both the metallic and ceramic components must be finely dispersed in each other. Nanocomposites with increased optical, electrical, and magnetic properties, tribological, corrosion resistance, and other protective properties came from these combinations [25].

3.2.3 Polymer matrix composites

The incorporation of nanoparticles to a polymer matrix can increase its performance when the nature and amount of NPs are selected appropriately. Then this material is better described as "nanofilled polymer composites". When the filler is well dispersed and thus has significantly different or better properties than those of the matrix, such as reinforcing a polymer matrix with much stiffer ceramic, clay, or carbon nanotube nanoparticles, this strategy is especially effective in producing high-performance composites [26].

3.3 Nano-finishing

Certain properties are infused into fabrics at the finishing step in the sports textile industry by modifying their appearance and/or enhancing their resistance to water, physical, chemical, biological, mechanical, and general wear. Nano-finishing has sparked fresh ideas for various applications that would be difficult to complete using traditional methods. It has enormous potential in a wide range of applications [27]. The utilization of nanoparticles on textiles is a growing area of interest in nano-finishing. By examining the properties of the nanoparticle, these are applied to fabrics with a wide range of practical performances. Nanocomposites are used to finish fabric materials, which will result in multifunctional textiles [28]. Nano-finishing is classified in two ways:

3.3.1 Ex-situ

The material is manufactured with the precursor of nanoparticles by several common finishing procedures. Here, traditional ex-situ finishing techniques such as exhausting, spraying, cushioning, foaming, and printing can be used. The ex-situ finishing can be divided into three common essential steps; dispersion of nanoparticle, then application on textile, and final fixing.

3.2.2 In-situ

The in-situ finishing of textiles with nanostructured materials is evolved as a possible alternative to complex and multi-step ex-situ methods. In this technique, the nanostructured materials are synthesized from their precursors in the presence of textile fabric as a facile one-step procedure. So, this technique eliminates the need for final fixation. Here, mechanical properties are improved through physical and/or chemical interactions between nanoparticles and fibres [29]. Many researchers have created various multifunctional textile materials by adding nanoparticles. For instance, Montazer and his research group investigated the in-situ synthesis of various nanoparticles, including silver, TiO_2, ZnO, TiO_2/Ag, and ZnO/Ag nanocomposites. Self-cleaning and anti-bacterial characteristics are among the major attributes of textiles made using these nanoparticles [30].

4. Nanomaterials and nanocomposites in sports equipment

Like any other revolutionary innovation in materials science, nanotechnology has had a significant impact on the level of competitiveness in sports. Nanocomposites have several unique qualities that make them extremely versatile with a broad array of everyday products and industrial applications [31]. Nanocomposites and nanostructures have gained popularity in athletic materials and equipment including sports clothes, golf, tennis balls, rackets, skiing, tennis balls, archery, cycling, etc. An overview of nanotechnology in sports is shown in Figure 2. Nanomaterials like silica nanoparticles (SNPs), nanoclays, carbon nanotubes (CNTs), and fullerenes have unique features such as high rigidity or stiffness, durability, lighter weight, or resilience to abrasion or wear. Table 1 outlines the application of different nanomaterials in the sports industry. The application of nanomaterials in different sports equipment is shown in Figure 3.

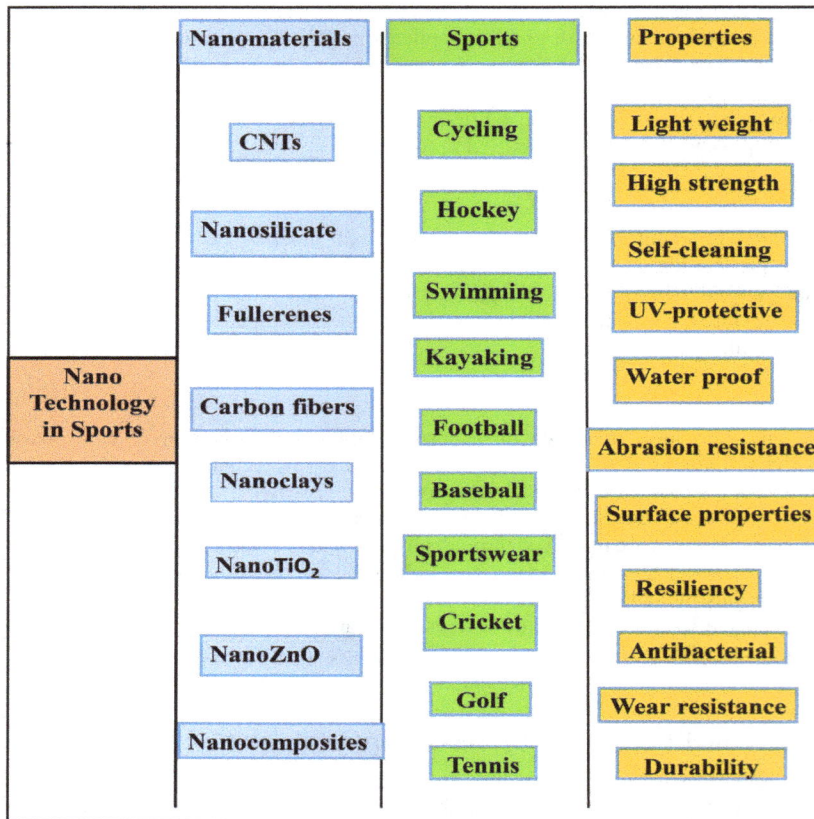

Figure 2 An overview of nanotechnology in sports.

4.1 Carbon nanotubes (CNTs)

Carbon nanotubes (CNTs) are the most widely utilized nanomaterial in sporting goods, comprising 14% of overall CNT use per year. Carbon nanotubes (CNTs) are extremely strong and stiff. They are stronger than steel, as stiff as diamond, yet six times lighter, making them ideal for high-strength, lightweight equipment such as ultra-lightweight bicycle frames and golf club shafts. Water sports like canoeing and boat racing have benefited from GO and Buckypaper (sheets of carbon nanotubes) (32). Glass fibre reinforced plastic modified with CNTs are being used in sports equipment for better mechanical properties [33]. In the sports of cycling, tennis, and pole vaulting, carbon fibre

Materials Research Forum LLC
https://doi.org/10.21741/9781644902295-4

reinforced epoxy composites are used. The thermal qualities, interfacial properties, mechanical properties, and fatigue resistance of sports equipment made of carbon fibre reinforced epoxy composite materials have increased significantly [34]. A flexible thermoplastic polyurethane/multi-walled carbon nanotubes (TPU/MWCNTs) strain sensor can detect an athlete's breathing type, frequency, and intensity [35]. Carbon nanotubes (CNTs) have long been utilized in rubber products and tire formulations to improve handling and grip, making them ideal for racing and sports vehicle tires [36].

4.2 Silicon nanoparticles

A waterproof polyester fabric coated with nanosilicate had developed to keep dry for two months of soaking and is popular for swimwear [7, 37]. There is a danger of UV-protection-related skin lesions in outdoor activities. To reduce this risk, nanoparticles that have UV protective properties, such as SiO_2, Al_2O_3, ZnO, and TiO_2, have been garnering greater attention due to their chemical durability, cheap cost, accessibility, and non-toxicity. Ultra-violet protection is based on UV light absorbance rather than reflection and/or scattering [11]. For instance, nanopaints containing self-cleaning, anti-bacterial, and anti-stain SiO_2 nanoparticles that clean the interior atmosphere, isolate the area, and shelter it from UV radiation from the sun are used to paint the stadium walls. Silica nanoparticles are added to fill voids between carbon fibres to strengthen fishing rods without adding weight.

4.3 Fullerenes

It is used in badminton and tennis sport to lower the weight and twisting of the racquets frame and control vibration. It is also used in cycling to reduce the weight and stiffness of bicycles. In archery sport, it provides better vibration control in arrows. In bowling, it reduces the chipping and cracking of balls. Nanomaterials like fullerenes, CNTs, cellulose nanofibres, SNPs, and nanoclays improve the inherent properties of various sporting equipment and materials [40].

4.4 Nanoclays

Nanoclays in shoe soles have the potential to alter lower extremity joint power flows during the stance phase of running [41]. Using nanoclay, an optimal anti-bacterial, anti-fungal cotton fabric with increased lightness, handling, softness, and surface properties was effectively achieved [42]. Nanoclays linings are used in sports equipment like tennis balls and footballs to function as a shield, maintaining consistent pressure inside the ball and enabling prolonged gameplay. In golf, nanoclay boosts the ball's resilience and bounce. It is also used in watercraft to lighten the load and increase the speed of the boat [43]. Nanohybrid like poly(ethylene terephthalate) (PET) and organically modified nanoclay are used in sports equipment. Due to its translucent nature, lightweight, and improved mechanical qualities, PET has various applications in numerous sectors, such as vehicle components, sports, and textiles [44].

4.5   Zinc-oxide nanoparticles

These are used to make T-shirts, sportswear, shoe insoles, socks, or facial cleansing wipes due to its anti-microbial property [54]. ZnO nanorods are used for self-cleaning, anti-microbial activity, and high tensile properties [55]. Single-directed nanofunctional shoes were created using a nanofibrous membrane of nanocomposite TiO_2 and ZnO material [59]. UV protection and self-cleaning qualities are also provided via a multilayered polydimethylsiloxane-ZnO-polydimethylsiloxane composite coating on cotton fabric [60]. A highly hydrophobic ZnO NPs treatment on sports fabric has the potential use as a composite in lowering drag in sportswear designs and vehicle movements on or underwater [56].

Table 1. Application of different nanomaterials in the sports industry.

Nanomaterials	Sport	Benefits	References
Carbon nanotubes	Cycling	It makes the cycle body frame lightweight with increased strength.	[32]
	Sportswear	Produces supercapacitors for self-powering textile energy.	[45]
	Hockey	It makes the stick lightweight with increased strength.	[46]
	Kayaking	Enhance abrasion in kayaks	[47]
	Car racing	Improve handling and grip.	[36]
Nano Silicate	Swimming	Waterproof polyester fabric	[7, 37]
	Outdoor games	UV-protective sportswear.	[11]
Fullerenes	Racket	Lighten weight and frame twisting	[39]
	Cycling	Lighter weight and rigidity of bicycles.	
	Archery	Provides better vibration control in arrows.	
	Bowling	Minimize chipping and cracking of balls	
Carbon Fibre	Fishing	Produce lighter and softer fishing rods.	[48]
	Vault poles	Increases flexibility with reduced weight and easy, flexible properties	[7]
	Baseball	Impart strong, lighter, and better-performing property	[49]
	Tennis racket	increase and improve energy damping	[50]

Nanoclays	Sport shoes	Lower extremity joint power flows during the stance phase of running	[41]
	Cotton fabric	Increases lightness, handling, softness, and surface properties were effectively achieved	[42]
	Football and cricket	Act as a barrier, enabling prolonged gameplay by maintaining steady pressure on the ball.	(41)
	Golf	Improve strength and bounce of the ball.	
Nano Titanium	Sportswear	Impart self-cleaning and anti-bacterial qualities	[28]
	Cotton fabrics	UV protection and self-cleaning capabilities	[51]
	Tennis	Provide superior elasticity and improve wear resistance	[52]
	Medical treatment	Repair of the tendon-bone tunnel	[53]
ZnO nanoparticles	Sportswear	Provide anti-microbial property	[54]
		ZnO Nano-rods are used for its self-cleaning, anti-microbial activity, and high tensile properties	[55]
		Lowering drag in sports	[56]
Nanocomposites	Sportswear	Impart anti-bacterial, acid resistant, and photocatalytic performance	[57]
	Weightlifting	Increase the durability and wear resistance of sports equipment and prevent sports injuries	[58]

4.6   Nanocomposites

In sportswear, nonmetal/metal doped hexagonally Ag/ZnO nanocomposites are used as nanofinishing on woolen fabric to impart anti-bacterial, resistant to acid and photocatalytic properties [57]. Deposition of $TiO_2/Fe_3O_4/Ag$ nanocomposites on a sonochemical process on polyester sports fabric surfaces was particularly successful in imparting anti-bacterial, photoactive, and high tensile strength to polyester fabrics [61]. Using pulse electrodeposition Ni/TiN- SiC nanocomposite was deposited on the dumbbell surface to increase sports equipment's durability and wear resistance and prevent sports injuries [58]. Fibre/epoxy composite was used as sports equipment due to its high impact resistance [62].

Emerging Applications of Nanomaterials   Materials Research Forum LLC
Materials Research Foundations 141 (2023) 75-100   https://doi.org/10.21741/9781644902295-4

5.   Analysis of nanomaterial's used in sports industries

In the sports industry, a wide range of nanomaterials are used, and the characterization of the materials requires very sophisticated analytical instruments. In this section, a brief overview of different characterization methods is discussed. Scanning electron microscope (SEM) is a versatile, sophisticated device with a high-energy electron, and the resulting electrons/X-rays are examined. These emitted electrons/X-rays provide information about a material's topography, morphology, composition, grain orientation, crystallographic information, and so on. Topography refers to an object's surface features, such as texture, smoothness, and roughness.

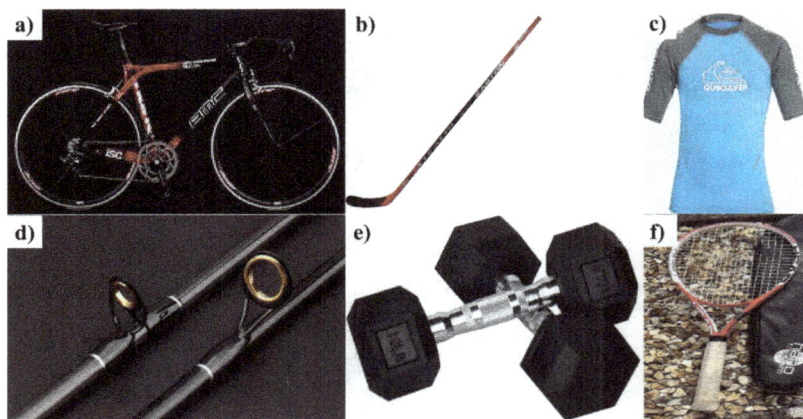

Figure 3. Application of nanomaterials in different sports equipment; a) CNTs bike, b) CNT grip composite stick, c) UV protective swimwear, d) carbon nanofibre fishing rod, e) Ni/TiN-SiC nanocomposite in dumbbell, f) nanotitanium-based tennis racket,

Morphology refers to an object's shape and size, but topography refers to its surface features or "how it looks" [63]. SEM is used in scanning the surface of fabric materials to find out what the surface is like, whether the surface is smooth or rough. For example, SEM analyzed PES fibres that underwent morphological alterations due to plasma for making anti-microbial sportswear. The smooth surface of the fibre may be seen by scanning the plasma treated PES fibres and untreated PES fibres. Due to plasma etching, the topography of the fibre was significantly altered. The fibre surface was knocked by plasma species that were energetic and extremely reactive, causing the fibre to erode. As a result, parallel to the fibre axis, unequal cracks, pits, and striations developed, increasing fibre surface roughness [64]. TiO_2-NPs are used to make high-performance sports fabric, and its deposition on fibre surface can be identify at higher resolution. The fibre surface topography can be observed after 1000 times magnification. For this unique feature of SEM, the unfinished sports polyester fibre surface is found smooth and sleek by comparing the images. In contrast, the finish on which TiO_2-NPs are deposited on the fibre surface varies randomly, with some aggregation formed due to TiO_2-NPs self-aggregating phenomenon. TiO_2-NPs were scattered randomly, resulting in an uneven surface [65].

In the X-ray diffraction technique, an X-ray and a high-energy electromagnetic wave are used to determine the degree of crystallinity of materials. Using closed tubes, synchrotron radiation sources, and spinning anodes X-rays can be generated. Sealed tubes and spinning anodes, both utilized in laboratory equipment, use the same principle to produce X-rays. After being formed by heating a tungsten filament in a vacuum chamber, electrons are accelerated by a high potential field and directed to a target, which creates X-rays. This X-ray collides with the material, and the diffracted beam is captured on a photographic film for later analysis [66]. Generally, XRD is used to identify the crystallinity nature of materials. For example, using XRD crystallinity of ZnO-NPs used in sports fabric to produce anti-microbial sportswear can be identified. The XRD patterns of the ZnO-NPs materials are quite similar to the primary diffraction peaks of zinc oxide's hexagonal crystalline structure. CuO-NPs produce small signals, suggesting the existence of the oxide rather than metallic copper. The crystallite diameter of CuO-NPs was calculated to be 30 nm on average [67].

The principle of UV-visible spectroscopy is based on the absorption of ultraviolet or visible light by chemical substances, leading to the production of various spectra. The interaction of light and matter is the basis of UV- Visible spectroscopy. It undergoes excitation and de-excitation absorbing medium light and producing a spectrum [68]. UV- visible spectroscopy shows results using the wavelength absorbed by the absorption medium. Different nanoparticles used in sports equipment absorb light of different wavelengths. For example, the absorption peak for ZnO-NPs produced with 5% soluble starch is 361 nm in the UV visible range. This ZnO-NPs is generally used for making functionalized sportswear [69].

Two phases are involved in the formation of a TEM image. The first involves a specimen scattering an incident electron beam. This scattered radiation is focused by an objective lens, which creates the primary image, which is then magnified further by additional lenses to create a highly magnified final image [70]. TEM is generally used to find out the particle size and shape. For example, the particle size and structural image of nanoparticles were determined using a High-Resolution Transmission Electron Microscope (HRTEM). It shows that the particle size of silica nanoparticles used to produce sports fabric with high physical strength is below 100 nm [71]. Different nanoparticles used in different sectors of sports equipment have different sizes. Like for MgO-7, MgO-6, MgO-5, MgO-4, MgO-3, MgO-2, MgO-1 the sizes are 69.1, 56.3, 47.3, 35.9, 26, 14.4, 7.6 nm respectively [72].

6. Methods for evaluating the characteristics of sportswear

Standard approaches are required to produce the various features of nanotechnology-enable sportswear. Table 2 summarizes the standard approach to investigate these properties briefly. The tests for waterproof breathable fabrics are carried out according to ISO 4920:2012 [73]. In this approach, a precise amount of distilled water is sprayed onto the fabric to test its resistance to surface wetting, compared to reference photographs. According to ISO 811: 1989 and BS EN 20811: 1992, the evaluation of water resistance based on hydrostatic pressure is also relevant [74, 75]. In this process, a constantly increasing water pressure is applied to one face of the cloth sample until penetration occurs. Water-vapor permeability resistance is an important property of sportswear and is investigated by standard test protocol [76]. Fabric weight change is linked to breathability in this way. A sample with a high breathability value is one with a value of more than 20,000 g/m^2 per day. The Bundesmann rainshower test was used to determine fabric water repellence in ISO 9865: 1991 and BS EN 29865: 1993. Other standard methods for waterproof breathable fabrics are AATCC 35-2006 and BS 5066: 1974 rain tests [78,79,80,81].

Table 2. The standard protocols for the assessment of nanotechnology enabled sportswear and shoe properties.

Property	Standard No.	Title
	ISO 811	Hydrostatic pressure test to determine resistance to water penetration.
	BS EN 20811	
	ISO 492	Fabric's resistance to surface wetting (spray test)
	ISO 15496	Determination of water vapor and permeability of textiles for quality control purposes.
	BS 5636	Determination of the permeability of fabrics to air
Waterproof and breathability	ASTM D737-04	
	BS EN ISO 9237	
	BS EN 29865	The Bandsman rainshower test is used to determine cloth water repellency
	ISO 9865	
	AATCC 35	Fabric's resilience to a fake shower test
	BS 5066	
	TM444	The construction of water-resistant footwear test
	STM 505	Water resistance test in footwear
	TM230	Water penetration test for footwear
	EN 13073	Water resistance for footwear
	TM77	Water penetration test
	AATCC 100	Textile materials with anti-bacterial treatments

Anti-bacterial	AATCC 147	Textile components with anti-bacterial treatments
	ISO 16187	Anti-bacterial activity for footwear
UV protection	AS/NZS 439 9	Evaluation and classification of sun protective apparel
	AATCC 183	The ability of fabrics to transmit or block erythemal weighted ultraviolet light.
	ASTM D6544	The ability of fabrics to transmit or block erythemal weighted ultraviolet light.
	TM160	The light fastness of footwear
Self-cleaning	ISO 27448	Semiconducting photocatalytic materials with self-cleaning properties.
	US20090313855A1	Self-cleaning outsoles for footwear
Hot and cold protection	ISO 20344	Personal protective equipment for footwear
	EN 12784	Thermal insulation test for the whole shoe
	ISO 15831	Clothing, physiological impacts, and the use of a thermal manikin to test thermal insulation
	STM 567 Endofoot	Testing thermal performance and moisture management
	TM436	Thermometric insulating value and cold rating of the entire shoe
Comfort	STD 223	The test methods and equipment available to assess footwear comfort
	SATRA STM 175	Permeability and Absorption Test
	SATRA TM183	Whole shoe cushion assessment test
	SATRA TM142	Falling mass shock absorption test
	SATRA TM194	Longitudinal stiffness of footwear test
	SATRA TM256	Torsional stiffness of footwear test
	SATRA TM190	Ground insulation index test

Color fastness to light is another most sought property of fabric, addressed by the TM160 method [81]. The method determines color resistance to the action of an artificial light source (Xenon lamp) that closely approximates natural daylight through the glass of higher intensity. The test for self-cleaning outsoles of shoes is described in US20090313855A1[82]. The instrument Endo-foot measures thermal insulation, moisture absorption, and breathability properties in conjunction with a hose to describe full footwear's thermal and sweat management performance by STM 567 [83]. The method TM436 is used for all forms of footwear with a closed upper and describes the assessment of real full shoe thermal insulation value and cold rating [84]. It also calculates full footwear thermal insulation value and cold rating with a hose.

Materials Research Forum LLC
https://doi.org/10.21741/9781644902295-4

The ground insulation index and absorption properties of footwear are determined by TM190 [85]. The STM175 absorption and permeability test equipment simulates the environment inside a shoe, which is 100% relative humidity at foot temperature [86]. The TM444 technique is responsible for non-flex induced water penetration resistance [87]. The method STM 505 determines how well footwear resists water infiltration while the wearer walks through it [88].

7. Current status of nanotechnology in sports

Although information technology is the wave of electricity, nanotechnology will certainly be the wave for the long term. Within the last few decades, nanotechnology has been evolving. Nanotechnology has numerous applications in every major industry, especially textiles, within the sportswear market. Several fabric companies have started to think about developing an oversized range of sports apparel by nanotechnology. A Swiss company named Scholler has developed a nano-based technology to produce clothing that is responsible for the self-cleaning property, wind and water resistance, air permeability, rain- and snow-repellent feature, and an optimal balance of comfort for extreme atmospheric conditions in sports such as mountaineering and skiing [89]. SoleFresh™ socks were developed by JR Nanotech, a UK-based company [90], and this was made with Ag-NPs to get rid of athlete's foot odor. Hyosung, a Korean corporation, produced nanosilver-based nylon fibres for everyday activities and sports, apparel, sports bags, and running shoes [91] . Nanotechnology gave footwear soles anti-slip qualities. In harsh climes, nanotechnology promotes wicking in sportswear, shielding mountaineers from the cold and rain and regulating breathable clothes' temperature [92]. The running shoes for MARS have been designed inspired by the SpaceX project. Each component of the running shoes has been carefully considered to ensure that it is space-ready. The upper is constructed of nylon and carbon nanotubes. The body's heat activates the nanotubes and contracts the nylon upper's fibres, allowing it to conform to the size of the foot gently. The rubber parts then enhance durability by strengthening the upper structure.

Various sporting equipment companies are developing innovative technologies/products, such as Wilson's Double Core™ and BLX™ (tennis), InMat's Air D-Fence™ and Nanolok™ (tennis), Holmenkol's nano-CFC® (skiing), Easton's N-FUSED™ (archery), St. Croix's NSi™ (fly-fishing), and Yonex's Nanopreme™ (golf) [93].

8. Concluding remarks and future aspects of nanotechnology in sports

Nanotechnology has improved many facets of sports engineering, including sports items, sports venues like a stadium, ground coverings, and sportswear. It has shown its vital role in the case of athlete's performance and sports venues, especially in the Olympic games. Nanotechnology-infused sportswear has multifunctional features such as resistance to water, heat, cold, bacteria, and unpleasant odors and being pliable and lightweight with high strength. Nanotechnology supports better performance and efficiency by improving the comfort of the wearer's clothing. Nanotechnology is used to treat athletes by developing biomimetic nano-biological ligament material used to repair tendon bone. It is also used in athletes' nutrition due to enhanced absorption and utilization. As nanomaterial-infused clothing is breathable, athletes may maintain high activity levels for longer periods. When compared to standard sports equipment, the sports sectors in which nanotechnology has brought benefits are improved athlete's comfort, better athlete's safety, higher athlete's performance, sportswear with greater strength and lightweight, sportswear with amazing properties such as anti-odor, water resistance, anti-microbial, anti-stain, thermal protection, UV protection, breathability, athletes nutrition, athletes medicare, improved sports equipment, improved sports venue etc. In recent years, there has been much study into nanotechnology to develop new ways and concepts for making sports items that utilize nanotechnology for higher performance. According to market research, sportswear has experienced tremendous development in recent years, with many creative goods being manufactured, like Nike's "Smart Shoes" through which an athlete can track his fitness. Consumers pay extra for these shoes because of their structural effectiveness and stylistic appeal, giving them a significant edge in pursuing their sport. The future strategy for developing improved performance items will be centered on addressing multifunctional criteria such as moisture control, abrasion resistance, temperature control, greater strength, lightweight, wear comfort, water resistance, as well as other aspects based on the kind of the sport, physical activity, and environmental conditions. Besides these advantages, nanotechnology in sports also has disadvantages. Nanoparticles can be harmful to the wearer's skin and environment. From the past to the present, only a few studies have been conducted on nanoparticles to determine their impact on health, safety issues, and long-term viability.

In comparison with nanocoated fabric nanofibres and nanocomposites are less toxic for humans. More studies need to focus on the bio-safety of high-performance sportswear that conforms with safety regulations, decreasing environmental issues associated with nanomaterials. Furthermore, after the possible hazard related to nano-enhanced sports items have been replaced, it is critical to closely manage their exit route, time of usage, and arrangement to avoid nanoparticles being discharged into the environment. Nanotechnology is a modern and exciting technology in a variety of fields that have altered many aspects of our daily lives, including sports. Nanoengineered sports equipment and materials are far greater strength, resistance, and durability than traditional sports equipment. Various businesses have been developing new goods and plan to incorporate nanotechnology into sports items. Many top athletes are opting for nano-enhanced athletic equipment to improve their performance. Even though nano-intensified sports equipment cannot be distributed to every athlete without ensuring that the athletes receive the same sports equipment, it relies on the human spirit and innate athletic potential to be faithful and tough to referee. As a result, the ultimate goal should be to put more faith in the frontier of execution than in investing in high-tech athletic equipment.

References

[1] Spending on Olympic medals pays off in swimming and track - Nikkei Asia.

[2] Ahmed M, Germano S. Tokyo Olympics Daily: Countries weigh medal hauls as Games draw to a close | Financial Times. 2021 Aug;

[3] De Bosscher V, Shibli S. TOKYO 2020 Evaluation of the elite sport expenditures and success in 14 nations. 2021.

[4] Ćibo M, Šator A, Kazlagić A, Omanović-Mikličanin E. Application and impact of nanotechnology in sport. In: IFMBE Proceedings. Springer, Cham; 2020. p. 349-62. https://doi.org/10.1007/978-3-030-40049-1_44

[5] Harifi T, Montazer M. Application of nanotechnology in sports clothing and flooring for enhanced sport activities, performance, efficiency and comfort: a review. J Ind Text [Internet]. 2017 Jan 28;46(5):1147-69. Available from: http://journals.sagepub.com/doi/10.1177/1528083715601512 https://doi.org/10.1177/1528083715601512

[6] Kim KS, Park CH. Thermal comfort and waterproof-breathable performance of aluminum-coated polyurethane nanowebs. Text Res J. 2013;83(17):1808-20. https://doi.org/10.1177/0040517512464288

[7] Chunyan L, Xingliang L, Sijin M, Yanfen X. Study on application and biosafety of nano-materials in sports engineering. In: Proceedings of the 2011 International Conference on Future Computer Science and Education, ICFCSE 2011. IEEE; 2011. p. 131-4. https://doi.org/10.1109/ICFCSE.2011.40

[8] Heine E, Knops HG, Schaefer K, Vangeyte P, Moeller M. Antimicrobial Functionalisation of Textile Materials. In: Springer Series in Materials Science. 2007. p. 23-38. https://doi.org/10.1007/978-3-540-71920-5_2

[9] Z.Yang, H.Peng, W.Wang, T.Liu, Crystallization behavior of poly (ε caprolactone) / layered double hydroxide nanocomposites. J Appl Polym Sci. 116(5)(2010)2658-67. https://doi.org/10.1002/app.31787

[10] Arvidsson R, Molander S, Sandén BA. Assessing the Environmental Risks of Silver from Clothes in an Urban Area. Hum Ecol Risk Assess. 2014;20(4):1008-22. https://doi.org/10.1080/10807039.2012.691412

[11] Yang H, Zhu S, Pan N. Studying the mechanisms of titanium dioxide as ultraviolet-blocking additive for films and fabrics by an improved scheme. J Appl Polym Sci. 2004;92(5):3201-10. https://doi.org/10.1002/app.20327

[12] Kathirvelu S, D'Souza L, Dhurai B. UV protection finishing of textiles using ZnO nanoparticles. Indian J Fibre Text Res. 2009;34(3):267-73.

[13] Veronovski N, Rudolf A, Smole MS, Kreže T, Geršak J. Self-cleaning and handle properties of TiO2-modified textiles. Fibers Polym. 2009;10(4):551-6. https://doi.org/10.1007/s12221-009-0551-5

[14] Liu K, Cao M, Fujishima A, Jiang L. Bio-inspired titanium dioxide materials with special wettability and their applications. Vol. 114, Chemical Reviews. 2014. p. 10044-94. https://doi.org/10.1021/cr4006796

Materials Research Forum LLC
https://doi.org/10.21741/9781644902295-4

[15] Shishoo R. Textiles in sport. Textiles in Sport. 2005. 1-364 p. https://doi.org/10.1533/9781845690885.1

[16] Cai Y, Ke H, Dong J, Wei Q, Lin J, Zhao Y, et al. Effects of nano-SiO2 on morphology, thermal energy storage, thermal stability, and combustion properties of electrospun lauric acid/PET ultrafine composite fibers as form-stable phase change materials. Appl Energy. 2011;88(6):2106-12. https://doi.org/10.1016/j.apenergy.2010.12.071

[17] Mondal S. Phase change materials for smart textiles - An overview. Appl Therm Eng. 2008;28(11-12):1536-50. https://doi.org/10.1016/j.applthermaleng.2007.08.009

[18] Pause B. Development of Heat and Cold Insulating Membrane Structures with Phase Change Material. J Ind Text. 1995;25(1):59-68. https://doi.org/10.1177/152808379502500107

[19] Okubo M, Saeki N, Yamamoto T. Development of functional sportswear for controlling moisture and odor prepared by atmospheric pressure nonthermal plasma graft polymerization induced by RF glow discharge. J Electrostat. 2008;66(7-8):381-7. https://doi.org/10.1016/j.elstat.2008.03.003

[20] Senthilkumar M, Sampath MB, Ramachandran T. Moisture Management in an Active Sportswear: Techniques and Evaluation-A Review Article. J Inst Eng Ser E. 2012;93(2):61-8. https://doi.org/10.1007/s40034-013-0013-x

[21] Huang ZM, Zhang YZ, Kotaki M, Ramakrishna S. A review on polymer nanofibers by electrospinning and their applications in nanocomposites. Compos Sci Technol. 2003;63(15):2223-53. https://doi.org/10.1016/S0266-3538(03)00178-7

[22] Ramakrishna S, Fujihara K, Teo W-E, Lim T-C, Ma Z. FRONT MATTER. In: An Introduction to Electrospinning and Nanofibers. 2005. p. i-xi. https://doi.org/10.1142/5894

[23] Tjong SC. Structural and mechanical properties of polymer nanocomposites. Vol. 53, Materials Science and Engineering R: Reports. 2006. p. 73-197. https://doi.org/10.1016/j.mser.2006.06.001

[24] Park BG, Crosky AG, Hellier AK. Material characterisation and mechanical properties of Al2O3-Al metal matrix composites. J Mater Sci. 2001;36(10):2417-26. https://doi.org/10.1023/A:1017921813503

[25] Curtin WA. Theory of Mechanical Properties of Ceramic-Matrix Composites. J Am Ceram Soc. 1991;74(11):2837-45. https://doi.org/10.1111/j.1151-2916.1991.tb06852.x

[26] Jesson DA, Watts JF. The interface and interphase in polymer matrix composites: Effect on mechanical properties and methods for identification. Vol. 52, Polymer Reviews. 2012. p. 321-54. https://doi.org/10.1080/15583724.2012.710288

[27] Montazer M, Morshedi S. Photo bleaching of wool using nano TiO2 under daylight irradiation. J Ind Eng Chem. 2014;20(1):83-90. https://doi.org/10.1016/j.jiec.2013.04.023

Materials Research Forum LLC
https://doi.org/10.21741/9781644902295-4

[28] Montazer M, Pakdel E. Functionality of nano titanium dioxide on textiles with future aspects: Focus on wool. Vol. 12, Journal of Photochemistry and Photobiology C: Photochemistry Reviews. Elsevier B.V.; 2011. p. 293-303. https://doi.org/10.1016/j.jphotochemrev.2011.08.005

[29] Montazer M, Amiri MM. ZnO Nano Reactor on Textiles and Polymers : Ex-Situ ZnO Nano Reactor on Textiles and Polymers : Ex-Situ and In-Situ Synthesis , Application and Characterization Majid Montazer *, Morteza Maali Amiri Textile Department , Center of Excellence in Textile , A. J Phys Chem [Internet]. 2013;B(118):6. Available from: https://pubs.acs.org/doi/abs/10.1021/jp408532r https://doi.org/10.1021/jp408532r

[30] Bușilə M, Mușat V, Textor T, Mahltig B. Synthesis and characterization of antimicrobial textile finishing based on Ag:ZnO nanoparticles/chitosan biocomposites. RSC Adv. 2015;5(28):21562-71. https://doi.org/10.1039/C4RA13918F

[31] Da Costa LP. Engineered nanomaterials in the sports industry. In: Handbook of Nanomaterials for Manufacturing Applications. INC; 2020. p. 309-20. https://doi.org/10.1016/B978-0-12-821381-0.00014-4

[32] Taylor-smith BK. Sport and Nanotechnology : Are the Big Sports Looking to Go Small ? 2018;1-4.

[33] Zhang X, Wang P, Neo H, Lim G, Malcolm AA, Yang EH, et al. Design of glass fiber reinforced plastics modified with CNT and pre-stretching fabric for potential sports instruments. Mater Des. 2016;92:621-31. https://doi.org/10.1016/j.matdes.2015.12.051

[34] Yang X, Gu F, Chen X. Performance improvement of carbon fiber reinforced epoxy composite sports equipment. In: Key Engineering Materials. 2020. p. 228-33. https://doi.org/10.4028/www.scientific.net/KEM.871.228

[35] Daňová R, Olejnik R, Slobodian P, Matyas J. The piezoresistive highly elastic sensor based on carbon nanotubes for the detection of breath. Polymers (Basel). 2020;12(3). https://doi.org/10.3390/polym12030713

[36] Shao HQ, Wei H, He JH. Dynamic properties and tire performances of composites filled with carbon nanotubes. Rubber Chem Technol. 2018;91(3):609-20. https://doi.org/10.5254/rct.18.82599

[37] Song Z, Cai Y. Application of nano-materials in sports engineering. In: Advanced Materials Research. 2013. p. 281-4. https://doi.org/10.4028/www.scientific.net/AMR.602-604.281

[38] Subic A. Materials in sports equipment. Materials in Sports Equipment. Woodhead Publishing; 2019. 1-588 p.

[39] Paper C. The Effect of Different Percentages of Nano Clay Particles in Sport Shoe Sole on Power Flow across Lower Limb Segments during ... In 2014.

[40] Maryan AS, Montazer M, Harifi T, Rad MM. Aged-look vat dyed cotton with anti-bacterial/anti-fungal properties by treatment with nano clay and enzymes. Carbohydr Polym. 2013;95(1):338-47. https://doi.org/10.1016/j.carbpol.2013.02.063

[41] Bissessur R. Nanomaterials applications. In: Polymer Science and Nanotechnology: Fundamentals and Applications. Elsevier Inc.; 2020. p. 435-53. https://doi.org/10.1016/B978-0-12-816806-6.00018-2

[42] Saxena D, Rana D, Bhoje Gowd E, Maiti P. Improvement in mechanical and structural properties of poly(ethylene terephthalate) nanohybrid. SN Appl Sci. 2019;1(11):1-11. https://doi.org/10.1007/s42452-019-1406-3

[43] Borda D'Água R, Branquinho R, Duarte MP, Maurício E, Fernando AL, Martins R, et al. Efficient coverage of ZnO nanoparticles on cotton fibres for antibacterial finishing using a rapid and low cost: In situ synthesis. New J Chem. 2018;42(2):1052-60. https://doi.org/10.1039/C7NJ03418K

[44] Aladpoosh R, Montazer M. The role of cellulosic chains of cotton in biosynthesis of ZnO nanorods producing multifunctional properties: Mechanism, characterizations and features. Carbohydr Polym. 2015;126:122-9. https://doi.org/10.1016/j.carbpol.2015.03.036

[45] Gharehaghaji AA. Nanotechnology in sport clothing. In: Materials in Sports Equipment. Woodhead Publishing; 2019. p. 521-68. https://doi.org/10.1016/B978-0-08-102582-6.00018-6

[46] Zhu T, Li S, Huang J, Mihailiasa M, Lai Y. Rational design of multi-layered superhydrophobic coating on cotton fabrics for UV shielding, self-cleaning and oil-water separation. Mater Des. 2017;134:342-51. https://doi.org/10.1016/j.matdes.2017.08.071

[47] Norouzi N, Gharehaghaji AA, Montazer M. Reducing drag force on polyester fabric through superhydrophobic surface via nano-pretreatment and water repellent finishing. J Text Inst. 2018;109(1):92-7. https://doi.org/10.1080/00405000.2017.1329646

[48] Pan S, Lin H, Deng J, Chen P, Chen X, Yang Z, et al. Novel wearable energy devices based on aligned carbon nanotube fiber textiles. Adv Energy Mater. 2015;5(4):1-7. https://doi.org/10.1002/aenm.201401438

[49] Rasheed A, Khalid FA. Fabrication and properties of CNTs reinforced polymeric matrix nanocomposites for sports applications. In: IOP Conference Series: Materials Science and Engineering. 2014. https://doi.org/10.1088/1757-899X/60/1/012009

[50] Understandingnano. Carbon nanotubes being used to strengthen kayaks.

[51] C60 Fullerene: When Nanotechnology Meets World of Sport and Fitness. Thailand land of smiles.

[52] Nano Fly Fishing Rods With Higher Quality - Try On A New Type Of Fishing Rod. MAXCATCH.

Emerging Applications of Nanomaterials   Materials Research Forum LLC
Materials Research Foundations 141 (2023) 75-100   https://doi.org/10.21741/9781644902295-4

[53] Harifi T, Montazer M. Application of nanotechnology in sports clothing and flooring for enhanced sport activities, performance, efficiency and comfort: a review. Vol. 46, Journal of Industrial Textiles. 2017. p. 1147-69. https://doi.org/10.1177/1528083715601512

[54] Hiremath N, Bhat G. High-performance carbon nanofibers and nanotubes. In: Structure and Properties of High-Performance Fibers. 2017. p. 79-109. https://doi.org/10.1016/B978-0-08-100550-7.00004-8

[55] Paper C. The Effect of Different Percentages of Nano Clay Particles in Sport Shoe Sole on Power Flow across Lower Limb Segments during ... 2014;(October).

[56] Sadeghian A, Montazer M, Harifi T, Mahmoudi M. Aged-look vat dyed cotton with anti-bacterial / anti-fungal properties by treatment with nano clay and enzymes. Carbohydr Polym. 2013;95(1):338-47. https://doi.org/10.1016/j.carbpol.2013.02.063

[57] Montazer M, Seifollahzadeh S. Enhanced self-cleaning, antibacterial and UV protection properties of nano TiO_2 treated textile through enzymatic pretreatment. Photochem Photobiol. 2011;87(4):877-83. https://doi.org/10.1111/j.1751-1097.2011.00917.x

[58] Chen YF, Wu J-H, Huang C-C. Experimental investigation into mechanical properties of nanomaterial-reinforced table tennis rubber. Adv Technol Innov. 2016;1(2):41-5.

[59] Liu X. Biomimetic nano scaffold for sports ligament injury therapy. Int J Nanotechnol. 2021;18(1-4):214-25. https://doi.org/10.1504/IJNT.2021.114226

[60] Behzadnia A, Montazer M, Rad MM. In situ photo sonosynthesis and characterize nonmetal/metal dual doped honeycomb-like ZnO nanocomposites on wool fabric. Ultrason Sonochem. 2015;27:200-9. https://doi.org/10.1016/j.ultsonch.2015.05.021

[61] Bai H, Li Q. Electrodeposited Ni/TiN-SiC Nanocomposites on the Dumbbell: Reducing Sport Injuries. Coatings. 2022;12(2). https://doi.org/10.3390/coatings12020177

[62] Harifi T, Montazer M. A robust super-paramagnetic TiO_2:Fe_3O_4:Ag nanocomposite with enhanced photo and bio activities on polyester fabric via one step sonosynthesis. Ultrason Sonochem. 2015;27:543-51. https://doi.org/10.1016/j.ultsonch.2015.04.008

[63] Yusup EM, Mahzan S, Kamaruddin MAH. Natural Fiber Reinforced Polymer for the Application of Sports Equipment using Mold Casting Method. In: IOP Conference Series: Materials Science and Engineering. 2019. https://doi.org/10.1088/1757-899X/494/1/012040

[64] Akhtar K, Khan SA, Khan SB, Asiri AM. Scanning electron microscopy: Principle and applications in nanomaterials characterization. In: Handbook of Materials Characterization. 2018. p. 113-45. https://doi.org/10.1007/978-3-319-92955-2_4

[65] Ilić V, Šaponjić Z, Vodnik V, Lazović S, Dimitrijević S, Jovančić P, et al. Bactericidal efficiency of silver nanoparticles deposited onto radio frequency plasma

pretreated polyester fabrics. Ind Eng Chem Res. 2010;49(16):7287-93. https://doi.org/10.1021/ie1001313

[66] Li H, Deng H, Zhao J. Performance Research of Polyester Fabric Treated by Nano Titanium Dioxide (N ano-TiO2) Anti-ultraviolet Finishing. Int J Chem. 2009;1(1):57-62. https://doi.org/10.5539/ijc.v1n1p57

[67] Epp J. X-Ray Diffraction (XRD) Techniques for Materials Characterization. In: Materials Characterization Using Nondestructive Evaluation (NDE) Methods. Elsevier Ltd; 2016. p. 81-124. https://doi.org/10.1016/B978-0-08-100040-3.00004-3

[68] Medina-Ramírez IE, Arzate-Cardenas MA, Mojarro-Olmos A, Romo-López MA. Synthesis, characterization, toxicological and antibacterial activity evaluation of Cu@ZnO nanocomposites. Ceram Int. 2019;45(14):17476-88. https://doi.org/10.1016/j.ceramint.2019.05.309

[69] Kaur N. UV-Visible spectroscopy Contents : 2018;(August):1-13.

[70] Yadav A, Prasad V, Kathe AA, Raj S, Yadav D, Sundaramoorthy C, et al. Functional finishing in cotton fabrics using zinc oxide nanoparticles. In: Bulletin of Materials Science. 2006. p. 641-5. https://doi.org/10.1007/s12034-006-0017-y

[71] Buseck PR. Chapter 1. PRINCIPLES OF TRANSMISSION ELECTRON MICROSCOPY. In: Minerals and Reactions at the Atomic Scale. 2018. p. 1-36. https://doi.org/10.1515/9781501509735-005

[72] Nallathambi G, Ramachandran T, Rajendran V, Palanivelu R. Effect of silica nanoparticles and BTCA on physical properties of cotton fabrics. Mater Res. 2011;14(4):552-9. https://doi.org/10.1590/S1516-14392011005000086

[73] Huang L, Li DQ, Lin YJ, Wei M, Evans DG, Duan X. Controllable preparation of Nano-MgO and investigation of its bactericidal properties. J Inorg Biochem. 2005;99(5):986-93. https://doi.org/10.1016/j.jinorgbio.2004.12.022

[74] ISO. ISO - ISO 4920:2012 - Textile fabrics - Determination of resistance to surface wetting (spray test) [Internet]. 2017. Available from: https://www.iso.org/standard/50706.html

[75] ISO - ISO 811:2018 - Textiles - Determination of resistance to water penetration - Hydrostatic pressure test [Internet]. Available from: https://www.iso.org/standard/65149.html

[76] BS EN 20811:11992. Resistance of fabric to penetration by water (hydrostatic head test).

[77] Iso BSEN. Textiles - Measurement of water vapour permeability of textiles for the purpose of quality control. Management. 2006;3:15496.

[78] ISO 9865:1991. Textiles, determination of water repellency of fabrics by the Bundesmann rainshower Test.

[79] Institution BS. BS EN 29865:1993, ISO 9865:1991: Textiles- Determination of water repellency of fabrics by the Bundesmann rain-shower test. 1993.

[80] AATCC Test Method 35-2006, Water resistance: rain test.

[81] Method of test for the resistance of fabrics to an artificial shower. 1974;

[82] SATRA TM160 Testing materials for light fastness-Footwear.

[83] US20090313855A1 Self cleaning outsoles for shoes.

[84] SATRA STM 567 Endofoot Testing thermal performance and moisture management.

[85] SATRA TM436 Determination of whole shoe thermal insulation value and cold rating.

[86] SATRA TM190:2002 Ground insulation index test.

[87] SATRA STM 175 Permeability and Absorption Test.

[88] SATRA TM444 Methods for the assessment of material properties used in the construction of water-resistant footwear.

[89] STM 505 Testing water resistance in footwear.

[90] Sawhney APS, Condon B, Singh K V., Pang SS, li G, Hui D. Modern Applications of Nanotechnology in Textiles. Text Res J. 2008;78(8):731-9. https://doi.org/10.1177/0040517508091066

[91] Geetika Jaiswal1 VS and ACP. Nanotechnology in The Driver's Seat of Sportswear Industry: A Review of Current Trends and Future Applications. https://lupinepublishers.com/.

[92] Mohapatra H. Nanotechnology in Fibres and Textiles [Internet]. International Journal of Recent Technology and Engineering (IJRTE). 2013. Available from: https://www.researchgate.net/publication/276062034

[93] Baydal-Bertomeu JM, Puigcerver SA, González JC, Gomez J, Perez-Fernandez M, Sempere-Tortosa JR. Nanotechnology can provide a real breakthrough in the anti-slip properties of safety footwear soles. Footwear Sci. 2015;7(June 2015):S58-60. https://doi.org/10.1080/19424280.2015.1038607

[94] http://www.nanotechbuzz.com.

Materials Research Forum LLC
https://doi.org/10.21741/9781644902295-5

Chapter 5

Nanocatalysts for the Photodegradation of Organic Pollutants

Md. Yeasin Pabel, Md. Fardin Ehsan, Muhammed Shah Miran, and Md. Mominul Islam*

Department of Chemistry, University of Dhaka, Dhaka 1000, Bangladesh

*mominul@du.ac.bd

Abstract

Industrial and other anthropogenic activities lead to increase in persistent organic pollutants (POPs) in natural water bodies. POPs are harmful to not only the aquatic ecosystem but also human health. Researchers around the globe have been working to develop efficient, cost-effective, environmentally-friendly methods for the treatment of contaminated waters. The chapter basically focuses on the removal of POPs by catalytic advanced oxidation processes (AOPs) using nontoxic metal oxides of d-block elements such as TiO_2, ZnO, MnO_2 and iron oxides nanomaterials and composites with different counterparts. An insight into the fundamentals of photocatalysis is highlighted. This chapter would bestow an opportunity to evaluate the underlying prospects of the methods involving AOPs used for the degradation of POPs dissolved in water.

Keywords

Advanced Oxidation Process, Oxide Catalysts, Persistent Organic Pollutants, Photocatalysis, Degradation, Dyes

Contents

1.   Introduction

Rapid growth of population and industrial development has severely damaged the environment especially the water quality leading to clean water crisis globally. Ground water can be contaminated in a number of ways, industrial water discharge containing synthetic organic dyes, agricultural pesticides, and pharmaceutical drugs are major sources of the pollution [1,2]. Most of these pollutants are termed as persistent organic pollutants (POPs). The consequences of water pollution by POPs are devastating, since the POPs contaminated waters affect the aquatic ecosystem and human health [1,2]. Hence, research to degrade this POPs is ceaseless that, as for example, may be realized from the summary of recent publication on the degradation of POPs by advanced oxidation processes (AOPs) (Fig. 1).

There are several methods available to remedy wastewater, for instances, adsorption, filtration, flocculation, electrocoagulation, and AOPs [3-5]. Adsorption is the transfer of pollutants dissolved in water to the surface of a solid adsorbent. The filtration method involves the physical separation of suspended pollutants by sieving and particle trapping mechanism. The flocculation technique introduces highly charged molecules into wastewater to destabilize the charged pollutants suspended in the water. The electrocoagulation method involves application of a direct current source between electrodes (anode and cathode) dipped in the contaminated water. AOPs involve the generation of highly reactive hydroxyl radical (OH$^{\bullet}$), which fragments targeted pollutants. The use of AOPs, such as Fenton and Fenton-like processes, catalytic and photocatalytic oxidation, electrochemical oxidation, photoelectrochemical method, and ozonation have been efficiently applied to treat water containing POPs. Among these methods, AOPs, especially photocatalytic oxidation is considered most convenient and efficient method. Different types of d-block elements and their nanocomposites are mostly used as photocatalyst for the treatment of water [6,7].

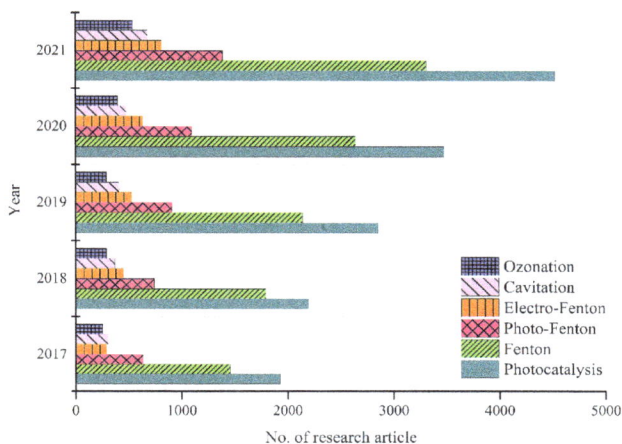

Figure 1. Number of articles published on POPs degradation by AOPs during 2017-2021. Data are collected from different digital sources.

This chapter mainly discusses the degradation of POPs with photocatalytic oxidation process. The fundamentals and photocatalytic activity of nontoxic oxides of *d*-block elements and their nanocomposites for the photodegradation of POPs are described. The critical parameters influencing photocatalytic degradation are also presented. Finally, the future prospect of the oxides of *d*-block elements as photocatalysts are pointed out.

2. Features of POPs family

POPs are organic molecules that are stable in air, water, and soil. Persistent molecules are defined as those have a half-life of more than 180, 60, and 2 days in soil, water, and air, respectively. POPs are basically hydrophobic, i.e., 'water-hating' or are lipophilic, i.e., 'fat-loving' chemicals. Thus, avoiding the aqueous phase they undergo partition to solids, notably organic matters present in aquatic system. These pollutants can get deposited away from the place of their origin, since they can be transported through migratory species, water, and air. Many POPs are currently used as solvents, pesticides, industrial chemicals (dyes) and pharmaceuticals. Chemical structures of some frequently used POPs are presented in Scheme 1. Although some POPs occur naturally, e.g., dioxins and dibenzofurans from volcanoes, most of them are chemicals used in different anthropogenic activities including agriculture as pesticides, e.g., dichlorodiphenyltrichloroethane (DDT), chlordane, heptachlor, mirex, hexachlorobenzene, aldrin, dieldrin, endrin, etc., in industry, e.g., perfluorinated compounds, polychlorobiphenyls, hexachlorobenzene, etc., or unintentionally produced, by-products of industrial processes or resulting from the combustion of certain materials (dioxins and furans). POPs are among the high-risk pollutants to human health and the environment [8]. They are highly toxic and can accumulate in living organisms.

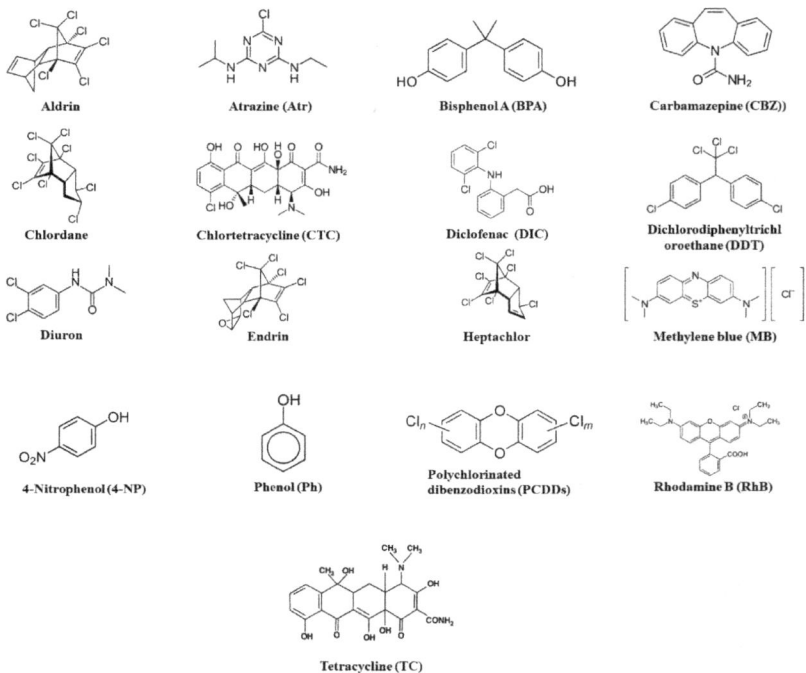

Scheme 1. Chemical structure of commonly studied POPs.

3.   AOPs used for water treatment

There are several methods practiced for the wastewater treatment including adsorption, filtration, precipitation, coagulation, electrocoagulation, flotation, membrane separation, chemical oxidation, chlorination, ozonation, and ultrafiltration [3,4]. Due to specific shortcomings of these methods, AOPs are well-established popular techniques. AOPs possess some specific advantages namely most convenient, high efficiency, simplicity, and minimum energy requirement. In this section, the chemistries of the AOPs-assisted degradation of POPs dissolved in water are briefly discussed.

Emerging Applications of Nanomaterials   Materials Research Forum LLC
Materials Research Foundations 141 (2023) 101-123   https://doi.org/10.21741/9781644902295-5

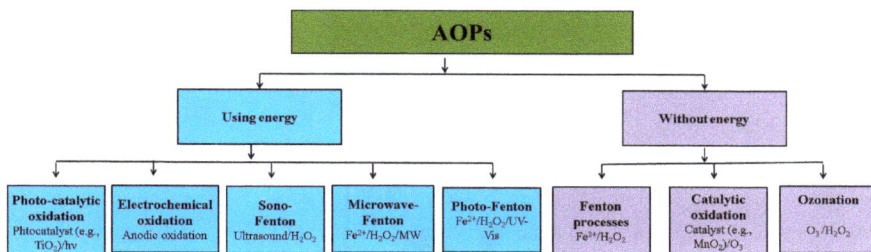

Figure 2. Classification of AOPs

Degradation is a process in which a large-sized organic molecule is broken down chemically into smaller ones or mineralized. For degrading organic species, AOPs are the most suitable since they lead to even mineralization of organic pollutants. AOPs, in fact, follow several chemical steps in removing pollutants from wastewater by oxidation with $OH^\bullet$ generated in situ. Since the $OH^\bullet$ radical is one of the strongest radical with an oxidation potential of 2.8 eV, once formed, it reacts unselectively with almost all organic species which quickly and efficiently are fragmented into smaller species including molecules and ions and inorganic molecules like CO_2 and H_2O. Steps of AOPs may be essentially divided into three parts: (i) Formation of $OH^\bullet$ species, (ii) Initial attacks on target molecules by $OH^\bullet$ and their breakdown to intermediate species and finally (iii) Further attacks on the thus-formed fragments by $OH^\bullet$ until mineralization. In situ generation of $OH^\bullet$ radical that is the working horse of the AOPs can be carried out by following different pathways. Depending on the scenario of energy consumption, the AOPs can be classified into different groups as depicted in Fig. 2. In the following sections, the photocatalytic pathways followed for degradation of POPs are comprehensively discussed.

4.   Photocatalytic degradation of POPs

4.1   Basic principles

Photocatalytic reaction is initiated when a photoelectron is promoted from the valence band (VB) of a photocatalyst (semiconductor), i.e., electrons promoted from the VB to the conduction band (CB) of the catalyst (e.g., metal oxide) upon illumination as depicted in Fig. 3.

Emerging Applications of Nanomaterials   Materials Research Forum LLC
Materials Research Foundations 141 (2023) 101-123   https://doi.org/10.21741/9781644902295-5

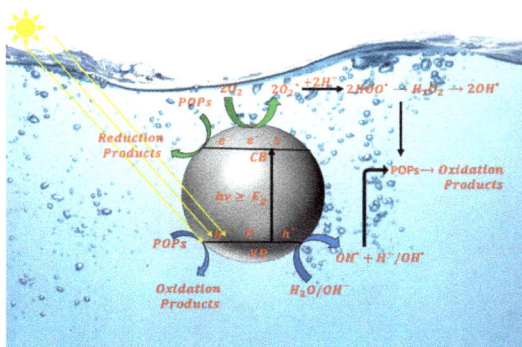

Figure 3. Mechanism of photocatalytic degradation of organic pollutant.

Such a transition of electron occurs only when the photon energy ($h\upsilon$) is either equal or greater than the band gap of the photocatalyst. Promotion of electrons from VB to CB leaves behind a hole (h^+) in the VB. The overall process is presented in Eq. 1.

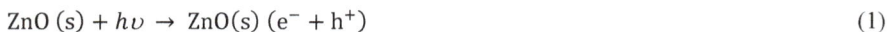

$$ZnO\ (s) + h\upsilon \rightarrow ZnO(s)\ (e^- + h^+) \tag{1}$$

Molecular oxygen (O_2) adhered to the surface of catalyst takes an electron from the CB and get reduced to superoxide radical ($O_2^{\bullet-}$) as:

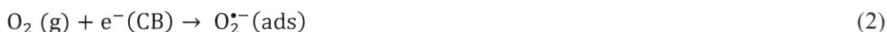

$$O_2\ (g) + e^-(CB) \rightarrow O_2^{\bullet-}(ads) \tag{2}$$

The $O_2^{\bullet-}$ species produced may get protonated forming hydroperoxyl radical ($HOO^{\bullet}$) which then, may produce H_2O_2 that ultimately dissociates into highly reactive $OH^{\bullet}$ radical as shown in Eqs. 3-5.

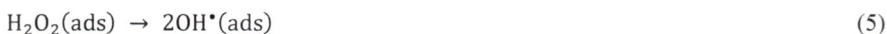

$$O_2^{\bullet-}(ads) + H^+(aq) \rightleftharpoons OOH^{\bullet}\ (ads) \tag{3}$$

$$2OOH^{\bullet}(ads) \rightarrow H_2O_2(ads) + O_2 \tag{4}$$

$$H_2O_2(ads) \rightarrow 2OH^{\bullet}(ads) \tag{5}$$

The thus-formed $OH^{\bullet}$ radicals react with organic species adsorbed on the catalyst surface or present at its vicinity non-selectively leading to form smaller fragments such as neutral molecules, or ions, and mineralized products (Eq. 6).

$$POPs + OH^\bullet(ads) \rightarrow CO_2(g) + H_2O(l) \tag{6}$$

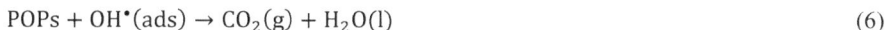

The generated $O_2^{\bullet-}$ species not only takes part in the oxidation process but also prevents the electron-hole recombination, which is very important to be an effective photocatalyst. On the other hand, the h^+ at the VB reacts with H_2O to produce $OH^\bullet$ radical (Eq. 7).

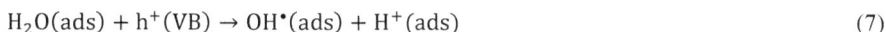

$$H_2O(ads) + h^+(VB) \rightarrow OH^\bullet(ads) + H^+(ads) \tag{7}$$

So far, it is clear that the photocatalyst used must have the ability to absorb photon to generate e^- and h^+ pairs when the catalyst is illuminated.

Some d-block metal oxides have wide band gap, e.g., the band gap of TiO_2 is around 3.2 eV that is equivalent to some energy of UV light. Thus, the efficiency of solar energy utilized for degradation of POPs is severely hindered by the metal oxides with wide band gap. Another limitation of using only metal oxides as photocatalyst is that there is a large probability of recombination of e^- and h^+ pairs. These limitations are considered to be a major problem in photocatalysis [9]. To overcome these two hurdles, natural polymers such as cellulose, chitosan, clay materials and so on have been used as supports. Moreover, support material of composites, i.e., matrix phase assists the catalysts particles to be dispersed as much as possible to result in a high surface area. Supportive materials have been introduced in extending the spectral response of metal oxides to visible region and to enhance the lifetime of spatial charge separation, forming in the composite, leading to reduced recombination probability [10]. On irradiation with visible light, supported catalyst absorbs the photons of light which assists the transfer of electrons from the highest occupied molecular orbital (HOMO) to the lowest unoccupied molecular orbital (LUMO). The excited electrons from the supported catalyst are transferred to the CB of metal oxides when the energy of CB of metal oxide is lower than that of HOMO of supported catalyst [10]. The electrons in the CB of metal oxide eventually react with O_2 (Eq. 2) present at the surface of solid catalyst to generate different types of highly reactive oxygen species (ROS), i.e., $O_2^{\bullet-}$, $OH^\bullet$, which are responsible for the degradation of the organic compound (Eq. 6) [11]. In the polymer-supported semiconductor metal oxide, polymer acts both as a good electron donor from its LUMO and an excellent hole acceptor in HOMO orbital from the VB of semiconductor metal oxide leading to substantially reduced recombination rate of light induced electron-hole pairs. The h^+ in the π orbital of polymer can migrate to the interface and react with H_2O to generate $OH^\bullet$.

4.2   Photocatalytic performance

4.2.1   3d-block transition metal oxides

Oxides of d-block elements are highly versatile in practical application and are termed as the "future materials". Beneficial properties of different metal oxides are the main reason for their increased applications in the interdisciplinary fields especially in heterogeneous photocatalysis. Nano-sized metal oxides are gaining a high popularity due to their unique thermal, mechanical, optical, magnetic and electronic properties. Nanoparticles of nontoxic oxides such as TiO_2, ZnO, Fe_2O_3 and MnO_2 have been used as pristine or in composites for photocatalytic degradation of POPs [12-15]. They have attractive characteristics because of their smaller size, and different morphologies and shapes. It is noteworthy that the structure of surface species is very different than the bulk species due to energetically different environments. The higher surface area of metal oxide nanoparticles contributes to bigger portion of their unique properties including good chemical reactivity, adsorption capability, enhanced mechanical stability and different electrical, magnetic and optical properties.

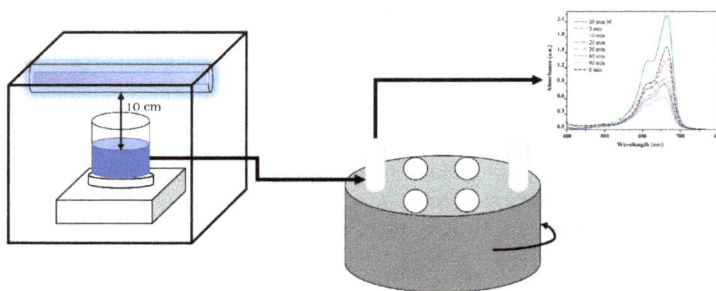

Scheme 2. A setup of photocatalytic degradation of dye

4.2.1.1 Titanium oxide

TiO_2 is the most explored d-block metal oxides used as photocatalysts owing to its inertness, stability, ease of preparation and environmentally benign [16]. TiO_2 is a semiconductor with a wide band gap in the range of 3.0–3.7 eV. When it is illuminated with light of wavelengths < 380 nm, the photogenerated exciton pair thereafter beginning redox reactions (Eqs. 2 and 7). The TiO_2 is of prime choice in applications especially as a photocatalyst because of its nontoxicity, chemical stability, and low cost etc. Most of the organic compounds present in aqueous solution get easily oxidized by the ROS resulting from illumination of TiO_2. A typical setup of photocatalytic degradation of organic dye is depicted in Scheme 2. The organic pollutants may be completely degraded called "mineralized", as a result of this oxidation events to form CO_2 and H_2O (Eq. 6) in the presence of illuminated TiO_2 with near-UV light. It has been revealed that nanosized TiO_2 has superior photocatalytic activity compared to bulk TiO_2. TiO_2 has three crystallographic phases: anatase, rutile, and brookite. The unit cell of orthorhombic brookite, tetragonal anatase, and tetragonal rutile contain eight, four, and two TiO_2 units, respectively. Due to its high photocatalytic effectiveness and adsorption affinity for organic molecules, anatase is considered as the most active among these phases.

Many organic pollutants present in wastewater such as pesticides, aromatic compound, azo dyes, and nitrite compound etc. (Scheme 1) is easily degraded by the TiO_2 nanoparticles under the exposure of UV light. On the contrary, TiO_2 cannot be activated by visible light because of its wide band gap. To enhance the activity in visible and infrared region the composite formation of TiO_2 with other semiconductor materials, doping etc. have been practiced. The catalytic activity of different forms of TiO_2 and their composites with various chemical entities for removing POPs are reviewed (Table 1). At a glance, the formation of composites generally exhibits better catalytic performance towards photo-induced degradation of POPs. The percent of removal of total organic carbon (TOC) by $PAni/TiO_2$ is more than two-fold larger than that of pristine TiO_2. The enhanced oxidative property of $PAni/TiO_2$ nanocomposite has been claimed to be due to the transfer of an electron from TiO_2 to PAni [17].

Table 1. *Photocatalytic degradation of POPs using some d-block metal oxides and its composites as photocatalysts.*

Photocatalyst	Contaminant(s)	Reaction Conditions			Efficiency [%]	Ref.
		Catalyst dose [mg/L]	Irradiation Source	pH		
Aeroxide P25	MCP	2668	UV	5.0	79	[18]
	CPS	15.72	300W halogen lamp (400-800 nm)	6.8	84	[19]
N/TiO$_2$	DDT	1000	150 W halogen lamp (> 420 nm, 14.38 W/m^2)	7.0	100	[20]
Fe$_3$O$_4$@TiO$_2$	BPA; DBP	1000	Hg-vapor lamp	7.0	100 (BPA; DBP);	[21]
MnO$_2$	4-CP	250	1 W UV-C lamp (258-262 nm, 1.45 mW/cm^2)	4	100	[12]
V-α-MnO$_2$	Ph	1250	UV lamp	6	76	[13]
MnO$_2$/NiO	TC	500	125 W UV irradiation (365 nm)	-	89	[14]
ZnO	2,4-D	-	UV lamp (365 nm)	6.8	100	[22]
Fe-ZnO	CPS	25	30 W UV lamps (374 nm)	-	93	[23]
ZnO-SnO$_2$	TCP	1000	15 W light source (300-400 nm)	-	99	[24]
Ultrathin α-Fe$_2$O$_3$ nanosheet	BPS	200	300 W Xe lamp (420 nm)	-	91	[25]
Co, Ni-Fe$_3$O$_4$	CF	8000	200 W Hg vapor lamp	8.0	98	[26]
Fe$_3$O$_4$@TiO$_2$@Au	2,4,6-TCP	500	150 W Xe lamp	10.0	98	[27]

2,4-D: 2,4-dichlorophenoxyacetic; 2,4,6-TCP: 2,4,6-Trichlorophenol; 4-CP: 4-Chlorophenol; BPA: Bisphenol A; BPS: Bisphenol S; CF: Carbol Fuchsin; CPS: Chlorpyrifos; DBP: Dibutyl phthalate; DDT: Dichlorodiphenyltrichloroethane; MCP: Methylcyclopropane; Ph: Phenol; TC: Tetracycline and TCP: Triclopyr

4.2.1.2 Manganese oxide

MnO_2 is actually a versatile material as it can be incorporated in various fields including catalysis. MnO_2 based materials have been received much attentions in the area of photocatalysis because of their narrow band gap (1–2 eV) that belong to the visible regions (1.8 -3.2 eV) of the solar radiation as well as due to their specific properties like multiple crystal structures and oxidation states, natural abundance, low-cost, and nontoxicity to the environment. In 1994, for the first-time photocatalytic activity of amorphous MnO_2 has been studied through the oxidation of 2-propanol [28]. However, the catalytic activity of MnO_2 mainly depends on its crystal form as well as morphology, porosity, specific surface area, and oxidation state. The physical and chemical properties of MnO_2 nanostructures largely depends on the nature of crystallographic structures (geometry, lattice parameter, and the size of tunnel and so on). For example, because of the binding force between the valance electron and the parent atom, the band gap is inversely proportional to particle size.

MnO_2 exhibits several crystallographic phases, which are known as α-, β-, γ-, δ-, and λ-forms. The catalytic performance of MnO_2 largely depend on its crystallographic nature. Amorphous form of MnO_2 have been reported to exhibit a better photocatalytic activity than the crystalline phase. This enhanced catalytic activity of amorphous MnO_2 has been explained in terms of surface defects. Moreover, the morphology also influences the activity. MnO_2 nanorods show outstanding performance towards the photodegradation process. A complete degradation of MB has been achieved within only 20 min using MnO_2 nanorods [29]. The performance of MnO_2-based materials towards the photocatalytic degradation of POPs are extensively reviewed and summarized in Table 1. The details of other factors affecting the catalysis such as pH, dose, initial concentration and so on are described below. Besides the advantages of using MnO_2-based catalysts, there are, of course, several specific limitations known. In some cases, these drawbacks have been mitigated through the formation of composites with other materials. For instance, the efficiency has been reported to improve drastically by using the heterostructure of MnO_2/NiO composites rather than single phase of MnO_2 [14]. Supporting system functions in two main roles during catalysis namely (i) to prevent the loss of MnO_2 photocatalyst and (ii) to delay the electron hole pair recombination in enhancing overall catalytic performance.

4.2.1.3 Zinc oxide

ZnO is a wide band gap (3.37 eV) semiconductor with high exciton binding energy of 60 meV. It absorbs a bigger portion of the UV spectrum and performs better than TiO_2 as a photocatalyst. ZnO are generally found in three types of crystal structures namely i) rocksalt, ii) wurtzite and iii) zinc blende structure. Among the three structures, ZnO is wurtzite is most common due to its excellent thermodynamic stability. ZnO wurtzite crystal structure is hexagonal at 25 °C and 1 atm of which two lattice parameters, a and c, have the values of 0.32960 and 0.52065 nm, respectively. The wurtzite ZnO is formed by sublattices of atoms that are stacked alternatively along c-axis in hexagonal close packing fashion. So, each Zn^{2+} sub-lattice contains four Zn^{2+} ions surrounded by four O^{2-} ions and vice versa, coordinated at the edges of a tetrahedron. The polar terminal faces of wurtzite form are the Zn terminated (0001) and O terminated (000$\bar{1}$) faces (c-axis oriented). The non-polar (11$\bar{2}$0) (a-axis) and (10$\bar{1}$0) faces both contain an equal number of Zn and O atoms. The physical and chemical properties of the polar faces are different than the non-polar faces. The electronic structures are slightly different for O terminated faces than the other three faces. The tetrahedral coordination is responsible for causing inherent polarity of the ZnO crystal. Interestingly, the Zn–O bond is strongly ionic in character as well. So, the properties of ZnO can be classified as in between the ionic and covalent compound.

The photodegradation of POPs are more effective with ZnO catalyst because of its mobility of electron (200–300 cm^2/Vs), which have higher values than that of TiO_2 (0.1-4.0 cm^2/Vs). Higher values of mobility results in rapid electron transfer resulting in higher quantum efficiency. The surface of ZnO nanoparticles is responsible for its luminescent properties especially its oxygen deficient-related defects or oxygen vacancies and their chemical states. XPS analysis reveals that the oxygen vacancies and defects are more prominent in undoped ZnO resulting in negligible green emission. Pure ZnO has a broad band at wavelength of about 369 nm in the absorption spectra. The position of the VB of ZnO is lower than that of the VB of TiO_2; as such, the oxidation potential of $OH^\bullet$ generated by ZnO is higher than that of $OH^\bullet$ produced by TiO_2. This is also a reason contributing to the higher photocatalytic performance of ZnO in degrading pollutants compared to TiO_2 [30, 31]. Mechanism of formation of $OH^\bullet$ species are the same as illustrated in Fig. 3 and Eqs. 2-5.

The sizes of ZnO nanoparticles dominate the optical properties and photocatalytic performance of ZnO. As particle size of ZnO decreases the band gap of ZnO increases. Like molecules, ZnO also consists of molecular orbitals but due to higher number of closely compact molecular orbitals form characteristic band structure. The major constraint of ZnO is its fast recombination rate of photogenerated electron-hole pairs. This phenomenon diminishes exciton pairs which hinders the photodegradation reaction. ZnO can also serve as anti-microbial agent. The highly reactive species generated while UV irradiation on the ZnO surface can react with the microorganisms such as bacteria. This not only can remove POPs but also can destroy pathogens in water making it disinfected. Utilization of ZnO and its nanocomposites used as photocatalyst for degrading POPs from aqueous solution are summarized in Table 1. It may be seen that the performance of degradation of POPs is increased by composites with basically other metal or metal oxides [23,24].

4.2.1.4 Iron oxide

Iron oxides are the most abundant oxides on the earth and have been used for various purpose including catalysis. Application of iron oxides and its nanocomposites as photocatalysts for degrading POPs are listed in Table 1. However, iron oxides are found in nature in different stoichiometric and crystalline structures. Generally, iron oxides are classified as würtzite (FeO), hematite (α-Fe$_2$O$_3$), maghemite (ν-Fe$_2$O$_3$), and magnetite (Fe$_3$O$_4$). Out of all the phases, α-Fe$_2$O$_3$ is the most stable state of iron oxides at ambient conditions. The structure of α-Fe$_2$O$_3$ is a hexagonal crystal system consisting of iron atoms surrounded by six oxygen atoms. In α-Fe$_2$O$_3$, the iron and oxygen atoms are arranged in a corundum structure, which is a trigonal-hexagonal scalenohedral with space group R-3c, lattice parameters a = 5.0356 nm, c = 13.7489 nm, and six formula units per unit cell. α-Fe$_2$O$_3$ is hexagonal closed-packed lattice where the oxide ions (O^{2-}) are arranged along the (001) plane. The cations (Fe^{3+}) in the (001) basal planes occupy two-thirds of the octahedral interstices. The tetrahedral positions remain vacant. So, pairs of FeO$_6$ octahedrons result from this cationic configuration. Three surrounding octahedrons in the same plane share the edges of octahedrons, and one face is shared with an octahedron in an adjacent plane in the (001) direction. The important distinctive characteristic for the Fe$_2$O$_3$ nanoparticles is its superparamagnetic behavior that contributes to its stabilization. As the size of the particles enters from micrometer to nanometer scale, drastic change in magnetic characteristics of Fe$_2$O$_3$ has been observed. Nanoparticles of 6-15 nm range exhibit superparamagnetic behavior, while microparticles behave like ferromagnetic material.

α-Fe$_2$O$_3$ is a promising candidate for the wastewater treatment under visible light since it is an n-type semiconductor with a band gap in the range of 2.1–2.3 eV [32]. In fact, α-Fe$_2$O$_3$ can absorb around 43% of the visible light. However, the photocatalytic activity of α-Fe$_2$O$_3$ is limited due to the low separation efficiency and fast recombination of photoinduced charge carriers. UV exposure promotes the generation of ROS that are produced from the defect sites of α-Fe$_2$O$_3$ nanoparticles or visible light electron-hole pairs. The electron-hole pairs that have been produced will help generate ROS such as O$_2^{\bullet-}$, OH$^{\bullet}$, etc. These highly reactive species can react indiscriminately with variety of organic compound (Eq. 6).

4.2.2 Supported Metal Oxides Nanocomposites

The application of sole metal oxides as photocatalysts in the degradation of POPs introduces several problems, for example, particle agglomerates, limited re-usability, and formation of secondary pollutant. The following limitations of unsupported photocatalysts are known:

1. *Recombination of electron and hole:* Recombination occurs due to the lack of charge separation.

2. *Excitation source:* Metal oxides with larger bands cannot be activated in radiation of visible range. They are generally active in UV light. Development of an effective photocatalyst, therefore, becomes difficult.

3. *Poor adsorption:* Adsorption of POPs on the photocatalyst is one of the main processes for the degradation reaction. But metal oxides are poor adsorbents.

4. *Re-usability:* It is very difficult to separate metal oxides from reaction medium that limits the repeated use of the catalysts.

5. *Secondary pollutants:* Leaching introduces heavy metal/metal ions in water and reduces photocatalytic efficacy of a photocatalyst.

6. *Photocorrosion:* The corrosion phenomenon of metal oxide photocatalyst (ZnO) hinders its photocatalytic efficiency.

To address these problems, metal oxides are immobilized with organic polymers or inorganic materials which acts as the support of the solid catalyst powder as well as good adsorbents. The chemical structures of some support materials are shown in Fig. 4. The synthetic polymers used as matrix are usually toxic, non- degradable, and non-sustainable. Different nontoxic organic/inorganic matrices, for instances, chitosan, cellulose, starch, natural clay minerals etc., are employed as a catalyst support. The support materials shown are non-toxic and have some unique properties namely hydrophilic in nature, physical and chemical resistance, thermally stable and so on. The metal oxides/polymer or clay nanocomposite possesses a high photodegradation activity towards POPs because of the enhanced migration and separation of e^- and h^+, the strong electrostatic interaction between the charged surface of the catalyst and the pollutants, and large surface area of the nanocomposite as described earlier. In some specific cases, the support materials assist efficient oxidation of water contaminants [33].

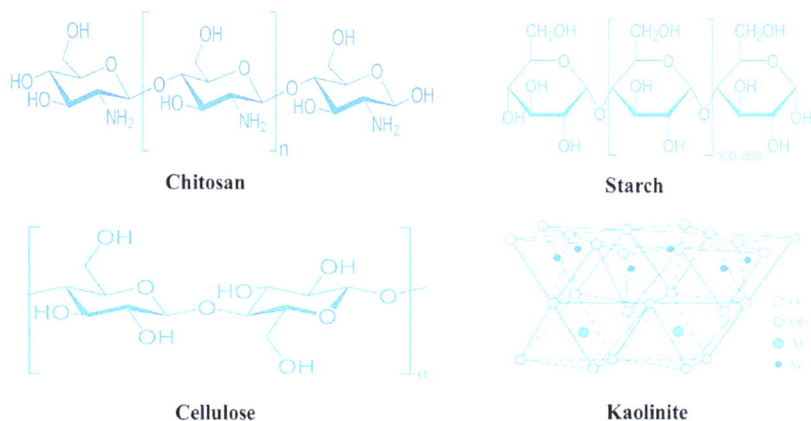

Chitosan

Starch

Cellulose

Kaolinite

Figure 4. Chemical structures of some nontoxic catalyst supports

Table 2. *Photocatalytic degradation of POPs using metal oxides and its composites as photocatalysts*

Support material (SM)	Photocatalyst (P)	Weight ratio of P to SM	Pollutant (s)	Remarks	Ref.
Cellulose	TiO_2	0.05:1	E1; E2; E3	99.5, 99.5, and 98.5% degradation for E1, E2, and E3 respectively	[34]
	ZnO	-	MO	complete degradation with 0.4 M composite	[38]
Cellulose paper	TiO_2-Ag_2O	-	An	97% degradation	[39]
Alginate/carboxymethyl cellulose	TiO_2	0.2:1	CR	99.7% degradation	[40]
Chitosan	N, S doped TiO_2; ZnO	0.5:1	TC	91% degradation in 20 min	[41]
Alginate	TiO_2	0.5:1	CHD; IBP ATL; CBZ;	CHD, ATL, IBP and CBZ can be removed up to 99, 85, 99 and 99 %	[42]
Sodium alginate	ZnO	-	TCS	high rates of degradation were observed (90% in 90 min)	[43]

An: Aniline; ATL: Atenolol; CBZ: Carbamazepine; CHD: Chlorhexidine; CR: Congo Red; E1: Estrone; E2: 17β-Estradiol; E3: Estriol; IBP: Ibuprofen; MO: Methyl orange; TC: Tetracycline and TCS: Triclosan

The advantages of immobilization of solid catalysts on the support materials may be noted as it reduces particles agglomeration, catalyst loss, metal ion leaching, and increases adsorption of POPs. The effectiveness of a nanocomposite is successful when there is good interaction between the matrix and the nanoparticle as well as good dispersion of nanoparticle in the matrix as described above. Assistance of the support material for enhancing the photocatalytic activities of different solid materials are summarized in Table 2. Various studies on the usage of different types of d-block metal oxides, especially TiO_2, and ZnO, have been carried out to synthesize support metal oxide composites [34,35]. Frequently used biopolymers for nanomaterials or composite synthesis are cellulose, chitosan, dextran, and starch [34-36]. Several methods such as *in situ* polymerization, sol-gel approach and direct blending/mixing have been followed to prepare supported metal oxide catalysts [37].

Making composites with cellulose has been reported to improve the activity by 1.3 to 3.5-times relative to pure cellulose matrix or metal oxide as a consequence of enhanced charge separation (described earlier). It is generally accepted that the synergistic interaction between both species, metal oxide and support material, is mandatory for increased efficiency. Indeed, the support material not only enhances the dispersion of metal oxide nanoparticles but also improves its adsorption of organic pollutants. Besides, the high surface area of the supported metal oxides has led to an increased harvesting of light for a superior activation of reactive species, hence boosted the photocatalytic reaction.

4.3   Factors influencing photocatalytic degradation

Experimental conditions such as initial concentration of pollutants, catalyst dose, solution pH, reaction temperature, reaction time, size, structure, and surface area of the photocatalyst, inorganic ions, light intensity, irradiation time, dissolved oxygen (DO) have been recognized to play substantial role on the efficiency of a photocatalytic method. Some of these factors that significantly alter the efficiency and rate of reaction are discussed briefly.

4.3.1   Initial concentration of pollutants

It is common that initial concentration of a reactant of any chemical reaction influence the rate of reaction. Heterogeneous catalytic degradation of POPs is associated with different steps as described earlier. The adsorption that is naturally an equilibrium process is a must since only the adsorbed species on the surface of catalyst participate in photocatalytic reaction. An increase in pollutants concentration shows positive effects for the adsorption up to a certain limit indeed. The efficiency of degradation generally decreases with increasing in the concentration of pollutants if other factors remain constant. Fig. 5 shows the effect of initial concentration of POPs on the rate and efficiency of photodegradation. In this case, both rate and efficiency decrease as the concentration of MCP increases. As the concentration of POPs is increased, it would be more difficult for the photons to reach the catalyst surface due to scattering effect. As a consequence, the rate of the reaction is decreased [44]. Another explanation is that at high initial concentration more species are adsorbed on the surface of catalyst that may lead to the decrease in the production of OH• radicals on the surface of catalyst, resulting in less percentage degradation of POPs [45].

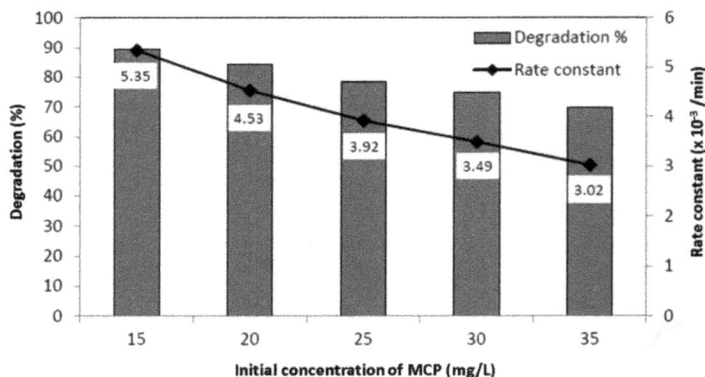

Figure 5. Effect of initial concentration of monocrotophos (MCP) on the photo degradation reaction. Experimental conditions: pH = 5; temp = 25 °C; light = UV; intensity of light = 30 W/m²; recirculation type photocatalytic reactor; recirculation rate = 400 mL/min [18]

4.3.2 Dose of catalysts

The optimum amount of photocatalyst is a very crucial factor to avoid its excess use. The determination of optimum dose of catalyst is sometimes difficult, since it depends on the various factors involved, e.g., type and concentration of targeted POPs, nature of light and its intensity, and the type and particle morphology of the catalyst used. The photodegradation increases with increasing the amount of catalyst up to a certain limit then decreases with increase in amount of catalyst. Initially the increase in loading of catalyst increases the number of active sites available for photodegradation reaction as can be seen in Fig. 6 [46]. However, beyond a certain amount of catalyst many probable incidents may happen. These may include (a) the transparency of solution decreases resulting in blocking of radiation to reach the catalyst surface and therefore percentage degradation starts decreasing, (b) interaction among the particles become significant which may result in the deactivation of activated species by collision with ground state of catalyst and (c) if the amount of pollutant molecules is small enough, the addition of more photocatalyst could not enhance the photodegradation efficiency.

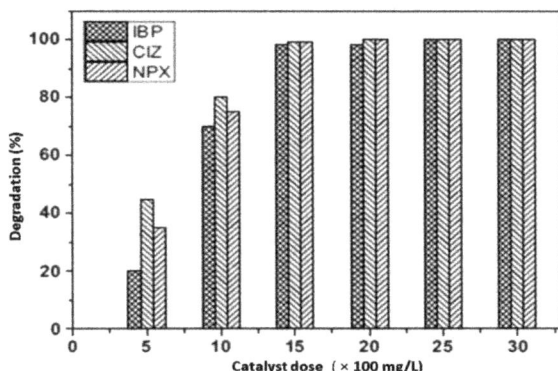

Figure 6. Effect of loading of surface functionalized titanium dioxide (TiO₂–NH₂) nanoparticles grafted into polyacrylonitrile (PAN)/multi-walled carbon nanotube composite nanofibers on efficiency. Experimental conditions: pH = 2; light source UV and irradiation time = 120, 40, and 25 min for IBP, NPX and CIZ, respectively [46]

4.3.3 Solution pH

The pH of medium is one of the most important factors affecting the photocatalytic degradation. A typical example is shown in Fig. 7. In this particular case, the efficiencies of degradations of IBP, NPX, and CIZ by TiO₂–NH₂ photocatalyst are improved with decreasing pH value. pH of solution in fact determines the nature surface charge of catalyst particles and alters the potential of catalytic reactions. The degree of adsorption of POPs on the surface of catalyst varies with amount and nature of the surface charge that eventually changes the reaction rate. This could be explained on the basis of point of zero charge (PZC) [44,47]. For example, the PZC for TiO₂ particles is $pH_{pzc} = 6.8$ [48]. At pH < 6.8, the surface of the catalyst gets positively charged and protonated according to the Eq. 9. Opposite phenomenon happens at pH > 6.8 according to the Eq. 10

$$pH = pzc \text{ (neutral)}: TiO_2 + H_2O \rightarrow Ti(OH)_4 + 2OH^- \tag{8}$$

$$pH < pzc \text{ (acidic)}: Ti(OH)_4 + H^+ \rightarrow Ti(OH)_4^+ \text{ (surface species)} \tag{9}$$

$$pH > pzc \text{ (alkaline)}: Ti(OH)_4 + OH^- \rightarrow TiO_2^- \text{ (surface species)} + H_2O \tag{10}$$

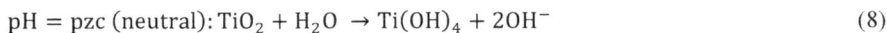

Thus, in acidic medium the TiO₂ surface is positively charged and the surface attracts anionic molecules and repel cationic species. Similarly, in alkaline medium the surface is negatively charged and the surface attracts cationic molecules and repel anionic species.

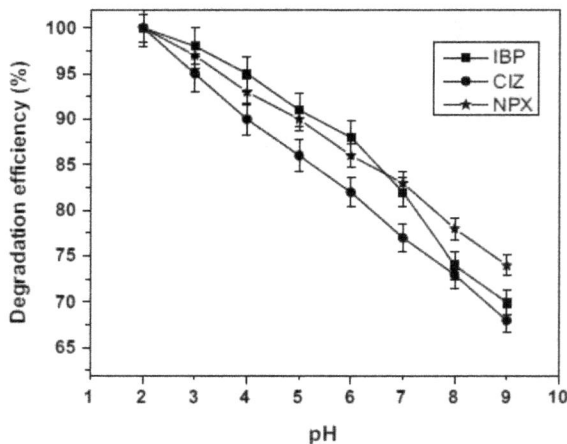

Figure 7. Effect of the solution pH on the degradation efficiency. Other conditions may be seen in Fig. 6 [46]

5.   Concluding remarks

The chapter describes the insight into the recent developments in the emerging field of photochemical wastewater treatment with nontoxic d-block metal oxides with the lights of importance, fundamental of catalysis, and application of catalysts for efficient degradation of POPs. From the recent scenario, it is explored that the AOP gains a significant popularity for removing POPs through photocatalytic means. It also reveals that among the various d-block metal oxides of which only a few of them are nontoxic namely TiO_2, MnO_2, Fe_2O_3 and ZnO are of focal point of photo-assisted AOPs. Set of methods for facile syntheses of metal oxides and nanocomposites catalysts with different microstates are available in literatures.

Anatase form of TiO_2 shows the best photocatalytic activity compared to its other crystallographic forms. While hexagonal wurtzite form of ZnO that is the most common exhibit a better catalytic efficiency in terms of quantum efficiency towards the photodegradation of POPs compared to TiO_2. MnO_2 and Fe_2O_3 and their composites are also assessed to exhibit good photocatalytic activities. Magnetic Fe_2O_3 is identified to be a unique support of nanocomposite that aides in easy separation of catalyst from wastewater after treatment via external magnetic field.

The loss of metal oxides photocatalyst, particle agglomerates, recombination of electron and hole, photocorrosion and so on that are genuine problems in practical uses have been evaluated to minimize by making composite with support materials. Various support materials are known, but most of them are synthetic and may lead to toxicity into the remediated water. Only the support materials those are biocompatible and nontoxic such as cellulose, chitosan and clay are preferred. The former two materials are biopolymers and can exhibit inherent adsorption capability due to their −OH linkage. Clay is aluminosilicate that can hold the metal oxide very strongly in composite by electrostatic interaction. In composites, the support materials lead to the formation of small particles with a high surface-to-volume ratio. Besides other advantages, the support materials that have also inherent adsorption capability thus offer a greater extent of adsorption of POPs that is prerequisite of a photocatalytic reaction.

References

[1] Hagetorn, C., Mc Coy, E. L., and Rahe, T. M. (1981). The potential for ground water contamination from septic effluents. Journal of Environmental Quality, 10(1), 1-8. https://doi.org/10.2134/jeq1981.00472425001000010001x

[2] Honeycutt, M., & Shirley, S. (2014). Dieldrin. Encyclopedia of Toxicology: Third Edition, 1100-1102. https://doi.org/10.1016/B978-0-12-386454-3.00132-9

[3] Sonkusare V.N., Chaudhary R.G., Bhusari G.S., Mondal A., Potbhare A.K., Mishra R.K., Juneja H.D., and Abdala A.A., Mesoporous octahedron-shaped tricobalt tetroxide nanoparticles for photocatalytic degradation of toxic dyes, ACS Omega.2020, 5, 7823-7835. https://doi.org/10.1021/acsomega.9b03998

[4] Hossain, M., Pabel, M., & Islam, M. (2022). Fenton-like processes for the removal of cationic dyes. In Advanced Oxidation Processes in Dye-Containing Wastewater (pp. 29-89). Springer, Singapore. https://doi.org/10.1007/978-981-19-0882-8_2

[5] Deb, A. K., Miran, M. S., & Mollah, M. Y. A. (2013). Active carbon prepared from vegetable wastes for the treatment of Pb (II) in aqueous medium. Bangladesh Journal of Scientific and Industrial Research, 48(2), 97-104. https://doi.org/10.3329/bjsir.v48i2.15739

[6] Sonkusare V.N., Chaudhary R.G., Bhusari G., Rai A.R., and Juneja H.D., Microwave-mediated synthesis, photocatalytic degradation and antibacterial activity of α-Bi2O3 microflowers/novel γ-Bi$_2$O$_3$ microspindles, Nano-Struct. Nano-Objects,2018,13, 121-131. https://doi.org/10.1016/j.nanoso.2018.01.002

[7] Sousa, M. A., Gonçalves, C., Pereira, J. H., Vilar, V. J., Boaventura, R. A., & Alpendurada, M. F. (2013). Photolytic and TiO$_2$-assisted photocatalytic oxidation of the anxiolytic drug lorazepam (Lorenin® pills) under artificial UV light and natural sunlight: A comparative and comprehensive study. Solar Energy, 87, 219-228. https://doi.org/10.1016/j.solener.2012.10.013

[8] J. Jacob (2013). A Review of the accumulation and distribution of persistent organic pollutants in the environment. International Journal of Bioscience, Biochemistry and Bioinformatics, 3(6), 657-661. https://doi.org/10.7763/IJBBB.2013.V3.297

[9] Yasmina, M., Mourad, K., Mohammed, S. H., & Khaoula, C. (2014). Treatment heterogeneous photocatalysis; factors influencing the photocatalytic degradation by TiO$_2$. Energy Procedia, 50, 559-566. https://doi.org/10.1016/j.egypro.2014.06.068

[10] Dehghani, M., Nadeem, H., Singh Raghuwanshi, V., Mahdavi, H., Banaszak Holl, M. M., & Batchelor, W. (2020). ZnO/cellulose nanofiber composites for sustainable sunlight-driven dye degradation. ACS Applied Nano Materials, 3(10), 10284-10295. https://doi.org/10.1021/acsanm.0c02199

[11] Zhang, K. L., Liu, C. M., Huang, F. Q., Zheng, C., & Wang, W. D. (2006). Study of the electronic structure and photocatalytic activity of the BiOCl photocatalyst. Applied Catalysis B: Environmental, 68(3-4), 125-129. https://doi.org/10.1016/j.apcatb.2006.08.002

[12] Vela, N., Calín, M., Yáñez-Gascón, M. J., el Aatik, A., Garrido, I., Pérez-Lucas, G., Fenoll, J., & Navarro, S. (2019). Removal of pesticides with endocrine disruptor activity in wastewater effluent by solar heterogeneous photocatalysis using ZnO/Na$_2$S$_2$O$_8$. Water, Air, & Soil Pollution, 230(6), 1-11. https://doi.org/10.1007/s11270-019-4185-y

[13] Ashkarran, A. A., Fakhari, M., Hamidinezhad, H., Haddadi, H., & Nourani, M. R. (2015). TiO$_2$ nanoparticles immobilized on carbon nanotubes for enhanced visible-light photo-induced activity. Journal of Materials Research and Technology, 4(2), 126-132. https://doi.org/10.1016/j.jmrt.2014.10.005

[14] Chouke, P.B.; Dadure, K.M.; Potbhare, A.K.; Bhusari, G.S.; Mondal A.; Chaudhary, K.; Singh, V.; Desimone, M.F.; Chaudhary, R.G.; Masram, D.T. Biosynthesized δ-Bi$_2$O$_3$ Nanoparticles from Crinum viviparum Flower Extract for Photocatalytic Dye Degradation and Molecular Docking, ACS Omega 2022, 7, 24, 20983-20993. https://doi.org/10.1021/acsomega.2c01745

[15] Sun, W., Meng, Q., Jing, L., Liu, D., & Cao, Y. (2013). Facile synthesis of surface-modified nanosized α-Fe$_2$O$_3$ as efficient visible photocatalysts and mechanism insight. The Journal of Physical Chemistry C, 117(3), 1358-1365. https://doi.org/10.1021/jp309599d

[16] Akpan, U. G., & Hameed, B. H. (2009). Parameters affecting the photocatalytic degradation of dyes using TiO$_2$-based photocatalysts: A review. Journal of Hazardous Materials, 170(2-3), 520-529. https://doi.org/10.1016/j.jhazmat.2009.05.039

[17] Jangid, N. K., Jadoun, S., Yadav, A., Srivastava, M., & Kaur, N. (2021). Polyaniline-TiO$_2$-based photocatalysts for dyes degradation. Polymer Bulletin, 78(8), 4743-4777. https://doi.org/10.1007/s00289-020-03318-w

[18] Sraw, A., Kaur, T., Pandey, Y., Sobti, A., Wanchoo, R. K., & Toor, A. P. (2018). Fixed bed recirculation type photocatalytic reactor with TiO$_2$ immobilized clay beads for the degradation of pesticide polluted water. Journal of Environmental Chemical Engineering, 6(6), 7035-7043. https://doi.org/10.1016/j.jece.2018.10.062

[19] Amiri, H., Nabizadeh, R., Martinez, S. S., Shahtaheri, S. J., Yaghmaeian, K., Badiei, A., Nazmara, S., & Naddafi, K. (2018). Response surface methodology modeling to improve degradation of chlorpyrifos in agriculture runoff using TiO_2 solar photocatalytic in a raceway pond reactor. Ecotoxicology and Environmental Safety, 147, 919-925. https://doi.org/10.1016/j.ecoenv.2017.09.062

[20] Ananpattarachai, J., & Kajitvichyanukul, P. (2015). Photocatalytic degradation of p, p′-DDT under UV and visible light using interstitial N-doped TiO_2. Journal of Environmental Science and Health, Part B, 50(4), 247-260. https://doi.org/10.1080/03601234.2015.999592

[21] Chalasani, R., & Vasudevan, S. (2013). Cyclodextrin-functionalized $Fe_3O_4@TiO_2$: Reusable, magnetic nanoparticles for photocatalytic degradation of endocrine-disrupting chemicals in water supplies. ACS Nano, 7(5), 4093-4104. https://doi.org/10.1021/nn400287k

[22] Topkaya, E., Konyar, M., Yatmaz, H. C., & Öztürk, K. (2014). Pure ZnO and composite ZnO/TiO_2 catalyst plates: a comparative study for the degradation of azo dye, pesticide and antibiotic in aqueous solutions. Journal of Colloid and Interface Science, 430, 6-11. https://doi.org/10.1016/j.jcis.2014.05.022

[23] Khan, S. H., Pathak, B., & Fulekar, M. H. (2018). Synthesis, characterization and photocatalytic degradation of chlorpyrifos by novel Fe:ZnO nanocomposite material. Nanotechnology for Environmental Engineering, 3(1), 1-14. https://doi.org/10.1007/s41204-018-0041-3

[24] Yadav, S., Kumar, N., Kumari, V., Mittal, A., & Sharma, S. (2019). Photocatalytic degradation of triclopyr, a persistent pesticide by ZnO/SnO_2 nano-composites. Materials Today: Proceedings, 19, 642-645. https://doi.org/10.1016/j.matpr.2019.07.746

[25] Shao, P., Ren, Z., Tian, J., Gao, S., Luo, X., Shi, W., Boyin, Y., Li, J., & Cui, F. (2017). Silica hydrogel-mediated dissolution-recrystallization strategy for synthesis of ultrathin α-Fe_2O_3 nanosheets with highly exposed (1 1 0) facets: a superior photocatalyst for degradation of bisphenol S. Chemical Engineering Journal, 323, 64-73. https://doi.org/10.1016/j.cej.2017.04.069

[26] Koli, P. B., Kapadnis, K. H., & Deshpande, U. G. (2019). Transition metal decorated Ferrosoferric oxide (Fe_3O_4): An expeditious catalyst for photodegradation of Carbol Fuchsin in environmental remediation. Journal of Environmental Chemical Engineering, 7(5), 103373. https://doi.org/10.1016/j.jece.2019.103373

[27] Choi, K. H., Min, J., Park, S. Y., Park, B. J., & Jung, J. S. (2019). Enhanced photocatalytic degradation of tri-chlorophenol by $Fe_3O_4@TiO_2@Au$ photocatalyst under visible-light. Ceramics International, 45(7), 9477-9482. https://doi.org/10.1016/j.ceramint.2018.09.104

[28] Cao, H., & Suib, S. L. (1994). Highly efficient heterogeneous photooxidation of 2-propanol to acetone with amorphous manganese oxide catalysts. Journal of the American Chemical Society, 116(12), 5334-5342. https://doi.org/10.1021/ja00091a044

[29] Kim, E. J., Oh, D., Lee, C. S., Gong, J., Kim, J., & Chang, Y. S. (2017). Manganese oxide nanorods as a robust Fenton-like catalyst at neutral pH: Crystal phase-dependent behavior. Catalysis Today, 282, 71-76. https://doi.org/10.1016/j.cattod.2016.03.034

[30] Shohel, M., Miran, M. S., Susan, M.H. & Mollah, M. Y. A. (2016). Calcination temperature-dependent morphology of photocatalytic ZnO nanoparticles prepared by an electrochemical-thermal method. Research on Chemical Intermediates, 42(6), 5281-5297. https://doi.org/10.1007/s11164-015-2358-x

[31] Hassan, F., Miran, M. S., Simol, H. A., Susan, M. H., & Mollah, M. Y. A. (2015). Synthesis of ZnO nanoparticles by a hybrid electrochemical-thermal method: Influence of calcination temperature. Bangladesh Journal of Scientific and Industrial Research, 50(1), 21-28. https://doi.org/10.3329/bjsir.v50i1.23806

[32] Mishra, M., & Chun, D. M. (2015). α-Fe$_2$O$_3$ as a photocatalytic material: A review. Applied Catalysis A: General, 498, 126-141. https://doi.org/10.1016/j.apcata.2015.03.023

[33] Crittenden, J. C., Zhang, Y., Hand, D. W., Perram, D. L., & Marchand, E. G. (1996). Solar detoxification of fuel-contaminated groundwater using fixed-bed photocatalysts. Water Environment Research, 68(3), 270-278. https://doi.org/10.2175/106143096X127703

[34] Mafra, G., Brognoli, R., Carasek, E., López-Lorente, Á. I., Luque, R., Lucena, R., & Cárdenas, S. (2021). Photocatalytic cellulose-paper: Deepening in the sustainable and synergic combination of sorption and photodegradation. ACS Omega, 6(14), 9577-9586. https://doi.org/10.1021/acsomega.1c00128

[35] Saad, A. M., Abukhadra, M. R., Ahmed, S. A. K., Elzanaty, A. M., Mady, A. H., Betiha, M. A., Shim, J., & Rabie, A. M. (2020). Photocatalytic degradation of malachite green dye using chitosan supported ZnO and Ce-ZnO nano-flowers under visible light. Journal of Environmental Management, 258, 110043. https://doi.org/10.1016/j.jenvman.2019.110043

[36] Banerjee, A., & Bandopadhyay, R. (2016). Use of dextran nanoparticle: A paradigm shift in bacterial exopolysaccharide based biomedical applications. International Journal of Biological Macromolecules, 87, 295-301. https://doi.org/10.1016/j.ijbiomac.2016.02.059

[37] Prasanna, S. S., Balaji, K., Pandey, S., & Rana, S. (2019). Metal oxide based nanomaterials and their polymer nanocomposites. In Nanomaterials and Polymer Nanocomposites (pp. 123-144). Elsevier. https://doi.org/10.1016/B978-0-12-814615-6.00004-7

[38] Hasanpour, M., Motahari, S., Jing, D., & Hatami, M. (2021). Numerical modeling for the photocatalytic degradation of methyl orange from aqueous solution using

cellulose/zinc oxide hybrid aerogel: Comparison with experimental data. Topics in Catalysis, 1-14. https://doi.org/10.1007/s11244-021-01451-y

[39] Sboui, M., Lachheb, H., Bouattour, S., Gruttadauria, M., La Parola, V., Liotta, L. F., & Boufi, S. (2021). TiO$_2$/Ag$_2$O immobilized on cellulose paper: A new floating system for enhanced photocatalytic and antibacterial activities. Environmental Research, 198, 111257. https://doi.org/10.1016/j.envres.2021.111257

[40] Thomas, M., Naikoo, G. A., Sheikh, M. U. D., Bano, M., & Khan, F. (2016). Effective photocatalytic degradation of congo red dye using alginate/carboxymethyl cellulose/TiO$_2$ nanocomposite hydrogel under direct sunlight irradiation. Journal of Photochemistry and Photobiology A: Chemistry, 327, 33-43. https://doi.org/10.1016/j.jphotochem.2016.05.005

[41] Farhadian, N., Akbarzadeh, R., Pirsaheb, M., Jen, T. C., Fakhri, Y., & Asadi, A. (2019). Chitosan modified N, S-doped TiO$_2$ and N, S-doped ZnO for visible light photocatalytic degradation of tetracycline. International Journal of Biological Macromolecules, 132, 360-373. https://doi.org/10.1016/j.ijbiomac.2019.03.217

[42] Sarkar, S., Chakraborty, S., & Bhattacharjee, C. (2015). Photocatalytic degradation of pharmaceutical wastes by alginate supported TiO$_2$ nanoparticles in packed bed photo reactor (PBPR). Ecotoxicology and Environmental Safety, 121, 263-270. https://doi.org/10.1016/j.ecoenv.2015.02.035

[43] Kosera, V. S., Cruz, T. M., Chaves, E. S., & Tiburtius, E. R. (2017). Triclosan degradation by heterogeneous photocatalysis using ZnO immobilized in biopolymer as catalyst. Journal of Photochemistry and Photobiology A: Chemistry, 344, 184-191. https://doi.org/10.1016/j.jphotochem.2017.05.014

[44] Kansal, S. K., Singh, M., & Sud, D. (2008). Studies on TiO$_2$/ZnO photocatalysed degradation of lignin. Journal of Hazardous materials, 153(1-2), 412-417. https://doi.org/10.1016/j.jhazmat.2007.08.091

[45] Manea, F., & Orha, C. (2018). Carbon-/zeolite-supported TiO$_2$ for sorption/photocatalysis applications in water treatment. In Photocatalysts-Applications and Attributes. IntechOpen. https://doi.org/10.5772/intechopen.80803

[46] Mohamed, A., Salama, A., Nasser, W. S., & Uheida, A. (2018). Photodegradation of ibuprofen, cetirizine, and naproxen by PAN-MWCNT/TiO$_2$-NH$_2$ nanofiber membrane under UV light irradiation. Environmental Sciences Europe, 30(1), 1-9. https://doi.org/10.1186/s12302-018-0177-6

[47] Behnajady, M. A., Modirshahla, N., & Hamzavi, R. (2006). Kinetic study on photocatalytic degradation of CI Acid Yellow 23 by ZnO photocatalyst. Journal of Hazardous Materials, 133(1-3), 226-232. https://doi.org/10.1016/j.jhazmat.2005.10.022

[48] Alaton, I. A., Balcioglu, I. A., & Bahnemann, D. W. (2002). Advanced oxidation of a reactive dyebath effluent: Comparison of O$_3$, H$_2$O$_2$/UV-C and TiO$_2$/UV-A processes. Water Research, 36(5), 1143-1154. https://doi.org/10.1016/S0043-1354(01)00335-9

Emerging Applications of Nanomaterials
Materials Research Foundations 141 (2023) 124-150

Materials Research Forum LLC
https://doi.org/10.21741/9781644902295-6

Chapter 6

Nanomaterials in the Automobile Sector

Mohammad Harun-Ur-Rashid[1] and Abu Bin Imran[2]*

[1]Department of Chemistry, International University of Business Agriculture and Technology, Dhaka-1230, Bangladesh

[2]Department of Chemistry, Bangladesh University of Engineering and Technology, Dhaka-1000, Bangladesh

* abimran@chem.buet.ac.bd

Abstract

The automobile sector is continuously putting efforts into ensuring the safety and comfort of passengers, intelligent traffic guidance systems, pollution reduction, and successful recycling processes. The constructive and successful utilization of nanomaterials is an effective and potential scientific strategy that simplifies the technologies to develop lighter, more durable and economical, faster, and smart devices by consuming less raw materials and optimum energy. Emerging nanomaterials in the automobile sector explores how nanomaterials and nanotechnology are utilized to boost the performance of materials and devices required for automobile application by designing and developing nanocomposites, nanoalloys, nanocoatings, nanocatalysts, nanolubricants, and nanoadditives. In addition to these, this chapter will focus on the application of green polymer nanocomposites in automobile sector to address the issues associated with nanotoxicity.

Keywords

Automobile, Nanomaterials, Nanotechnology, Nanoalloys, Nanocoatings, Nanocatalysts, Nanolubricants, Nanoadditives, Green Polymer Nanocomposites, Nanotoxicity

Contents

1.   Introduction

Free movement is the basic need of people and the fundamental requirement for the development of modern society. Automobiles play an important role in this regard. The automobile sector ceaselessly works on passenger safety and comfort, smart traffic guidance systems, pollution reduction, animal and human health safety, eco-friendliness, and eventually successful recycling processes. Day by day, the conventional materials have been replaced by the smart nanomaterials such as nanocomposite gels [1-3], nanocomposite polymer films [4-6], structural colored nanomaterials [7-9], molecular machines [10], nanomaterial based biosensors [11], and so on. Considering all these factors, the utilization of nanomaterials is an effective and prospective scientific strategy that put forward the technologies to design and develop lighter, smaller, safer, economical, durable, faster, and intelligent devices through the consumption of less amount of raw materials and optimum energy. The productive, constructive, and functional applications of nanomaterials through the incorporation of operative technology are termed nanotechnology. Nanomaterials with the help of potential nanotechnology may offer stronger but lighter engine and body materials, fuel cells, self-healing and scratch-resistant coatings, self-cleaning windshields, improved tires, quality lubricants, better catalytic convertors, nanoporous filters, eco-friendly corrosion protection, smart sensors and vision system, durable tires, and so on [12].

Fuel economy regulation and customer expectations, battery-electric vehicle (BEV) technology, advanced driver assist systems (ADAS), mobility-as-a-service (MaaS), safety, innovative manufacturing technologies, and material cost are the major driving forces that influence the material change in automobile sectors [13]. The utilization of nanomaterials in a different format for automobile construction has been greatly dominated by seven fundamental reasons depicted in Fig. 1. Customers are expecting that their new vehicles would be more fuel-efficient than their previous one. Greenhouse gas (GHG) emission regulations, corporate average fuel economy (CAFE), and customers' expectations have pressured automobile producers to improve vehicle efficiency. In this context, manufacturing lightweight vehicles is the key to meet imposed regulations and consumer expectations. Involvement of nanotechnology, nanomaterials, and nanocomposites are capable of producing lightweight construction materials without compromising other desirable properties and features [14]. The next corresponding driving force for material

Emerging Applications of Nanomaterials

Materials Research Foundations 141 (2023) 124-150

Materials Research Forum LLC

https://doi.org/10.21741/9781644902295-6

change is the prospective utilization of battery electric vehicle (BEV) technology. Batteries used in electric vehicles are substantially heavier than internal combustion engines (ICE) with comparable performance. In particular, the weight of Chevrolet Bolt (electric vehicle) 136 kg more than Volkswagen Golf of similar size [15]. Similarly, the new Mini EV (2020 model) of BMW is 130 kg heavier than a base Mini Cooper. For this reason, BEVs are coercing automakers to attempt to find novel and creative ways to reduce vehicles' weight. The third major driver is ADAS which includes infotainment, comfort, and productivity features for consumers. The thrust for automatic vehicles will lead to more sensors and additional supporting devices that lead to a heavier vehicle of 130-180 kg extra weight. The automakers plan to increase the consumers' safety and comfort by including more and more automatic features almost every model year [16]. This trend is likely to continue for personal vehicles and may become more prevalent for shared vehicles. ADAS and users' demand for smarter vehicles with more content will need lightweight construction materials. Vehicles used for ride-sharing are projected to run 5-7 times more miles per year than personal vehicles. As the components (suspension systems, doors, seats, and so on) of shared vehicles are used extensively, shared vehicles require improvements in robust joint design, structural durability, better thermal control, higher fatigue targets, and vehicle end-of-life cycle. These requisites demand high-performance materials that can achieve engineering targets under harsher conditions. Automobile producers are always trying to provide safer vehicles to a consumers where selecting effective material is crucial. It has been believed that the lighter vehicle is more vulnerable than a heavier one in a crash; however, it is not universally true. Developments in vehicle design, application of smart technologies, and selection and optimization of novel construction materials with a high strength-to-weight ratio can result in lightweight vehicles with expected safety. Manufacturing technology acts as both an obstacle as well a trigger to creating novel materials. Huge capital investments in present manufacturing procedures (injection and stamping molding) reduce the fast transformation to novel materials and innovative processes. However, the adaptation of Industry 4.0 catalyzes the creation of modern materials and groundbreaking manufacturing processes. Last but not the least driving force is the price of the material. The overall cost involved in processing and optimizing materials is a major variable in the automotive sector. In this highly competitive business, consumers are always looking for value for their money. As the world will face the scarcity of existing fuel very soon, the automakers are trying to introduce different technologies for producing fuel-efficient motor vehicles. The details of low fuel consumption technologies found by automakers have been summarized in Fig. 2.

Emerging Applications of Nanomaterials   Materials Research Forum LLC
Materials Research Foundations 141 (2023) 124-150   https://doi.org/10.21741/9781644902295-6

Figure 1. The major driving forces that influence the material change in automotive sectors.

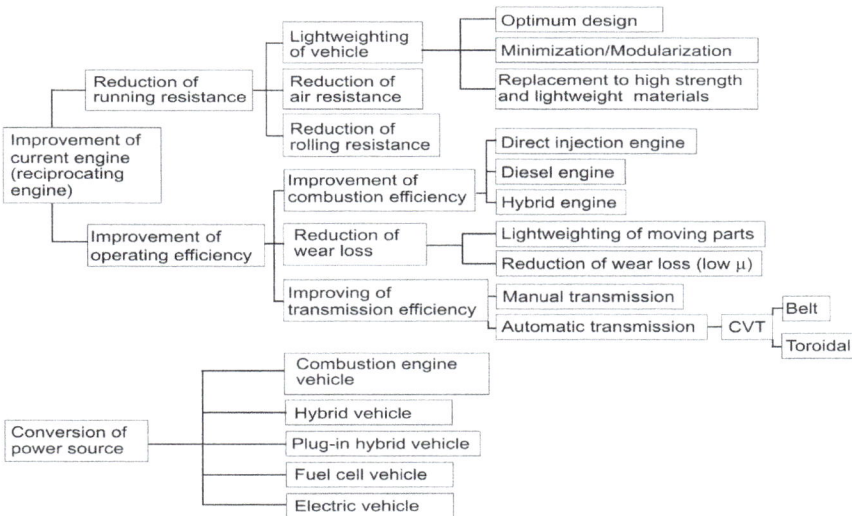

Figure 2. Details of low-fuel-consumption technologies. The figure has been reproduced with the permission from [17].

To date, steel is the primary material for constructing automotive structures. Recently, different high-strength steels (HSSs) have been developed and utilized to construct a lightweight automotive body. To attain more lightweight body, better HSSs are required. Nanostructured steels or ultrafine-grained (UFG) steels or nanosteels are highly promising to become one of the new HSSs for better performances. For example, high manganese-containing twinning induced plasticity (high-Mn TWIP) steels having austenite microstructures are classified as second generation advanced high strength steels (AHSSs) or Gen-2 steels which have of superior tensile strength and elongation. Next generation or third-generation AHSSs (Gen-3 steels) are steels containing less alloy elements than TWIP steels with moderate elongation between HSS and TWIP. Fig 3 shows the projection of utilization of the materials that are going to be decreased by 2040 while Fig 4 represents the projection of the utilization of materials that are going to be increased by 2040 in automobile sectors. The utilization of mild, HSS, AHSS, and hot formed steel (HF) is projected to be reduced in future automobile industries. On the contrary, the consumption of Gen-3 steel, aluminum alloys of different grades (Al alloy 5xxx and 7xxx grades), polymers, and polymer nanocomposites (PNCs) will increase for future production of automobiles. Al alloys of different grades are suitable for automotive doors, roof panels, lift gates, hoods, fenders, and body sides. PNCs find their potential applications in automobile construction due to their outstanding and novel physical and chemical properties. Present and future applications of nanomaterials in automobile construction are given in Table 1.

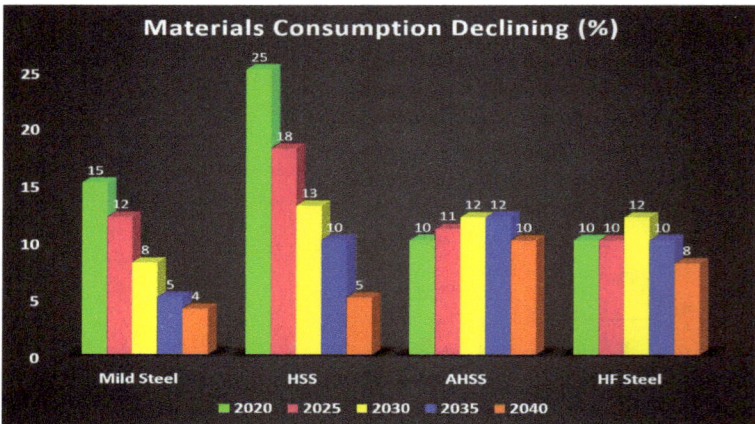

Figure 3. The projection of materials utilization that is going to be decreased by 2040 in automotive sectors [13].

Materials Research Forum LLC
https://doi.org/10.21741/9781644902295-6

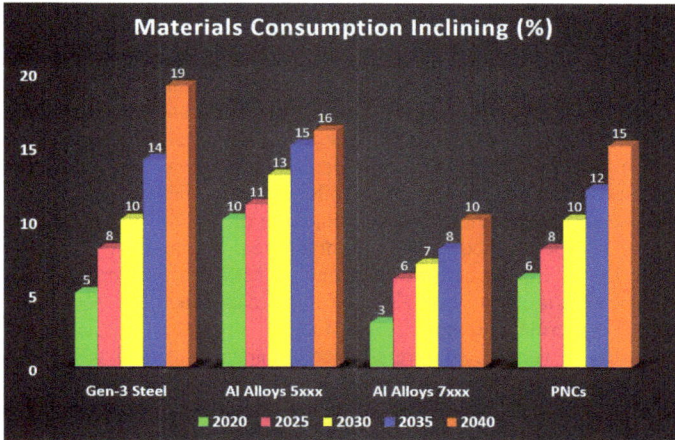

Figure 4. The projection of materials utilization that is going to be increased by 2040 in automotive sectors [13].

Table 1. Present and future applications of nanomaterials in automobile construction. The future applications have been represented by italic letters.

Application /Effect	Functionalities	External Components	Car body	Interior	Chassis and tires	Electronics	Engines
Mechanical	Hardness, friction, tribological features, breaking resistance	Nanocoating *Polymer glazing*	*Nanosteel*		Carbon black in tires *Nanosteel*		*Low-friction aggregate components*
Geometric	Large surface-to-volume ratio, porosity		*Gecko effect*	Nano filler *Gecko effect*		Super caps Fuel cells	
Electro-magnetic	Size dependent electro-magnetic		*Gluing on command*		*Switchable materials*	GMR sensors *Solar cells*	Piezo injectors
Optical	Color, fluorescence, transparency	*Electro chromatic layers*		Antiglare coatings			
Chemical	Reactivity, selectivity, surface properties	Care and sealing systems	Corrosion resistance	Dirtprotection *Fragrance in the cabin*			Catalysis Fuel additives

Emerging Applications of Nanomaterials   Materials Research Forum LLC
Materials Research Foundations 141 (2023) 124-150   https://doi.org/10.21741/9781644902295-6

2. Polymer nanocomposites (PNCs)

The PNCs, where the nanomaterials are utilized as a reinforcer, owing very large interfacial area per volume, and the gaps between the fillers and polymer matrix are significantly short. As a result, molecular interactions between the fillers and matrix will endow PNCs novel features that the conventional polymers or common macrocomposites do not have. The application of PNCs in automobile parts and systems is intended to escalate manufacturing speed, environmental sustainability, thermal stability, recycling, and reduce body weight. The successful and effective utilization of PNCs for structurally noncritical parts of automobiles such as body panels, front and rear fascia, valve/timing covers, cowl vent grills, interior, and truck beds could reduce several hundred millions kilograms of automobile body weight every year [18]. PNCs and nanofillers composed of automobile parts provide better thermal and dimensional stability, strength, noise reduction, improved modulus, corrosion resistance, cost versus performance ratio, and reliability than their counterparts. General Motors (GM), Montell USA, and Toyota have developed polypropylene (PP)/clay hybrid PNCs of better mechanical strength for automobile applications [19, 20]. The first commercial application of nanocomposites for automotive exterior components was accomplished by GM in 2002. They introduced a nanoclay/thermoplastic olefin nanocomposite running board to their GMC Safari and Chevrolet Astro vans [21]. Clay-based nanocomposites are applied to manufacture automobile's side doors, molding, cargo bed bridge, console, panels, engine cover, seat backs, and trim. Mercedes-Benz, Honda, Ford, Pontiac Bonneville, and Porsche have developed and utilized thermoplastic polyolefin resin based PNCs reinforced by silicon carbide (SiC) nanoparticles and PP for making their automobiles' interior and exterior components [22]. The appropriate application of PNCs in the automobile industry may reduce the amount of commonly used 900 kg of various metals to approximately 600 kg. Some of the commercially available PNCs developed by various companies, including Toyota, GM, InMat LLC, and Ube, have been summarized in Table 2. The application of PNCs for making automobile parts is shown in Fig. 5.

Table 2. Commercially available PNCs.

Polymer Matrix	Nanofiller	Applications
Polyamide 6	Exfoliated clay	Timing belt cover
Thermoplastics olefin	Exfoliated clay	Exterior step assist
Polyisobutylene	Exfoliated clay	Tires
Polyamides nylon 6, 66, 12	Exfoliated clay	Auto fuel systems
Polysulfone	Halloysite clay	Ultrafiltration membranes
Epoxy	Clay	Automobile body parts
Polyurethane	Clay	Automobile body parts

Green PNCs, produced from eco-friendly polymers reinforced with nanofillers, hold huge potential for constructing automobiles' body and engine parts without creating any harmful and hazardous impact on the environment and health [24]. Green PNCs are free from the risk of any health hazards and are expected to be environmentally friendly and biodegradable. Green PNCs are fabricated by fortifying the eco-polymer matrices such as cellulose, natural rubber, polysaccharides, chitin, collagen, polyhydroxyalkanoate, poly(ethylene oxide), poly(vinyl alcohol), polyethylene glycol, polyamide, polycarbonate, polyanhydride, polycaprolactone, and polylactic acid through the incorporation of eco-friendly nanofillers such as cellulose nanomaterials (nanofibers, nanoparticles, nanocrystals, etc.), natural fibres (jute, bamboo, zein, etc.), silica, clay, starch nanocrystals, halloysite nanotubes, nanocarbon, and carbonaceous nanomaterials. Many automobile companies such as BMW, Mercedes, Proton, Audi, Volkswagen, Daimler, Ford, Opel, and Cambridge Industry utilize different natural fibers-based PNCs in automobile sectors [25]. The applications of natural fiber-reinforced polymer nanocomposites in automotive industry have been condensed in Table 3.

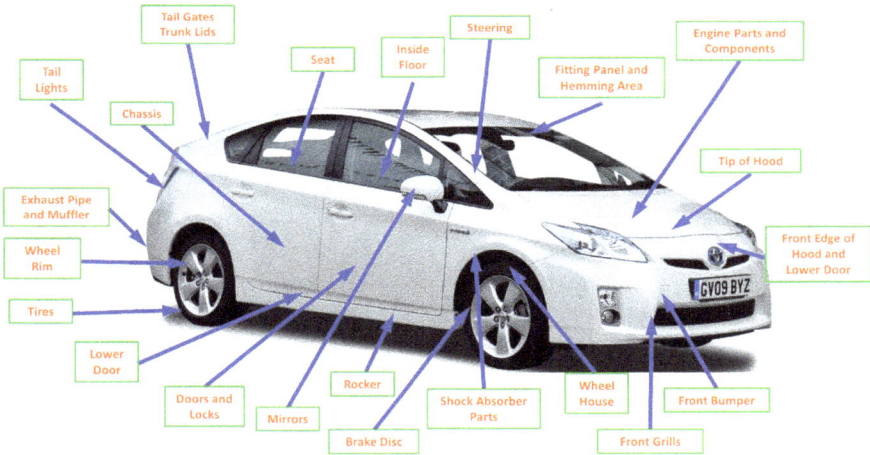

Figure 5. Prospective areas and components of an automobile for the appropriate utilization of green polymer nanocomposites for manufacturing parts. The figure has been reproduced and modified with permission from [23].

Table 3. The application of natural fiber composites in automotive industry.

Manufacturer	Model	Applications
Audi	A2, A3, A4, A4 Avant, A6, A8, Roadstar, Coupe	Bootliner, hat rack, side and back door panel, spare tirelining, and seatbacks
BMW	3, 5, and 7 series and other	Seatback, molded footwell linings, headliner panel, bootlining, door panels, and noise insulation panels
Citroen	C5	Interior door paneling
Daimler	A, C, E, and S classes	Pillar cover panel, door panels, car
Chrysler	EvoBus (exterior)	windshield/car dashboard, and a business table
Ford	Mondeo CD 162, Focus	Floor trays, B-pillar, bootliner, door inserts, and door panels
Fiat	Punto, Brava, Marea, Alfa Romeo 146, 156, 159	Door panel
GM	Cadillac De Ville, Chevrolet Trail Blazer	Seat backs, cargo area floor mat
Lotus	Eco Elise (July 2008)	Body panels, interior carpets, spoiler, and seats
Mercedes Benz	C, S, E, and A classes	Glove box (cotton fibers/wood molded, flax/sisal), instrument panel support, insulation (cotton fiber), molding rod/apertures, seat backrest panel (cotton fiber), trunk panel (cotton with PP/PET fibers), door panels (flax/sisal/wood fibers with epoxy resin/UP matrix), and seat surface/backrest (coconut fiber/natural rubber)
	Trucks	Internal engine cover, bumper, wheel box, engine insulation, sun visor, interior insulation, and roof cover
Mitsubishi		Cargo area floor, instrumental panel, and door panels
Opel	Vectra, Astra, Zafira	Door panels, instrumental panel, headliner panel, and pillar cover panel
Peugeot	406	Seatbacks, parcel shelf front, and rear door panels,
Renault	Clio, Twingo	Rear parcel shelf
Rover	2000 and others	Rear storage shelf/panel and insulations
Saab	9S	Door panels
Saturn	L300	Package trays and door panel
Toyota	Raum, Brevis, Harrier, Celsior,	Spare tire cover, door panels, floor mats, and seatbacks
VAUXHALL	Corsa, Astra, Vectra, Zafira	Interior door panels, pillar cover panel, headliner panel, and instrument
Volkswagen	Passat Variant, Golf, A4, Bora	Door panel, boot-lid finish panel, seatback, and bootliner
Volvo	V70, C70	Natural foams, seat padding, and cargo floor tray

PNCs utilized in the automobile industry for making tires. Precipitated carbon black (70-500 nm) blended with amorphous silica has been used for decades as nanofillers for the rubber compound for automobile tire manufacturing [26]. Modern advancement in nanotechnology has enabled the application of carbon nanotubes, rubber nanoparticles, and nanoclay for enhancing durability, rolling resistance, and wet grip of automobile tires. Most of automobile tires are made of rubber mixed with different nanomaterials as fillers. PNCs of nanosilica, carbonaceous nanomaterials, silicate, clay, nanocalcium carbonate, and others blended with diverse polymers such as styrene butadiene rubber (SBR), epoxidised natural rubber (NR), butyl rubber (BR), and so on. The major tire industries use epoxidised NR/organo-clay nanocomposite, SBR/clay nanocomposite, BR/clay nanocomposite, nanorubber, NR/BR/SBR-nanocalcium carbonate composite for manufacturing automobile tires.

Epoxy nanocomposites (ENCs) exhibit outstanding chemical, mechanical, thermal and electrical properties, high strength, and corrosion resistance, which are essential in automobile manufacturing. Numerous ENCs such as epoxy-inorganic NCs, epoxy-carbon fiber NCs, epoxy-clay NCs, epoxy-carbon nanotube NCs (ECNTNCs), epoxy-graphene NCs, epoxy-titanium NCs (ETNCs), epoxy-glass fiber NCs are highly potential to be used in automobile manufacturing industries [27]. The automotive brake system functions by transforming kinetic energy into thermal energy. To construct a perfect braking system, combining brake pad and disc of different specifications is necessary. Potassium titanate based nanomaterials could be an effective and suitable replacement in this regard. However, a phenolic resin based braking pad gets decomposed at high energy braking conditions. Modified epoxy resin can be an excellent candidate to overcome this limitation [28]. ENCs have been extensively used as coatings for automobile surfaces due to their excellent chemical, mechanical and thermal characteristics, corrosion resistance, and effective adhesion quality [29-31]. ETNCs can be used to manufacture automotive headlights, windows, and mirrors with antifogging and anti scratch properties [32]. Effective employment of EPNs can successfully protect automobile users from a severe accident occurred by fuel tank leakage by improving corrosion resistance properties, design variability, and safety at an optimum cost [33].

3.   Nanoalloys

Nanostructured steels or nanosteels are considered as Gen-2 and Gen-3 steels. Grain purification towards nanometer grain sizes is one of the major microstructural controlling approaches that will result in Gen-3 steels. Usually, the parts in an automobile body are divided practically into two segments having different functions. Components of the forepart and back frame section deform heavily during any collision to absorb impact belonging to the one segment, while the parts of the passengers' cabin, which deform less to protect the occupants from injury during a collision. Gen-3 steels with tensile strength of over 1000 MPa are suitable for making up the passengers' cabin [34].

Titanium (Ti) is an alluring metal for transportation equipment in automotive industries because of its high specific strength, heat, and corrosion resistance compared with other lightweight materials, such as Al, Mg alloys, and carbon fiber-reinforced polymer composites. Ti and its alloys have been utilized in the automobile sector to curtail the vehicles' weight [35]. Over the last few years, Ti nanoalloys, Ti-6Al-4V and Ti-Al-Fe, have been used for making engine valves, exhaust systems, and connecting rods of sports motor vehicles. In addition, many Ti-alloys of Cu, Al, Si, and Nb have been developed to enhance mechanical strength, deformation characteristics, corrosion, and oxidation resistance at higher temperatures. At present, Ti-alloys have found more applications in the automobile sector. Fuel tanks, fracture-split connecting rods, and fuel cells for automotive are constructed from Ti nanoalloys [36]. Nissan and Toyota have used different Ti nanoalloys commercially to make their automobiles exhaust systems. For example, Ti-1Cu was used for NISSAN GTR Spec-V in 2009 and for NISSAN GT-R in 2016, and Toyota used Ti-0.5Al-0.45Si-0.2Nb for TOYOTA LEXUS LFA in 2009. Many fuel cell-driven vehicles use polymer electrolyte fuel cells (PEFCs), where a separator is considered one of the essential components. Ti foils are used to construct separators employed in the fuel cell-driven vehicle "MIRAI" produced by Toyota Motor Corporation in 2014.

Fig. 6 depicts the chronological development of Ti nanoalloys automotive parts since 1990. Honda Motor Company introduced a connecting rod made of Ti-3Al-2V RE (Rare earth) in their sports automobile. In1998, Toyota released ALTEZZA, the first mass-produced automobile with Ti nanoalloy-based intake and exhaust valves. Volkswagen AG introduced the LUPO with Ti-6.8Mo-4.5Fe-1.5Al (or low-cost beta, LCB) alloys-based coil springs in 2000. The CORVETTE Z06 from Chevrolet, manufactured by GM, USA, was manufactured with Ti alloy based (Cp-TiGr.2) exhaust system in 2001 to improve the performance of vehicles by reducing body weight. To become the third most versatile automotive construction material after Fe and Al, Ti and its alloy must defeat the limitation of the present average cost of mass-produced parts. It is essential to practice new smelting methods to reduce the production cost. Moreover, remarkable improvement in performance can extend the utilization of low automotive parts of Ti nanoalloys through the replacement of iron parts. It is necessary to recognize the potentialities and find parts and structures that can effectively and successfully utilize Ti nanoalloy's performance from environmental conditions around the automobile industry.

Low cost fuel cell-driven vehicles require cheap and durable cathode catalyst for oxygen reduction reaction (ORR). Previously, Pt nanoparticles were modified with another metal support fuel cell with lower operation cost. In this context, a limited electrochemical dissolution Cu nanoparticles produces Pt-Cu nanoalloy based catalysts capable of maintaining activity even under the harsh automobile drive-cycles to be applied in fuel cells driven motor vehicles [37]. Pt-Cu core-shell nanocatalyst is a significant development for making cathode catalysts suitable for proton-exchange membrane fuel cells (PEMFCs). It is highly potential to be used in the automobile sector to support motor vehicle activity under usual driving conditions at continuous circuit voltage (Fig. 7).

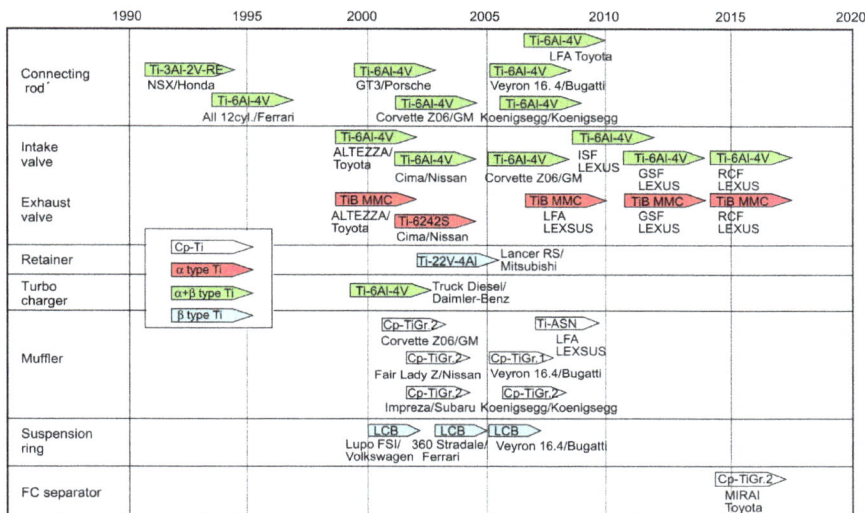

Figure 6. Chronological development of typical practical Ti parts in the automobile industry. The figure has been reproduced with permission from [17].

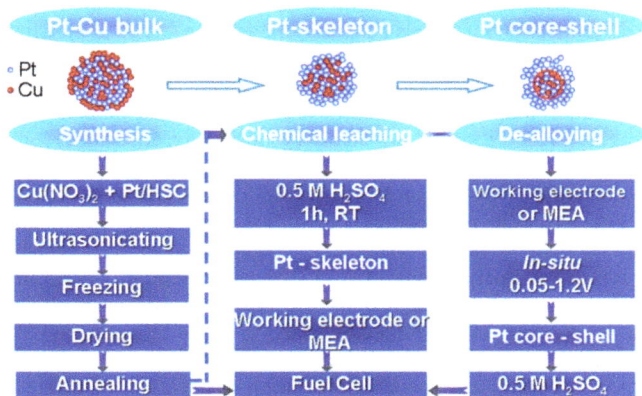

Figue 7. Profiles of consecutive steps for the preparation of three different systems Pt-Cu bulk alloy, skeleton and core–shell. The figure has been reproduced with the permission from [37].

Alloy nanoparticles composed of Cu and Zn can improve automobile engine performance and emission characteristics when mixed with fuels [38]. Cu-Zn nanoalloy particles are highly promising to be used as fuel additive for diesel engines. Nanoscale ceramics are used in automobile motors and transmissions to provide higher heat and wear resistance features. For instance, nanostructured ceramics based zirconium (Zr) and alumina nanoalloys have been selected for making cylinder of an engine. Nanoalloy composed of silicon carbide and nitride nanocrystals is used for manufacturing valve springs and ball bearings.

4. Nanolubricants

The objectives of using a lubricant is to reduce the friction, wear, and tear of automobile engines when two facets of engine components come in contact with each other. An appropriate and sufficient amount of lubricant supports the components of engines or other system to run fluently and incessantly. In addition, the lubricant prevents unexpected pressure or vibrations at bearings, minimize corrosion, provide sealing, and transport contaminants to filter for cleaning purpose. Inadequate and inappropriate utilization of lubricant reduces the engine life and may cause engine damage, fire, local welding, and overall engine failure. The application of lubricants is referred as lubrication which might be classified as boundary lubrication, mixed lubrication, hydrodynamic lubrication, and elastohydrodynamic lubrication [39]. Lubricants are commonly composed of base oil such as esters, hydrogenated polyolefins, silicones, fluorocarbons, etc. (90%) and additives (10%). Dry graphite, graphene, carbon nanomaterials, tungsten disulfide, and molybdenum disulfide are some examples of commonly used non-liquid powder lubricants. Nanomaterials are used as lubricant additives to enhance the tribological properties of lubricants. Addition of nanomaterials decreases the friction of two contatcing surfaces and increases the thermal conductivity of lubricants. The concentration and size of nanomaterilas are major factors that determine the tribological featurs of lubricants. The friction and wear of the worn surfaces is a principal cause of energy dissipation in automobile engines. Nanomaterials are highly promising for utilizing as lubricants, which are called nanolubricants, in automobile sectors since nanolubricants are capable of creating superlubricity [40-43]. In the automobile lubrication system, the boundary lubrication region is a highly concerned area to be taken care of due to the high friction coefficient. The boundary friction coefficient is higher in piston and camshaft regions. If the lubricant film is not thick enough to meet the requirement, grooves, microcracks, and valleys will be appeared on the rubbing surfaces due to the adhesive and abrasive wear. This undesirable situation can be handled by applying nanolubricants, which will help from nanoscale dimension by making the tribopair surfaces smooth and improving antifriction features in automotive engines. The nanolubrication mechanisms of carbon nanotube and graphene have been shown in Fig. 8 and Fig. 9, respectively [44].

Materials Research Forum LLC
https://doi.org/10.21741/9781644902295-6

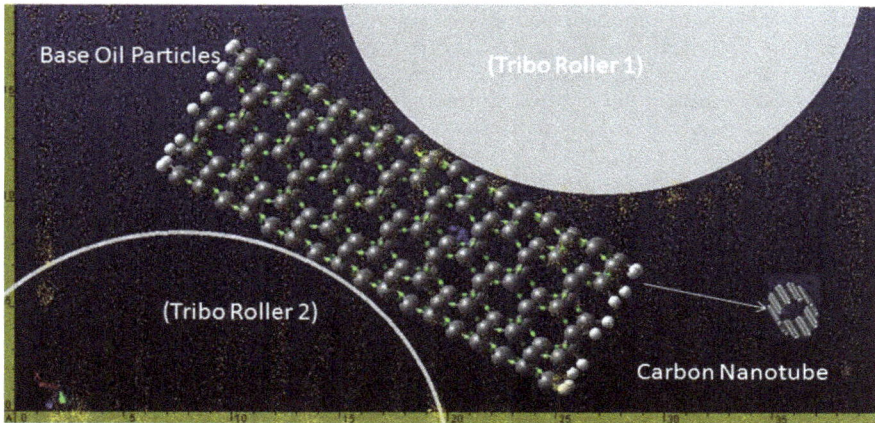

Figure 8. Nanolubricant mechanism of carbon nanotube in automobile engines. The figure has been reproduced with the permission from [44].

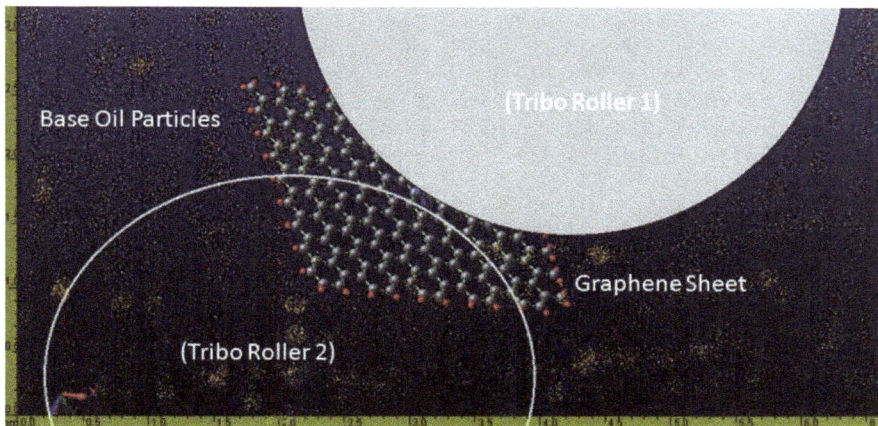

Figure 9. Nanolubricant mechanism of graphene in automobile engines. The figure has been reproduced with the permission from [44].

Graphene-based nanolubricants with anti wear and antifriction features can minimize exhaust emissions and save energy by improving the tribological behavior of automobile engines [45]. The loss of power during engine operation is due to the friction and wear that result in higher fuel consumption and emissions. Lubricant, loaded with graphene nanoaditive, contributes to saving the energy of automobile engines through the self-healing mechanism of the worn surfaces of ean engine. Application of graphene based nanolubricant enhances the antifriction feature by 29-35% and anti-wear feature by 22-29%. In addition, the engine performance is improved by 7-10% while the fuel consumption and emission (CO_2, HC, and NOx) are decreased by 17% and 5.42%, respectively. The combination of copper and graphene (Cu/Gr) nanomaterials also improves the properties of automobile lubricants [46]. The addition of Cu/Gr nanolubricants lowers the friction coefficient by 26.5-32.6% and wear rate by 25-30%. A homogeneous mixture of silica and graphene (SiO_2/Gr) can enhance water-based lubricant's rolling properties and tribological performances [47]. The addition of SiO_2/Gr nanoadditive brings down the friction coefficient by 48.5% and wear rate by 79%.

5. Nanocatalyst and nanoadditives

A catalytic converter, installed in an automobile vehicle, is an emission control device employed to reduce the toxic emissions from an ICE. The noble metal catalysts are generally used as catalytic converters in an automobile. Among platinum group metals, palladium (Pd) blended with suitable metal and composite oxides is an effective catalyst for the oxidation of carbon monoxide (CO) at low temperature. Pd metal nanocatalyst is the best possible catalyst for automobile applications [48]. Small particle size and greater dispersibility make the Pd nanocatalyst highly active. Other metals belonging to platinum group such as rhodium is utilized as a reduction nanocatalyst and Pt is utilized as reduction-oxidation nanocatalyst in catalytic converter. Structural formation after calcination, chemisorptions of CO, and the regeneration of PdO nanocatalysts have been represented graphically in Fig. 10, Fig. 11, and Fig. 12, respectively.

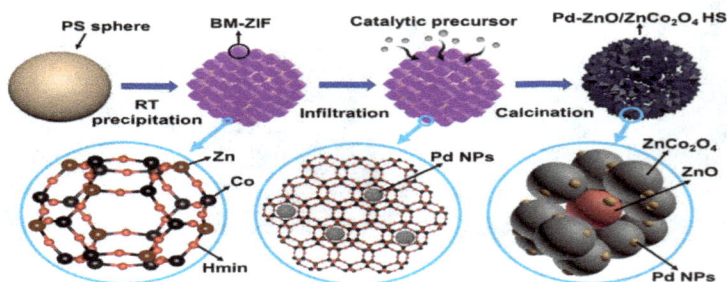

Figure 10. Structural formation of palladium oxide (PdO) after calcination. The figure has been reproduced with the permission from [48]

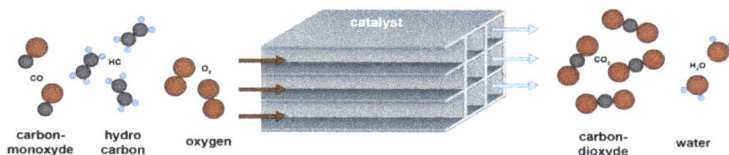

Figure 11. Chemisorptions of CO over PdO nanocatalysts. The figure has been reproduced with the permission from [48]

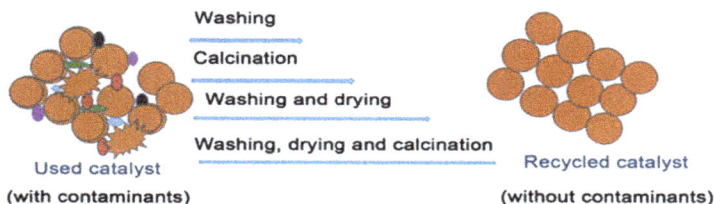

Figure 12. Regeneration of PdO nanocatalysts. The figure has been reproduced with the permission from [48].

Biofuels are receiving extensive attention to be used as fuel for automobile engines. The effect of the addition of nanomaterials such as the oxides of Al, Ce, Cu, Fe, Mg, Mn, Ni, Pd, Pt, Si, Ti, Zn, Zr, and graphene nanoparticles, carbon nanotube, and many more in biofuel has a great impact on the performance of automotive engines [49]. The biofuels are mixed with titanium dioxide (TiO_2) at different concentrations and examined for the performance and emission properties of automobile engines. The addition of TiO_2 with biofuel improves the performance of automobile engines such as power, torque, exhaust gas temperature (EGT), brake specific fuel consumption (BSFC), and brake thermal efficiency (BTE). In addition, fuel nanoadditives minimize the emission of GHGs and particulate matters and curtail the specific fuel consumption (SFC) [50]. Al oxide nanoadditives significantly reduce the ignition delay (ID) and combustion time (CD), while enhance the peak pressure and heat release rate (HRR) at the highest load and cylinder pressure. The nanoadditive is capable of minimizing hydrocarbon and CO emissions by 26.72% and 48.43%, respectively [51]. Addition of carbon nanotube to automobile fuel at 30, 60, and 90 ppm proportions enhances the power, BTE, and BSFC of diesel engines by 3.67%, 8.12%, and 7.12%, respectively [52]. Graphene oxide (GO) nanoaditive at different concentrations can improve automobile engine performance remarkably [53]. Cerium oxide (CeO_2) nanoparticle addition to fuel enhances the performance and emission characteristics of automotive engines [54-56]. Moreover, the addition of nanoadditives can extend the fuel droplet life during fuel injection and combustion process (Fig. 13 and Fig.

Emerging Applications of Nanomaterials
Materials Research Foundations 141 (2023) 124-150

Materials Research Forum LLC
https://doi.org/10.21741/9781644902295-6

14). The delay in the evaporation process may reduce the loss of fuel and improve fuel efficiency significantly.

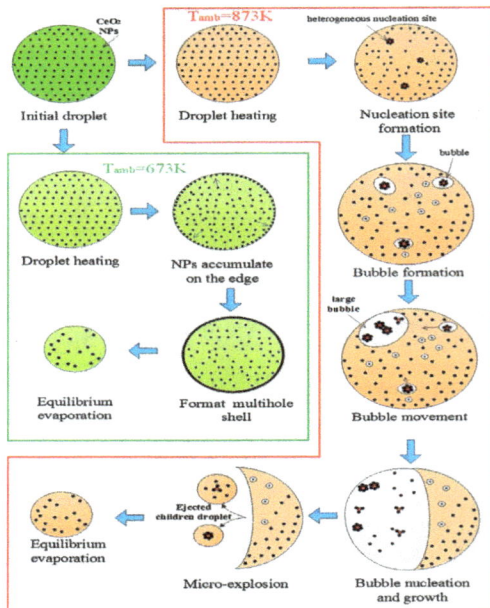

Figure 13. Regeneration of PdO nanocatalysts. The figure has been reproduced with the permission from [55].

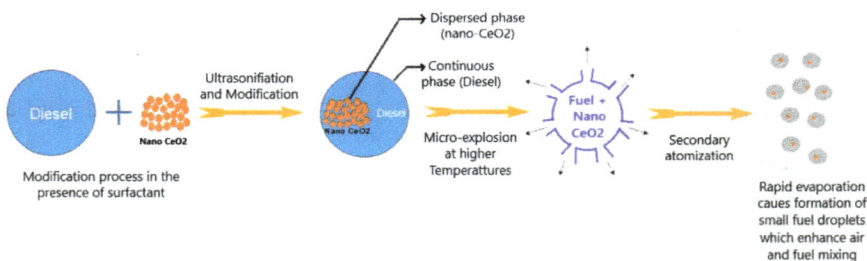

Figure 14. Step-by-step process of micro-explosion when cerium oxide (CeO₂) nanoparticle is blended with diesel. The figure has been reproduced with the permission from [54.].

Emerging Applications of Nanomaterials   Materials Research Forum LLC
Materials Research Foundations 141 (2023) 124-150   https://doi.org/10.21741/9781644902295-6

Nanomaterials to be used as nanoaddtives should have essential properties such as greater surface area to volume ratio, micro-explosion, oxygen buffer, enhanced thermal conductivity, antiwear, corrosion inhibition, and catalytic activity (Fig. 15). Oxygen buffer properties ensure the availability of oxygen for complete combustion and curtail the amount of unburnt emission. Higher surface area to volume ratio enhances the catalytic performance by providing greater surface area for fuel particles to interact. Nanoaddtive should be capable of supporting micro-explosion of the automotive fuel at the nozzle, ensuring better atomization by mixing air and fuel properly. Antiwear and corrosion inhibition characteristics of nanoadditives will keep the fuel tank and delivery system clean and corrosion free. Nanoaddvitives having better thermal conductivity act as heat sink and minimize the engine temperature and emissions. Improved catalytic performance of nanoadditive increase the fuel combustion rate and lower engine emissions. Since late 1990, CNTs blended with nylon can also be found in the automobile fuel system to avoid static charge build-up.

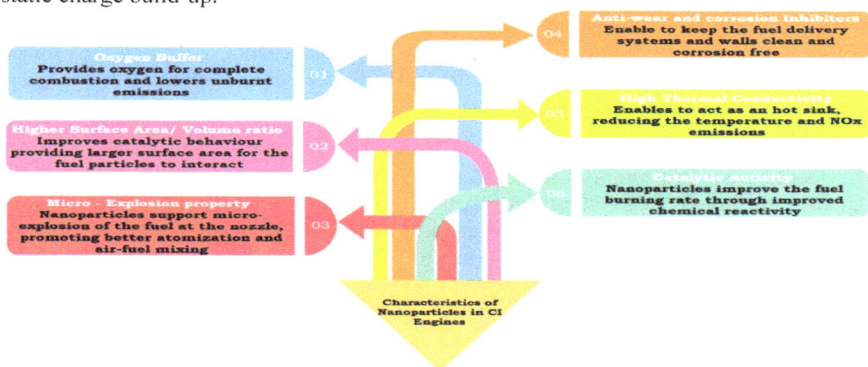

Figure 15. Nanoparticles with desirable properties in CI engines to enhance performance and emissions simultaneously. The figure has been reproduced with the permission from [54.].

6. Nanocoating

Nanocoating is the formation of nanolayer thickness on the surfaces of materials to be protected from adverse conditions such as wearing, chemicals, corrosion, heat, radiation, friction, dirt, and fire. Nanotechnology continuously attempts to improve its outcomes by applying nanomaterials to meet the ever-increasing expectations on modern advanced nanocoatings. Carbon nanotube based composite provides automobile exterior body panels that supports the application of electrostatic paint lines [57]. Ceramic nanoparticles are added to auto-paints to improve their scratch resistance ability and protect their initial gloss for an extended period of time. Self-cleaning nanocoatings employed on the exterior and interior body parts provide antifogging windshield and mirror. Polymer coatings have been extensively used in the automobile industry to inhibit the corrosion of metallic surfaces. The corrosion protection sometimes fails due to the repeated exposure of the coated surfaces to harsh chemicals or strong electrolytes. To improve the defensive mechanism of surface coatings, 2-aminothiazole (AT) modified TiO_2-ZrO_2 nanoparticles are incorporated to polyurethane (PU). The PU-AT/TiO_2-ZrO_2 PNC coated surface shows excellent corrosion inhibition during seawater exposure [58]. The synthesized PNC coating, PU-AT/TiO_2-ZrO_2, can provide outstanding hydrophobic, mechanical, and barrier attributes that reduce the degradation and extend the life of the coated surfaces. A self-cleaning hydrophobic nanocoating (SCHN) is highly prospective in the automobile sector [59]. Ni-SiC nanocoatings are suitable for using on the surface of automobile cylinder liners [60]. Fire and flame retardant nanocoatings can be used to coat automobile surfaces to protect from fire and to make fire alarms [61-64]. Flame retardant nanocoating composed of carbon black (CB), polyvinyl alcohol (PVA), and carbon nanotube can be used as a smart coating, which is capable of creating short time alarm at 350 °C and in the case of fire [61]. The experimental results show that the flame retardant nanocoating can be used as a fire-warning sensor. There are varieties of nanocoatings prepared from metal, ceramic, and PNCs capable of protecting the metal surfaces from corrosion [65-68].

7. Major challenges and possible solutions

The adoption and utilization of nanomaterials in automobile sectors can be afflected by numerous factors, including the risk of parts failure, demand and supply chain, recycling at the end of a product life cycle, environmental issues, cost-performance and risk-benefit ratio, and consumers' expectations. There is a huge chance of failure when newly developed materials are utilized commercially for the first time. There is the risk of vehicles recall from the automobile sector due to the failure of components made of newly developed materials. Many automakers have set up their own R&D sections to examine the novel materials and conduct feasibility surveys for effective and successful utilization to avoid such a scenario. These processes are expensive, time-consuming, and create obstacles to using innovative nanomaterials. Usually, newly developed materials are tested in the appropriate place at a low-volume vehicles. For example, BMW has used carbon fiber reinforced polymers (CFRPs) extensively in its "i-series" vehicles. Irregularity in demand and supply chain is another factor affecting the usage of nanomaterials in automobile sectors. Due to unpredictable conditions, the suppliers may not be able to maintain the supply chain for providing the required amount of nanomaterials with desirable qualities ceaselessly. This unexpected interruption can result in a delay in vehicle production. To avoid these undesirable circumstances, automakers rely on multiple suppliers rather than a single one and ensure that the required materials have a strong supply chain globally before initiating the use of nanomaterials in massive production. Manufactures deploy their investments globally to ensure continuous production by maintaining a quality supply chain of materials and developing infrastructure to drive innovation. Considering the health and environmental impact of nanotechnology products, it is essential to implement a life-cycle-based method associated with risk assessment for identifying potential problems and adopting safer manufacturing processes that are harmless to health and environmental components. Nanomaterial based products are synthesized by either top-down or bottom-up methods which generate a great deal of wastage. The manufacturing of nanotechnology products requires the highest clarity, which is involved with post-processing or reprocessing stages, low yield, usage of toxic acidic or basic chemicals and organic solvents, a requirement of extreme operation conditions, the consumption of utilities, and the generation of GHGs that eventually produce waste materials. The excessive use of resources and energy ultimately creates severe and long-term impacts on the environment and human health due to the accidental or unintentional exposure of nanomaterials. So, ensuring the recyclability of nanomaterials is a principal factor in material qualification for achieving a greater risk-benefit ratio. Although the incorporation of nanomaterial can enhance the functionalities of bulk material, the production and processing cost may leave-behind the benefits. For instance, the price of carbon fiber is approximately ten times more than a similar amount of steel. The automobile sector is very competitive in terms of manufacturing cost, since the customers have many options, and several fragments of the automotive market are price-sensitive. Therefore, the selection, optimization, and manufacturing technologies of nanomaterials should be conducted very carefully to attain a maximum cost-performance ratio. Among multiple market forces prevailing in automobile sector, customers responsiveness always plays an important role in material selection, specifically for materials used for interior construction. For example, if leather seats are preferred by the customers, an automobile manufacturer may be reluctant to go for new material even though the new one is better but cheaper than the conventional one. The automakers can address this type of issue by optimizing the appearances and qualities of newly developed materials by adopting nanotechnologies that are state of the art. The utilization of artificial intelligence and innovative manufacturing approaches such as thin-wall casting, additive manufacturing, and resin transfer molding will create the opportunities for nanomaterial-based construction materials for automobile industries.

8.   Future prospects

The automobile sector demands the rapid and practical development of nanotechnology, which can greatly impacting the expectations of consumers to improve their vehicles. Applying appropriate nanomaterials through effective nanotechnologies will generate smart automobile sectors and trends. Electronic control of automobile components is one of the quick-changing aspects of advanced and future automobiles. The electronically regulated antilock braking system, fuel injection, headlamp brightness control, exhaust emission, automated air conditioning, sensor-based driver's seat adjustment, and automatic steering control are some of the instances where the application of nanomaterials could be the best solution. Nanotechnology implementation may enable artificial intelligence for safe driving by reducing personal failures. Nanomaterials can be utilized to develop and fabricate all automotive components and subsystems. Advanced nanoparticles can be used as nanofillers for smart nanocomposites, nanodevices, nanocoatings, nanoelectronics, nanobatteries, nanofuelcells, nanosensors, and so on for cutting-edge automobiles.

Conclusion

Nanomaterials have greatly influenced the advancement of the automobile industry since the physical and chemical properties of the bulk material can be modified and controlled at the nanoscale level. The chapter has focused on some of the crucial and significant applications of nanomaterials in the automobile sector. Many of the nanotechnology applications for automobile purposes are limited to the research level because of the lack of successfull initiatives for commercialization. Nanomaterial selection, processing, optimization, fabrication, modification, manufacturing cost, recycling, and hazards are some of the limitations that hamper the commercial nanotechnology applications in automobile sector like other nanotechnology based sectors. The utilization of nanomaterials through nanotechnology is yet an incompletely developed process in any of the industries, but it is strongly believed that the nanotechnology is going to rule the automobile sector to a substantial range and will create products with novel characteristics which are unimaginable today.

Acknowledgements

A.B.Imran gratefully acknowledges the support of the Grant of Advanced Research in Education (GARE) (PS2016239) from the Ministry of Education, the Peoples Republic of Bangladesh. The author is also thankful to Committee for Advanced Studies and Research (CASR) in BUET for funding.

References

[1] M. Harun-Ur-Rashid, T. Seki, Y. Takeoka, Structural colored gels for tunable soft photonic crystals, Chem. Rec. 9 (2009) 87-105. https://doi.org/10.1002/tcr.20169

[2] M. Harun-Ur-Rashid, A. B. Imran, Superabsorbent hydrogels from carboxymethyl cellulose", in Ibrahim H. Mondal (ed.) Carboxymethyl Cellulose. Volume I: Synthesis and Characterization, Nova Science Publishers, New York, 2019, pp. 159-182.

[3] M.R. Karim, Harun-Ur-Rashid, A.B. Imran, Highly stretchable hydrogel using vinyl modified narrow dispersed silica particles as cross-linker, ChemistrySelect, 5 (2020) 10556-10561. https://doi.org/10.1002/slct.202003044

[4] M. Harun-Ur-Rashid, T. Foyez, A. B. Imran, Fabrication of stretchable composite thin film for superconductor applications. In Sensors for Stretchable Electronics in Nanotechnology, CRC Press, 2021, pp. 63-78. https://doi.org/10.1201/9781003123781-5

[5] R. Bagade, A. K. Potbhare, R. G. Chaudhary, A. Mondal, M Desimone, R.Mishra, H.D. Juneja. Microspheres/Custard-Apples Copper (II) Chelate Polymer: Characterization, Docking, Antioxidant and Antibacterial Assay, ChemistrySelect, 4 (2019) 6233-6244. https://doi.org/10.1002/slct.201901115

[6] A-N. Chowdhury, J. Shapter, and A. B. Imran, Innovations in Nanomaterials, Nova Science Publishers, Inc., NY, USA, 2015

[7] M. Harun-Ur-Rashid, A. B. Imran, T. Seki, Y. Takeoka, M. Ishii, H. Nakamura, Template synthesis for stimuli-responsive angle independent structural colored smart materials, Trans. Mater. Res. Soc. 34 (2009) 333-337. https://doi.org/10.14723/tmrsj.34.333

[8] M. Harun-Ur-Rashid, A. B. Imran, T. Seki, M. Ishii, H. Nakamura, Y. Takeoka, Angle-independent structural color in colloidal amorphous arrays, ChemPhysChem, 11 (2010) 579-583. https://doi.org/10.1002/cphc.200900869

[9] Y. Takeoka, S. Yoshioka, M. Teshima, A. Takano, M. Harun-Ur-Rashid, M., T. Seki, Structurally coloured secondary particles composed of black and white colloidal particles, Sci. Rep. 3 (2013) 1-7. https://doi.org/10.1038/srep02371

[10] A. B. Imran, M. Harun-Ur-Rashid, Y. Takeoka, Polyrotaxane Actuators. In Soft Actuators, Springer, Singapore, 2019, pp. 81-147. https://doi.org/10.1007/978-981-13-6850-9_6

[11] M. Harun-Ur-Rashid, T. Foyez, I. Jahan, K. Pal, A. B. Imran, Rapid diagnosis of COVID-19 via nano-biosensor-implemented biomedical utilization: a systematic review, RSC Advances, 12 (2022) 9445-9465. https://doi.org/10.1039/D2RA01293F

[12] A.K. Chandra, N.R. Kumar, Polymer nanocomposites for automobile engineering applications. In Properties and Applications of Polymer Nanocomposites. Springer, Berlin, Heidelberg, 2017, pp. 139-172. https://doi.org/10.1007/978-3-662-53517-2_7

[13] S. Modi, A. Vadhavkar, Technology Roadmap: Materials and Manufacturing, Michigan USA, 2019.

[14] Q. Wang, S. Xiao, S.Q. Shi, L. Cai, Effect of light-delignification on mechanical, hydrophobic, and thermal properties of high-strength molded fiber materials. Sci. Rep. 8 (2018) 1-10. https://doi.org/10.1038/s41598-018-19623-4

[15] UBS. (2017, August). Retrieved from https://www.cargroup.org/wp-content/uploads/2017/08/Langan.pdf

[16] S.M. Zoepf, Automotive features: mass impact and deployment characterization, Doctoral dissertation, Massachusetts Institute of Technology, 2011.

[17] T. Furuta, Automobile applications of titanium. In Titanium for consumer applications, Elsevier, 2019, pp. 77-90. https://doi.org/10.1016/B978-0-12-815820-3.00006-X

[18] J.M. Garces, D.J. Moll, J. Bicerano, R. Fibiger, D.G. McLeod, Polymeric nanocomposites for automotive applications, Adv. Mater. 12 (2000) 1835-1839. https://doi.org/10.1002/1521-4095(200012)12:23<1835::AID-ADMA1835>3.0.CO;2-T

[19] A. Usuki, M. Kato, A. Okada, T. Kurauchi, Synthesis of polypropylene-clay hybrid, J. Appl. Polym. Sci. 63 (1997) 137-138. https://doi.org/10.1002/(SICI)1097-4628(19970103)63:1<137::AID-APP15>3.0.CO;2-2

[20] C.J. Chirayil, J. Joy, H.J. Maria, I. Krupa, S. Thomas, (2016). Polyolefins in automotive industry. In Polyolefin Compounds and Materials, Springer, Cham, 2016, pp. 265-283. https://doi.org/10.1007/978-3-319-25982-6_11

[21] D. Rosato, Automotive and nanocomposites. SpecialChem February 9 (2007) 4.

[22] N. Kakarala, S. Shah, SPE Automotive TPO Global Conference 2000 (2000) 147-158.

[23] NK. Akafuah, S. Poozesh, A. Salaimeh, G. Patrick, K. Lawler, K. Saito, Evolution of the automotive body coating process-A review. Coatings, 6 (2016) 24. https://doi.org/10.3390/coatings6020024

[24] M. Harun-Ur-Rashid, A.B. Imran, M.A.B.H. Susan, (2022). Green Polymer Nanocomposites in Automotive and Packaging Industries, Curr. Pharm. Biotechnol. 6 May (2022). https://doi.org/10.2174/1389201023666220506111027.

[25] A. B. Imran, M. A. B. H Susan, Natural fiber-reinforced nanocomposites in automotive industry. In Nanotechnology in the Automotive Industry, Elsevier, 2022, pp. 85-103. https://doi.org/10.1016/B978-0-323-90524-4.00005-0

[26] H. Vasiliadis, Nanotechnology in automotive tyres. ObservatoryNANO, 4 (2011) Briefing no 23.

[27] M.N. Sakib, A.A. Iqbal, Epoxy based nanocomposite material for automotive application-a short review, Int. J. Automot. Mec. Eng. 18 (2021) 9127-9140. https://doi.org/10.15282/ijame.18.3.2021.24.0701

[28] DSEA Chan, G.W Stachowiak, Review of automotive brake friction materials. Proc. Inst. Mech. Eng. Part D J. Auto. Eng. 218 (2004), 953-966. https://doi.org/10.1243/0954407041856773

[29] F. Lei, J. Yang, B. Wu, L. Chen, H. Sun, H. Zhang, D. Sun, Facile design and fabrication of highly transparent and hydrophobic coatings on glass with anti-scratch property via surface dewetting. Prog. Org. Coat. 120 (2018) 28-35. https://doi.org/10.1016/j.porgcoat.2018.03.008

[30] R.K. Mishra, K. Verma, R.G. Chaudhary, T.L. Lambat, K. Joseph, An efficient fabrication of polypropylene hybrid nanocomposites using carbon nanotubes and PET fibrils. Mater. Today: Procs. 29 (2020), 794-800. https://doi.org/10.1016/j.matpr.2020.04.753

[31] A.U. Chaudhry, S.P. Lonkar, R.G. Chudhary, A. Mabrouk, A.A. Abdala, Thermal, electrical, and mechanical properties of highly filled HDPE/graphite nanoplatelets composites. Mater. Today: Procs. 29 (2020) 704-708. https://doi.org/10.1016/j.matpr.2020.04.168

[32] S. Kugler, K. Kowalczyk, T. Spychaj, Transparent epoxy coatings with improved electrical, barrier and thermal features made of mechanically dispersed carbon nanotubes. Prog. Org. Coat. 111 (2017) 196-201. https://doi.org/10.1016/j.porgcoat.2017.05.017

[33] L.N. Shafigullin, A.M. Sotnikov, N.V. Romanova, E.S. Shabaeva, D.R. Sarimov, Development of a polymeric fuel tank with high barrier properties. In IOP Conference Series: Materials Science and Engineering, IOP Publishing, 570 (2019) pp. 012088. https://doi.org/10.1088/1757-899X/570/1/012088

[34] A. Asghari, A. Zarei-Hanzaki, M. Eskandari, Temperature dependence of plastic deformation mechanisms in a modified transformation-twinning induced plasticity steel. Mater. Sci. Eng. A 579 (2013) 150-156. https://doi.org/10.1016/j.msea.2013.04.106

[35] H. Singh, G.S. Brar, H. Kumar, V. Aggarwal, A review on metal matrix composite for automobile applications. Mater. Today: Proc. 43 (2021) 320-325. https://doi.org/10.1016/j.matpr.2020.11.670

[36] K. Takahashi, K. Mori, H. Takebe, Application of titanium and its alloys for automobile parts. In MATEC Web of Conferences, EDP Sciences 321 (2020) 02003. https://doi.org/10.1051/matecconf/202032102003

[37] A. Marcu, G. Toth, R. Srivastava, P. Strasser, Preparation, characterization and degradation mechanisms of PtCu alloy nanoparticles for automotive fuel cells. J. Power Sources, 208 (2012) 288-295. https://doi.org/10.1016/j.jpowsour.2012.02.065

[38] A. Gangwar, A. Bhardawaj, R. Singh, N. Kumar, Enhancement in performance and emission characteristics of diesel engine by adding alloy nanoparticle (No. 2016-01-2249). SAE Technical Paper (2016). https://doi.org/10.4271/2016-01-2249

[39] P. Kumar, R.K. Upadhyay, Nanomaterials Lubrication for Transportation System. In V. Kumar, A.K. Agarwal, A. Jena, R.K. Upadhyay R.K. (eds) Advances in Engine Tribology. Energy, Environment, and Sustainability, Springer, Singapore, 2022. https://doi.org/10.1007/978-981-16-8337-4_7

[40] J. Li, Y. Peng, X. Tang, Q. Xu, L. Bai, Effect of strain engineering on superlubricity in a double-walled carbon nanotube. Phys. Chem. Chem. Phys. 23 (2021) 4988-5000. https://doi.org/10.1039/D0CP06052F

[41] B. Wang, K. Gao, Q. Chang, D. Berman, Y. Tian, Magnesium silicate hydroxide-MoS2-Sb2O3 coating nanomaterials for high-temperature superlubricity. ACS App. Nano Mater. 4 (2021) 7097-7106. https://doi.org/10.1021/acsanm.1c01104

[42] H. Wang, Y. Liu, Superlubricity achieved with two-dimensional nano-additives to liquid lubricants. Friction 8 (2020) 1007-1024. https://doi.org/10.1007/s40544-020-0410-3

[43] W. Zhai, K. Zhou, Nanomaterials in superlubricity. Adv. Funct. Mater. 29 (2019) 1806395. https://doi.org/10.1002/adfm.201806395

[44] A. Kotia, K. Chowdary, I. Srivastava, S.K. Ghosh, M.K.A. Ali, Carbon nanomaterials as friction modifiers in automotive engines: Recent progress and perspectives. J. Mol. Liq. 310 (2020) 113200. https://doi.org/10.1016/j.molliq.2020.113200

[45] M.K.A. Ali, H. Xianjun, M.A. Abdelkareem, M. Gulzar, A.H. Elsheikh, Novel approach of the graphene nanolubricant for energy saving via antifriction/wear in automobile engines. Tribol. Int. 124, (2018) 209-229. https://doi.org/10.1016/j.triboint.2018.04.004

[46] M.K.A. Ali, X. Hou, M.A. Abdelkareem, Anti-wear properties evaluation of frictional sliding interfaces in automobile engines lubricated by copper/graphene nanolubricants. Friction, 8 (2020) 905-916. https://doi.org/10.1007/s40544-019-0308-0

[47] H. Xie, S. Dang, B. Jiang, L. Xiang, S. Zhou, H. Sheng, T. Yang, F. Pan, Tribological performances of SiO2/graphene combinations as water-based lubricant additives for magnesium alloy rolling. App. Surf. Sci. 475 (2019) 847-856. https://doi.org/10.1016/j.apsusc.2019.01.062

[48] S. Dey, G.C. Dhal, Highly active palladium nanocatalysts for low-temperature carbon monoxide oxidation. Polytechnica 3 (2020) 1-25. https://doi.org/10.1007/s41050-019-00018-x

[49] J. Lv, S. Wang, B. Meng, The effects of nano-additives added to diesel-biodiesel fuel blends on combustion and emission characteristics of diesel engine: a review. Energies 15 (2022) 1032. https://doi.org/10.3390/en15031032

[50] X. Zhang, N.T.L. Chi, C. Xia, A.S. Khalifa, K. Brindhadevi, Role of soluble nano-catalyst and blends for improved combustion performance and reduced greenhouse gas emissions in internal combustion engines. Fuel 312 (2022) 122826. https://doi.org/10.1016/j.fuel.2021.122826

[51] M.E.M. Soudagar, N.-N. Nik-Ghazali, M.A. Kalam, I.A. Badruddin, N.R. Banapurmath, M.A. Bin Ali, S. Kamangar, H.M. Cho, N. Akram, An investigation on the influence of aluminium oxide nano-additive and honge oil methyl ester on engine performance, combustion and emission characteristics. Renew. Energy 146 (2020) 2291-2307. https://doi.org/10.1016/j.renene.2019.08.025

[52] S.H. Hosseini, A. Taghizadeh-Alisaraei, B. Ghobadian, A. Abbaszadeh-Mayvan, Performance and emission characteristics of a ci engine fuelled with carbon nanotubes and diesel-biodiesel blends. Renew. Energy 111 (2017) 201-213. https://doi.org/10.1016/j.renene.2017.04.013

[53] S.S. Hoseini, G. Najafi, B. Ghobadian, M.T. Ebadi, R. Mamat, T. Yusaf,. Performance and emission characteristics of a CI engine using graphene oxide (GO)

nano-particles additives in biodiesel-diesel blends. Renew. Energy 145 (2020) 458-465. https://doi.org/10.1016/j.renene.2019.06.006

[54] K. Nanthagopal, R.S. Kishna, A.E. Atabani, A.H. Al-Muhtaseb, G. Kumar, B. Ashok, A compressive review on the effects of alcohols and nanoparticles as an oxygenated enhancer in compression ignition engine. Energy Convers. Manag. 203 (2020) 112244. https://doi.org/10.1016/j.enconman.2019.112244

[55] X. Wang, M. Dai, J. Wang, Y. Xie, G. Ren, G. Jiang, Effect of ceria concentration on the evaporation characteristics of diesel fuel droplets. Fuel 236 (2019) 1577-1585. https://doi.org/10.1016/j.fuel.2018.09.085

[56] V. Sajith, C.B. Sobhan, G.P. Peterson, Experimental investigations on the effects of cerium oxide nanoparticle fuel additives on biodiesel. Adv. Mech. Eng. 2 (2010) 581407. https://doi.org/10.1155/2010/581407

[57] P.I. Dolez, Nanomaterials definitions, classifications, and applications. In Nanoengineering, Elsevier, 2015, pp. 3-40. https://doi.org/10.1016/B978-0-444-62747-6.00001-4

[58] J.R. Xavier, Electrochemical and dynamic mechanical properties of polyurethane nanocomposite reinforced with functionalized TiO2-ZrO2 nanoparticles in automobile industry. App. Nanosci. (2022) 1-16. https://doi.org/10.1007/s13204-022-02393-x

[59] S. Maharjan, K.S. Liao, A.J. Wang, K. Barton, A. Haldar, N.J. Alley, H.J. Byrne, S.A. Curran, Self-cleaning hydrophobic nanocoating on glass: A scalable manufacturing process. Mater. Chem. Phys. 239 (2020) 122000. https://doi.org/10.1016/j.matchemphys.2019.122000

[60] Z. Zhaowei, D.U. Quanbin, (2019). Study on magnetic field assisted electrodeposition of Ni-SiC nanocoatings on the surface of automobile cylinder liner. J. Func. Mater./Gongneng Cailiao, 50 (2019) 03081-03089.

[61] L. Xia, Y. Lv, Z. Miao, L. Luo, W. Luo, Y. Xu, C. Yuan B. Zeng, L. Dai, A flame retardant fabric nanocoating based on nanocarbon black particles@ polymer composite and its fire-alarm application. Chemical Engineering Journal, 433 (2022) 133501. https://doi.org/10.1016/j.cej.2021.133501

[62] B. Palen, T.J. Kolibaba, J.T. Brehm, R. Shen, Y. Quan, Q. Wang, J.C. Grunlan, J. C. Clay-filled polyelectrolyte complex nanocoating for flame-retardant polyurethane foam. ACS Omega, 6 (2021) 8016-8020. https://doi.org/10.1021/acsomega.0c05354

[63] F. Yang, B. Yuan, Y. Wang, X. Chen, L. Wang, H. Zhang, Graphene oxide/chitosan nanocoating with ultrafast fire-alarm response and flame-retardant property. Polym. Adv. Technol. 33 (2022) 795-806. https://doi.org/10.1002/pat.5556

[64] H. Xie, X. Lai, H. Li, J. Gao, X. Zeng, Skin-inspired thermoelectric nanocoating for temperature sensing and fire safety. J. Colloid Interface Sci. 602 (2021) 756-766. https://doi.org/10.1016/j.jcis.2021.06.054

Materials Research Forum LLC
https://doi.org/10.21741/9781644902295-6

[65] C.M.P. Kumar, A. Lakshmikanthan, M.P.G. Chandrashekarappa, , D.Y. Pimenov, K. Giasin, Electrodeposition based preparation of Zn-Ni alloy and Zn-Ni-WC nanocomposite coatings for corrosion-resistant applications. Coatings 11 (2021) 712. https://doi.org/10.3390/coatings11060712

[66] A.A. Farag, (2020). Applications of nanomaterials in corrosion protection coatings and inhibitors. Corros. Rev. 38(1), 67-86. https://doi.org/10.1515/corrrev-2019-0011

[67] D.H. Abdeen, M. El Hachach, M. Koc, M.A. Atieh, A review on the corrosion behaviour of nanocoatings on metallic substrates. Materials 12 (2019) 210. https://doi.org/10.3390/ma12020210

[68] J.D. Maeztu, P.J Rivero, C. Berlanga, D.M. Bastidas, J.F. Palacio, R. Rodriguez, Effect of graphene oxide and fluorinated polymeric chains incorporated in a multilayered sol-gel nanocoating for the design of corrosion resistant and hydrophobic surfaces. App. Surf. Sci. 419 (2017) 138-149. https://doi.org/10.1016/j.apsusc.2017.05.043

Emerging Applications of Nanomaterials
Materials Research Foundations 141 (2023) 151-168

Materials Research Forum LLC
https://doi.org/10.21741/9781644902295-7

Chapter 7

Nanotechnology in Defence and Security

B.P. Choudhary[1*], Bhuvnesh Kumar [2], S. Sharma[3], A.K. Sharma[1], R. Karmakar[4],
N.B. Singh[5]

[1]Department of Applied Science, Chandigarh Engineering College, Jhanjeri, Mohali, India

[2]Research Development Cell, Sharda University, Greater Noida, India

[3]Department of Computer Application, Chandigarh School of Business, Jhanjeri, Mohali, India

[4]Department of Applied Sciences & Humanities, Roorkee Institute of Technology, Roorkee,
India

[5]Department of Chemistry and Biochemistry, Sharda University, Greater Noida, India

*pratapbhasker6@gmail.com

Abstract

Nanotechnology, an indispensable branch of technology is presently being used in many specific technologically advanced domains. Technology can offer possibilities far beyond previously known limits, which means a debate between hope and anonymity. The potential of nanotechnology to harvest its strategic advantages or cause a disaster depends on a nation's level of preparedness. The integration of nano-materials and nanodevices into the defence systems is aimed at improving their performance in terms of defensive and/or offensive capabilities, individual safety, and reliability of systems/ subsystems. Major military powers around the world have been working diligently to improve their understanding of this emerging technology area. The earliest developments in this area involved the production of organic explosives in the form of fine powders with a particle distribution from the submicron to nano-meter range. The challenge ahead is to move from powder to integration of energy nanomaterials into the production system. According to the published literature, the subject studies are in the early stage on this specific domain, wherein numerous techniques are recently reported for making products from nano-thermites. This chapter is aimed to discusses various means and ways to protect the population from attack and to prevent industrial accidents, disaster management and also the potential applications of nanotechnology in warfare.

Keywords

Defence, Quantum cryptography, Nanotechnology, Weapons, War

Materials Research Forum LLC
https://doi.org/10.21741/9781644902295-7

Contents

1.   Introduction

Nanotechnology, which deals with particles in the size range of 1 nm to 100 nm is expected to play a vital role in various sectors including defence and security due to exotic properties of nanomaterial quite different from those of bulk materials. It is exciting to be aware of the historic evidential information that the present day advanced human being has been continually in the pursuit of studying increasingly various substances from mineral and plant origin for manufacturing equipment for his safety [1]. Materials have become a lot more substantial that the successive intervals have been diagnosed through them-which include not only the Stone Age, but also the Bronze Age, and the Iron Age. Each of the ages persevered for a millennium. The Iron Age, in fact, prolonged until the 20th century and taken innovative adjustments, now no longer most effective with inside the residing requirements of the human society, however additionally with inside the development of struggle generation. The technology of bows, arrows, gadas (stone or steel-hooked up timber logs) and so on, gradually replaced through swords to spears and to axes, etc [2]. By the end of the 16th century, our understanding of creating explosives caused the generation of small firearms and artillery weapons due to assault and siege guns. Together with those arms, metal shields got here into life for safety towards such guns. In the beginning of the 20th century, tanks and lethal chemical guns were advanced and used at some stage in World War I [3]. In the inter–World Wars, length of extra than a 1/2 of decade, there has been an exponential increase with improvement of many new substances and varieties of deadly guns and equipment, produced from metals and alloys. Examples of such guns consist of rifles, and other weapons including the tanks, ships, submarines,

Emerging Applications of Nanomaterials   Materials Research Forum LLC
Materials Research Foundations 141 (2023) 151-168   https://doi.org/10.21741/9781644902295-7

aircraft, and so on, which have been drastically used at some stage in World War II. To have an edge on the adversaries, the growing race even nowadays is to develop extra superior substances for struggle and safety technologies [3].

In technological development after World War II, the invention and amalgamation of a lot of polymers or an aggregate of various styles of natural monomers have introduced a revolution in substances for various technology-based programs, which not only include fitness care but also the programs in agriculture, human comforts, home items, security, defence, etc [4-6]. The length has witnessed the evolution of multifunctional polymers which include insulating, semiconducting, accomplishing polymers, composites having mechanical energy near the ones of metals at a whole lot decrease density, fibres of ultra-excessive energy, adhesive for bonding multiple styles of substrates, coatings to guard surfaces from corrosion, erosion, and so on. All such awesome traits of polymers, merchandise primarily based totally on them, and extra importantly the scope of amendment of their traits to shape in a specific requirement make the existing technology, due to the Polymer Age, in its personal right [7-9]. The creation of nanotechnology on the flip of the remaining century giving substances of precise physicochemical traits, collectively with as reinforcing factors in polymer, ceramics, or steel matrices, has similarly superior the functionality of substances for unique human endeavours, thereby making Nano Age as an extension to the Polymer Age from the angle of substances [10-14]. Nanotechnology is predicted to unexpectedly unfold the inside of future and expected that this trend of nanotechnology growth will continue [15].

Overall goal of defence nanotechnology in respect of research and development program in any nation for security purpose is to assess all the aspects of nanotechnology applications in defence sector be it Air Force, Navy and the Army. This aspect is important to make the necessary investments in order to understand the probable technological breakthrough in future.

The improvement and usage of nanotechnology-empowered guns and structures will convey the required modifications in well-known abilities of defence forces [2-5]. Specifically, the land forces will discover giant effect on engagement functionality with the aid of using manner of lighter, unique and lengthy-staying power guns and structures. Usage of nanofibers, nanocomposites and CNTs will convey down the load of the gadget carried with the aid of using the soldier. Usage of light weight helmet, frame suit, more effective light weight bulletproof jacket, adaptive camouflage, discount in electromagnetic signatures, and nanoscale biosensors will substantially strengthen the safety degree of armed forces. Navigational aids enabled through NEMS, quantum dots and CNTs will further strengthen the navigational and manoeuvring functionality of armed forces [16]. Nanotechnology stimulated encryption and compression of facts may also have giant effect at the selection-making cycle. Nanotechnology has the potential to improve warfare for individual soldiers by enhancing their capabilities. Military organizations around the world might choose to implant or inject their soldiers with small, self-replicating machines in order to gain a variety of benefits [2-4]. The potential of having these nanomachines in soldiers' bodies would be the implementation of a system that monitors and records every soldier engaged in combat in real-time. Nanoelectronics help increase the performance of information systems as well as making virtual reality systems more powerful and less expensive. Virtual reality systems are important for training or for battlefield reconstruction. Nanomaterials can be used in manufacturing the guns lighter, which means there is more ammunition for them. Together, these new technologies can lead to guns that can target and fire automatically if an enemy is detected, thanks to self-guided bullets [6, 7]. Nanorobots can be designed to attack and destroy materials that could be used to create weapons, such as metals, or other materials. Bio-nanorobots could be designed to kill specific individuals, based on his genetic profile. Nanomedicine can provide improved wound healing in the battlefield, which will help keep soldiers healthy. This chapter describes various aspects of nanotechnology in defence and security.

2. Applications of nanotechnology in defence and security

Nanotechnology is now being used in different industries starting from medical applications to military usage. In fact, nanotechnology is the next big thing that has multiple applications in the military domain. All military systems miniaturized would give a significant strategic advantage over the enemy. Various applications of Nanotechnology in defence & security are given in Fig.1.

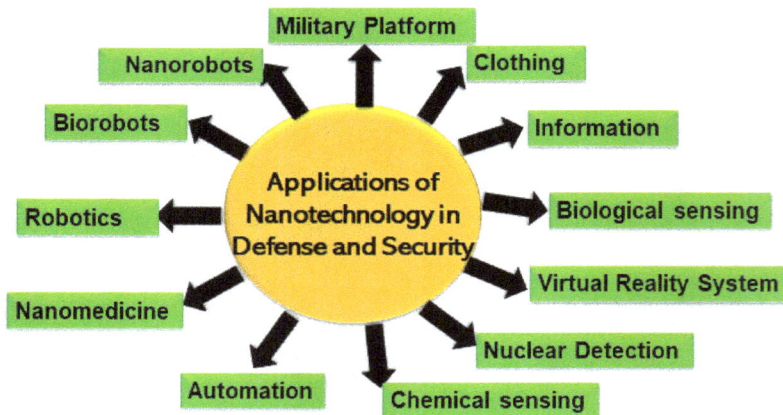

Figure 1. Applications of Nanotechnology in Defence & Security

2.1 Nanotechnology in the information system

Nanotechnology can be used to authenticate individuals. Some methods used in forensics, such as fingerprints, may also have its application to help identify people using fingerprint, iris, retinal, facial, and hand characteristics. It is also important to protect the related informhfxkation and its communication [17-20]. Nanocomposites can be used as a medium for recording personal information, such as fingerprints, imaging etc. Opal-based nanocomposites are used in defence technology, such as authentication. As far as anti-counterfeiting is concerned, the optical fiber cables can be used to protect brand clothing or pharmaceuticals. There is a growing need to understand the path of an object from its manufacture to its users. This is important for products as diverse as pharmaceuticals to aero-engine components. Positioning and localization techniques help ensure the quality of a product and help to track down any problems. It can be helpful in detecting counterfeit items. RFID tags are utilized to identify objects or goods. Metal nanoparticles or carbon nanotubes have been found to improve their characteristics [21]. Quantum cryptography can help to securely transmit information between two locations. It is based on quantum mechanics and uses entanglement effects to achieve its effects. The intensification in smartness of a soldier is being attained by incorporating the new phenomenon and technologies such as Real-Time Physiological Status Monitoring (RT-PSM) and Internet of Things (IoT). Such amalgamation has led to development of wearable systems not only limited to Soldier Wearable Integrated Power Equipment System (SWIPES) but also to Tactical Assault Light Operator Suit (TALOS) and Advanced Soldier Sensor Information Systems Technology (ASSIST). Such type of advanced systems contains a blend of components, such as smart glasses, sensor-enabled rings, smart textiles, smart watches, smart helmets, and smart uniforms, etc. They are supposed to be developed by using advanced materials such as conducting polymers, nanomaterials and/ or combination of both. [22-27]. Different components using nanomaterials in defence are shown in Fig.2.

Figure 2. Armed soldier with different component [28]

2.2 Nanotechnology in biological sensing

Bio-sensing holds a great future for nanotechnology applications especially for warning and an early detection of bio- threat including both the natural and manmade outbreaks of various biological agents like viruses, bacteria, and the toxins that can be deliberately spread among the population, livestock, or crops, with the goal to cause illness or death. There are large number of bacteria, viruses and toxins, which could be used as biological weapons in military /terrorist attacks. Nucleic acid and immunological techniques-based bio-sensors have been developed for detection of biological threats under lab conditions (Point detection). However, detection of such threats from a distance (Standoff detection) is in vogue and thus needs focussed attention.

For better biosensing, nanomaterials can be functionalized with appropriate functions depending on the chemical composition. In some cases, direct functionalization is possible, or via coating with functional polymers without affecting their specific properties [29]. Such functionalization increases the biocompatibility of the materials.

Nanomaterial application as chemical agent detection in real time offers an important benefit for the investigators and responders to strengthen preparedness for public health protection. Nanomaterial signatures of organic agent surrogates (Bacillus thurengensis or bacterial spores including Bacillus globigii and pathogenic Escherichia coli) had been received with the aid of using more than one agency in 2003. Principal aspect of evaluation turned into the use for preliminary discrimination research [30-32]. Discrimination of these three traces of one stress of mold, E. coli, and one stress of Candida albicans yeast was confirmed with the aid used in the chemometric approach discriminant feature evaluation, that is just like predominant aspect evaluation [33-37]. The complete LIBS spectra have been used for the evaluation and suitable separation of the traces turned into a great achievement. The identity of organic agent surrogates (consisting of the anthrax simulant a Gram-terrible bacterial biothreat simulant E. coli, B. atrophaeus, an organic toxin simulant α-hemolysin from Staphylococcus aureus, and a smallpox simulant MS-2 bacteriophage) turned into confirmed on stainless metallic (SS), polycarbonate, and aluminum foil substrate coupons [30].

2.3 Nanotechnology in nuclear detection

The dispersal of radioactive material in a densely populated area is a terrible threat. Nuclear separation or synthesis weapons are rather a problem of attacks by a sovereign power, although improvised bombs are also possible. Detecting nuclear materials with existing technologies can be very sensitive, but they are often bulky and expensive. Nanoscale materials have the potential to be used to detect radiation. They could be used to replace existing gamma ray detection methods or be used in scintillators to increase light output. Nanowire arrays are necessary for imaging techniques and high-rate counting. As the high-resolution germanium detector used to measure gas light needs to operate at very low temperatures, alternative mobile solutions based on nanomaterials that operate at room temperature are helpful in cases of nuclear accidents where measurements need to be done quickly outside the laboratory.

Areas of Nuclear Nano-Technology (NNT) deals with performance and safety in nuclear energy systems. The use of engineered-nanomaterials in nuclear energy systems has opened doors for improving the performance and safety of nuclear power. NNT deals with the use of the engineered-nanomaterials for future nuclear energy applications.

The danger of nuclear terrorism is of serious challenge for native land protection and demands new detection equipment, given that now no longer all gadgets that detects radiation can be differentiated among radioactive isotopes, a few aren't able to measure very excessive radiation costs, and others require pattern training in a laboratory. Laser Induced Breakdown Spectroscopy (LIBS) has been confirmed to find out radiological and nuclear isotopes, despite the fact that one number one drawback is the want for line-of-sight measurements (hence prescribing the ability packages). A standoff LIBS machine turned into used to illustrate the detection of uranium below diverse conditions like consisting of in soil and on surfaces including aluminium, plastic and ceramic tiles [38]. In addition, the approach of laser ablation molecular isotopic spectrometry has enabled using LIBS to decide isotope ratios for lighter factors including zinc, strontium, carbon and boron [39, 40]. LIBS has additionally been confirmed for diverse packages associated with the nuclear enterprise due to its potential to remotely verify fabric composition in regions now no longer correctly on hand with the aid of using employees (e.g., with a fibre optic machine or robot platform) and due to the potential to carry out speedy, in situ evaluation without a pattern training. These packages consist of the characterization of nuclear-associated substances including metallic [41-43], thorium and uranium is contained in depleted powders [44], beginning potential of yellow cake samples (powdered uranium listen used because the first step with inside the fabrication of nuclear fuel) [45], and established order of detection limits for common nuclear fission merchandise including cesium, strontium, and cerium (a common surrogate for radioactive plutonium) [46]. For the detection of nuclear radiation, we have classified nuclear radiation detector into two head viz. Ion collection based and non-ion collection-based detectors which directly or indirectly based on density of ions of the material used [47].

3.   Nanotechnology in virtual reality system

It is a regarded reality that generation is extraordinarily crucial for army effectiveness for any nation. Induction of recent and rising technology guarantees qualitatively advanced army pressure able to addressing each traditional in addition to uneven threats. The army ideas just like the revolution in army affairs are dynamic in nature and cope with advent of technological and doctrinal modifications with inside the armed forces. Recent revolution with inside the data and communique generation zone has added in induction of diverse new RMA technology with inside the militaries. Modern day armies are located inducting offshoots of diverse improvements in technology including nanotechnology, biotechnology, robotics, cognitive sciences, and few others into their scheme [19, 20].

Military technology also uses the diverse laptop simulations of structures allowing them to carry out operations at the simulated machine and demonstrating the consequences in actual time. Training is one region wherein digital fact technology is locating wider applicability with inside the defence. The motives, which have pushed the army to discover and appoint digital fact strategies of their education are essentially to lessen publicity to risks and to boom stealth. Simulation in fact is vital for the militaries for an easy purpose that cannot be 'on activity education' as a long way as battle is concerned. At times, it additionally becomes extraordinarily hard to illustrate the actual-lifestyles intellectual and bodily demanding situations of army lifestyles. Under such circumstances, the want arises to fugue the fact in a shape. This is wherein the digital fact packages mainly designed for the army may want to play a first-rate role. One crucial issue of digital fact - primarily based on simulation is that it gives an area to adopt diverse volatile and complex manoeuvres, which might be hard to carry out in actual lifestyles at the actual gadget. Digital fact makes it viable to simulate a gadget malfunction or horrific climate or any surprising scenario. All this will become viable without inflicting any harm both to the people or to the gadgets. Virtual fact machine allows to teach pilots to deal with emergencies.

4.   Nanotechnology in chemical (explosive) sensing

100 categories of explosives have been reported in the literature and detecting explosives in luggage, vehicles, cargo, planes, or human-carried explosives is a major security challenge. Several methods have been developed to detect explosives [8, 15,48-50]. The first is to detect traces of volatile compounds released into the atmosphere from explosive compounds. The second uses spectroscopy to check for the presence of compounds associated with explosives. The third method uses multiple detection methods at the same time to improve detection efficiency.

Most techniques rely on the fact that the absorption of volatile molecules found in explosives alters the physical or chemical properties of nanomaterials. It applies to electrochemical sensors, mass-based sensors, fibre optic sensors, photoluminescence-based detection, surface-enhanced Raman scattering (SERS), nanosensors, nanowires, and biosensors. Sensors are usually nano-sized materials with specific properties. Other methods that do not use nanotechnology, such as spectroscopy using wavelengths between microwaves and infrared rays and terahertz detection, are also possible. This technology has great potential not only in defence and security, but also in nano-catalysis [15].

Explosive-based weapons are simple, easy to install, and can cause enormous damage in recent years, the reliable and accurate detection of explosives has become one of the most important issues of international concern [51]. There are number of traditional methods for the detection and quantification of explosives but nanotechnology has played a pivotal role in production of nanosensors with low cost, portability, specificity, and ability in rapid identification [52].

5. Nanotechnology in nanomedicine

Nanoparticles may be used to correctly supply antidotes to humans at risk (antibiotics, anti-inflammatory drugs, anti-infective medicines, and vaccines) earlier than any capacity assault. Nanoparticles can enhance the steadiness of drugs, convey large concentrations, and offer a better-focused shipping. Depending on their structure, which may be designed on demand, nanoparticles can convey hydrophilic or hydrophobic molecules. It is vital to make accurate analysis and, relying at the result, to supply the applicable drug to the patient. This is subject of theragnostic, a mixture of drug shipping and diagnostics [11, 12]. It is feasible to increase arrays of biosensors with the potential to come across a massive quantity of viruses in parallel (as much as nearly a hundred), to understand proteins, DNA and RNA fragments, and to become aware of pollutants and microorganisms launched throughout an assault. In this subject, carbon nanotubes, nanoparticles, or quantum dots may be notably utilized in detection or imaging. As some distance as radioactive substances are concerned, just a few researches primarily based totally on nanoparticles were carried out. For example, magnetic nanoparticles can soak up low degree radioactive material. This is the case for nano magnetite composite debris which could soak up radioactive caesium atoms. Other examples difficulty remediation of nuclear waste. Caesium and strontium radioactive substances are regularly blended together. Nanostructured sodium silico titanate can selectively do away with Caesium ions whilst nanostructured sodium titanate can do away with strontium ions. In case of a radiological assault, a foam received through spaying polymer gels in suspension in water can lure radioactive debris, which bind to nanoparticles contained with inside the gel. Nanoparticles may be covered with proteins responding to the particular pollutants generated in a radiological or nuclear assault. These nanoparticles, injected into the blood of the tainted person, can detoxify the blood. In order that they may be now no longer attacked through the white blood cells, they may be organized in any such manner that they may be invisible to them: they may be "stealthy" with admire to the immune system. In the case of an explosive assault, many humans may be critically injured. Self-meeting strategies may be used to construct a nanofiber barrier to forestall bleeding without requiring cauterization, pressure, or adhesives. Nanotechnology is likewise beneficial in surgery, wound dressing, and implants. Nano-gadgets along with titanium nanotubes, nanocrystals, nanofibers, and nanoparticles may be used to deal with accidents precipitated through an explosive assault or incident.

After an accident or terrorist attack, it is often necessary to decontaminate the environment from chemical, biological, or radioactive products. With the help of nanotechnology, water and air filtration membranes can be developed. Metal oxides of aluminium, titanium, or cerium in the form of nanoparticles can be used to destroy bacteria and viruses [53-54]. Macro crystals are much less surface reactive and are not useful in this regard. Magnesium oxide nanocrystals can chemically absorb organophosphorus compounds at room temperature. They can be useful in chemical weapons attacks. Bacteria such as Escherichia coli, Bacillus cereus, and Bacillus globigii can be decontaminated in minutes using nanopowder made from calcium oxide and magnesium oxide. Carbohydrate-coated magnetic glicon nanoparticles can be used for rapid decontamination of anthrax.

Emerging Applications of Nanomaterials   Materials Research Forum LLC
Materials Research Foundations 141 (2023) 151-168   https://doi.org/10.21741/9781644902295-7

Photocatalytic action based on TiO_2 nanoparticles can be used to destroy a variety of organic pollutants and toxins. TiO_2 is an inexpensive material that can be manufactured with nanoparticles with a size of 3 to 5 nm. Photocatalysts can be expanded with photocatalytic nanowires or membranes. In the latter case, the membrane can be used as a protective mask as it allows oxygen and nitrogen to pass through and blocks toxic gases. Another example is a nanosponge made from mesoporous silica that can remove toxins from water. Those 6 nm nanopores can be filled with a self-assembled layer that allows them to trap mercury or other chemicals. Filters using nanocrystalline silver are also used, for example, to decontaminate water from Legionella pneumophila [15, 55]. Nanotechnology has opened up new opportunities in implantable delivery system. Nanoparticle can be made from bio-compatible and bio-degradable materials. The drug loaded in the nanoparticles is released in the targeted organ from the matrix by diffusion, swelling or degradation. Inhalation, implantation and injections are the other prominent methods in bio-medical field followed for drug delivery.

6.   Nanotechnology in automation and robotics

Many technologies have been developed to protect products and consumers from counterfeiting. More and more are based on nanotechnology. Researched as part of the Observatory Nano project, related to trademark theft and intellectual property theft prevention. The goal is to obtain cost-effective smart materials and packaging that enable accurate traceability and automatic identification. It is argued that everyone branches of battle (aerial, land and naval) may be converted with the aid of using the growing proliferation of unmanned structures. Military robots will to begin with take over all of the 'risky, stupid and grimy' responsibilities and could function below near human supervision. But someday – perhaps as quickly as 10-15 years – robots may be capable of functioning autonomously and will in large part update human infantrymen with inside the battlespace. With machines turning into smarter and extra capable, there may be a lesser want for people going into movement, or maybe of people being in non-stop manage of unmanned structures. Increasingly superior Artificial Intelligence will permit robots to function extra autonomously and they may expand many 'humanlike abilities', with a view to permit them to expand powerful techniques and functional behaviour. Computers can also additionally by no means emerge as universally clever, however they might without difficulty be capable of outsmart people in lots of domain names which are applicable for battle. The distinction is handiest that the generation is now maturing and that clever guns which can function effectively with little want for human supervision are actually technically viable. The armed Predator drones are handiest a demonstration of this widespread fashion and additionally evidence that robots have now reached the factor wherein they may be truly militarily beneficial. Military robots and different, probably self-reliant robot structures including unmanned fight air motors (UCAVs) and unmanned floor motors (UGVs) may want to quickly be delivered to the battlefield. Look similarly into the destiny and we can also additionally see self-reliant micro- and nanorobots armed and deployed in swarms of heaps or maybe millions. This developing automation of battle can also additionally come to symbolize a first-rate discontinuity with inside the records of battle: people will first be eliminated from the battlefield and can sooner or later also be in large part excluded from the selection cycle in destiny excessive-tech and excessive-pace robot battle. Although the present day technological problems will no question be overcome, the best barriers to computerized guns at the battlefield are probable to be felony and moral worries. Armin Krishnan explores the technological, felony and moral problems linked to fight robotics, inspecting each the possibilities and boundaries of self-reliant guns. He additionally proposes answers to the destiny law of army robotics thru worldwide regulation [56-58].

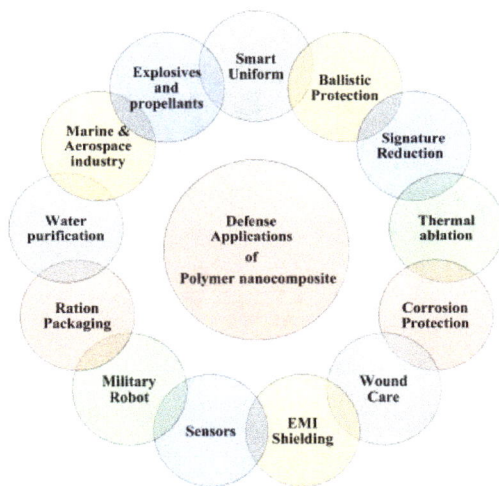

Figure 3 Applications of PNC in Defence [62]

7.   Nanotechnology in clothing

Protecting people before, during, and after the incident or attack is an equally important issue for civilians and military personnel. It involves clothing, equipment, and various protective measures against chemical, biological, or radioactive materials and explosives. Protective clothing contributes to the protection of individuals. Impermeable and selectively permeable clothing is one of the solutions [59]. Protective clothing should be lightweight and durable. This is not the case when using charcoal, which is used in many protective clothing today. Magnesium oxide nanoparticles loaded into nanofibers are more efficient than activated carbon. In general, nanofibers have very specific domains and can contain active chemistry as well as other functions [60, 61]. They can provide better protection against aerosols. Protective equipment is used daily by security guards who may face dangerous situations. They must be lightweight while providing a high level of protection against a variety of chemical or biological factors. Nanocomposites are increasingly being used to manufacture them. You also need protection from explosives, bullets, etc. Carbon nanotube fabrics provide better shielding than steel-based fabrics and are significantly lighter [61].

Polymer nanocomposites (PNCs) are used in different sectors for defence purposes (Fig.3) [62].

PNCs are utilized to make bulletproof vests, intelligent fabrics, helmets, gloves, and footwear for the protection of troops (Fig.4) [62].

Figure 4 Bullet proof dresses for Army personal by PNCs [62]

8. Nanotechnology in military platform

Buildings can be damaged or destroyed by explosives and fires. The electronic infrastructure that is essential for communication, data processing, and device control can be overridden by electromagnetic attacks or disrupted by malicious intrusions. Infrastructure, such as buildings and houses, can be destroyed or severely damaged by attacks and serious natural phenomena. Their mechanical resistance can be increased by incorporating nanomaterials. How can nanomaterials be used to improve the properties of concrete? Similarly, nano-meter-sized deposits increase the resistance of steel alloys to shock waves, and metal foams provide a high level of protection against ballistic projectiles. Other nanoobjects such as carbon nanotubes and inorganic fullerenes can be used to strengthen the structure against explosions, earthquakes, etc. Nano-coatings and nano-additives can significantly improve fire resistance. Nano-titanium or nano-silicon dioxides improve fire resistance of coating materials. This is also the case of nano-clays for acrylate and polymer nanocomposites. It turns out that buckyball nanocomposites are also flame retardants. Furthermore, carbon nanotube membranes are used to reduce flammability of glass fibre composites. We now exploit electromagnetic radiation far more than in the past. Frequencies in the 15 GHz range are radiated from many everyday devices (smartphones, WLANs, transmitter antennas, etc.). This wide variety of electromagnetic sources can interfere with critical electronic devices. Electromagnetic interference is a serious problem. Electromagnetic shielding can be provided by a conductive polymer or a nanocomposite containing a conductive polymer. Nanocrystalline silver particles can also be used. Finally, carbon nanofibers or carbon nanotubes can be used, but their cost is high. You also need to protect your electronics from electromagnetic bombs (E-bombs). This allows you to completely disable all electronic devices in the area of the explosion.

Conclusions

Nanotechnology offers immense opportunities for its applications in the security sector. As discussed above, nanotechnology and nanomaterials have a great potential to protect men and materials from dangers and risks. As with other advanced technologies, nanotechnology also has some ethical and social implications, which are not new. These issues need to be addressed appropriately for nanotechnology-based products and processes. Thus, there are suitable motives to examine the issues and to consider preventive measures at an early stage. Required work in this domain needs acceleration to carry out the needful, and, a lot of work is under progress world wide. Therefore, country wide guidelines and worldwide controls need to be coordinated closely. Likewise, verification of compliance might be the need to check whether the framed guidelines are legitimately followed. In bio-nanotechnology, application of nanotechnology in manipulation of gene and gene transfer may require specific approval in addition to the ethical clearance. Some software of nanotechnology might in general serve protecting safety, consisting of safety towards terrorist assaults. How to check the access of terrorist groups to such software and devices needs attention of all concerned. In a long run, one cannot brush aside the idea that containing the dangers of emerging technologies including genetic engineering, pervasive laptop networks, microsystems and nanotechnology. Reliably, stopping misuse and correctly making sure the compliance can additionally require a long way-attaining limits, extensive verification and a powerful machine for crook prosecution of transgressors, just like what has been evolved and emerged in civilian society. Long-time protection, therefore, requires strengthening of worldwide crook regulation. Scientifically, the emphasis should be given for interdisciplinary and multidisciplinary research and development work for a wider application of nanotechnology for its application in security sector with spin off to civil populace.

References

[1] N. Phukan, Nanotechnology and its Military Applications. Academic Journal of Nanotechnology (2019).

[2] C. Ngo, M.H. Van-de-Voorde, Nanotechnology for Defence and Security: Nanotechnology in a Nutshell, (2014), Springer. https://doi.org/10.2991/978-94-6239-012-6

[3] N.L. Fishher, Iron Man: Roman Masculinity and The Roman Military Dagger, Ph.D. Thesis, Cornell University, USA (2017).

[4] US Department of Defence Director of Defence and Engineering (2009). Defence Nanotechnology Research and Development Program.

[5] E. Edwards, Overview of Nanotechnology in Military Aerospace Applications. Nanotechnology Commercialization: Manufacturing Processes and Products. (2018) Wiley. https://doi.org/10.1002/9781119371762.ch5

[6] B.J. Buchanan (ed.), Gunpowder, Explosives and the State: A Technological History (University of Bath, Ashgate Publishing Limited, England, Aldershot, 2006).

[7] J.B.A. Bailey, Field Artillery and Firepower (Naval Institute Press, Annapolis, 2004)

[8] A. Kim Coleman History of Chemical Warfare, (Palgrave Macmillan, 2005). https://doi.org/10.1057/9780230501836

[9] Ulf. Schmidt, Secret Science: A Century of Poison Warfare and Human Experiments. (Oxford University Press, Oxford, New York, 2015).

[10] A. Molnar, V. Gerasimov, A. Badidova, Implementation of Biological Sources of Energy in the System of "Smart Clothes" Acta Mechanica Slovaca, 23 (3) (2019) 30 - 35 https://doi.org/10.21496/ams.2019.019

[11] G. Nichol, Applications of Nanotechnology in Military Medicine: From the Battlefield to the Hospital and Beyond. HDIAC Journal. 3 (2016), 32-35.

[12] C. Shipbaugh, P. Anton, G. Bloom, B. Jackson, R. Silberglitt, Nano-enabled Components and Systems for Biodefence. Biomedical Nanotechnology, 114-139, (2005) Taylor and Francis.

[13] D. Paul, L. Kelly, V. Venkayya, T. Hess, Evolution of U.S. Military Aircraft Structures Technology. Journal of Aircraft, 39, (2002) 18-29. https://doi.org/10.2514/2.2920

[14] D. Avery, Pathogens for War: Biological Weapons, Canadian Life Scientists, and North American Biodefence (University of Toronto Press, Toronto, 2013) https://doi.org/10.3138/9781442664982

[15] C. Zafer, Nanotechnology, Society and National Security. The Journal of Security Sciences, 10 (1) (2021) 193 - 216.

[16] A. Ahmed, M Mohsin, Z.A.S. Muhammad. Survey and technological analysis of laser and its defence applications. Defence Technology (2020). https://doi.org/10.1016/j.dt.2020.02.012.

[17] C. Cameron, Military-grade augmented reality could redefine modern warfare, (2010).

[18] C.E. Howard. Department of Defence invests in delivering augmented reality technology to foot soldiers, (2007).

Materials Research Forum LLC
https://doi.org/10.21741/9781644902295-7

[19] P. Herrero, A. de Antonio, Intelligent virtual agents keeping watch in the battlefield, (2005). https://doi.org/10.1007/s10055-004-0148-7

[20] A. Lele, Virtual reality and its military utility, Journal of Ambient Intelligence and Humanized Computing (2011). DOI 10.1007/s12652-011-0052-4. https://doi.org/10.1007/s12652-011-0052-4

[21] D. Srikanth, A. Reddy, V. Praveen, S. Yenneti, Optimizing the Nano Technology in Defence System for the Future War Figther, IOSR Journal of Electrical and Electronics Engineering, 9 (2) (2014) 51-55. https://doi.org/10.9790/1676-09245155

[22] J. Gubbi, R. Buyya, S. Marusic, M. Palaniswami, Internet of Things (IoT): a vision, architectural elements, and future directions, https://arxiv.org/ftp/arxiv/papers/1207/1207.0203.pdf

[23] I. Lee, K. Lee, The Internet of Things (IoT): applications, investments, and challenges for enterprises. Bus. Horiz. 58 (2015) 431-440. https://doi.org/10.1016/j.bushor.2015.03.008

[24] K.E. Friedl, M.J. Buller, W.J. Tharion, A.W. Potter, G.L. Manglapus, R.W. Hoyt, Real time physiological status monitoring (Rt-Psm): accomplishments, requirements and research roadmap, USA Technical Note TN (2016) 16-02.

[25] E. Fairbrass, L. Genders, G. Perez-Ortega, C. Swisher, V. Mittal, Value modeling and trade off analysis of the tactical assault light operator suit. Industrial and Systems Engineering Review, 5 (2017) 116-122. https://doi.org/10.37266/ISER.2017v5i2.pp116-122

[26] https://www.army-technology.com/features/featuresensor-sensibility-future-of-soldier-worn-systems/

[27] C. Hurley, Future vest technology empowering the future soldier. SoldierMod, 18 (2017) 33.

[28] https://www.reddit.com/r/ImaginaryTechnology/comments/d8c49z/terran_federation_standard_military_kit_by/

[29] M. Holzinger, A.L. Goff, S. Cosnier, Nanomaterials for biosensing applications: a review, Frontiers in Chemistry, 63 (2014) 1-10. https://doi.org/10.3389/fchem.2014.00063

[30] J.L. Gottfried, F.C. De Lucia Jr., C.A. Munson, A. W. Miziolek, Double-pulse standoff laser-induced breakdown spectroscopy for versatile hazardous materials detection, Spectrochimica Acta Part B: Atomic Spectroscopy, 62 (2007) 1405-1411. https://doi.org/10.1016/j.sab.2007.10.039

[31] J.L. Gottfried, Discrimination of biological and chemical threats in residue mixtures on multiple surfaces, Analytical and Bioanalytical Chemistry, 400 (2011) 3289-3301. https://doi.org/10.1007/s00216-011-4746-4

[32] J.D. Hybl, G.A. Lithgow, S.G. Buckley, Laser-induced breakdown spectroscopy detection and classification of biological aerosols, Applied Spectroscopy, 57 (2003) 1207-1215. https://doi.org/10.1366/000370203769699054

[33] S. Morel, N. Leone, P. Adam, J. Amouroux, Detection of bacteria by time-resolved laser-induced breakdown spectroscopy, Applied Optics, 42 (2003) 6184-6191. https://doi.org/10.1364/AO.42.006184

[34] A.C. Samuels, F.C. De Lucia Jr., K.L. McNesby, A.W. Miziolek, Laser-induced breakdown spectroscopy of bacterial spores, molds, pollens, and protein: initial studies of discrimination potential, Applied Optics, 42 (2003) 6205-6209. https://doi.org/10.1364/AO.42.006205

[35] J. Diedrich, S.J. Rehse, S. Palchaudhuri, Escherichia coli identification and strain discrimination using nanosecond laser-induced breakdown spectroscopy, Applied Physics Letters, 90 (2007) 163901-163903. https://doi.org/10.1063/1.2723659

[36] N. Kumar, A. Dixit, Nanotechnology for Defence Applications, (2019). https://doi.org/10.1007/978-3-030-29880-7. https://doi.org/10.1007/978-3-030-29880-7

[37] B. Unal, S. Aghlani, Use of Chemical, Biological, Radiological and Nuclear Weapons by Non-State Actors, Emerging trends and risk factors; (Lloyd's Emerging Risk Report-2016), Chatham House, The Royal Institute of International Affairs.

[38] R.C. Chinni, D.A. Cremers, L.J. Radziemski, M. Bostian, C. Navarro-Northrup, Detection of uranium using laser-induced breakdown spectroscopy, Applied Spectroscopy, 63 (2009) 1238-1250. https://doi.org/10.1366/000370209789806867

[39] F.R. Doucet, G. Lithgow, R. Kosierb, P. Bouchard, M. Sabsabi, Determination of isotope ratios using laser-induced breakdown spectroscopy in ambient air at atmospheric pressure for nuclear forensics, Journal of Analytical Atomic Spectrometry, 26 (2011) 536-541. https://doi.org/10.1039/c0ja00199f

[40] A.A. Bolshakov, X.L. Mao, J.J. Gonzalez, R.E. Russo, Laser ablation molecular isotopic spectrometry (LAMIS): current state of the art, Journal of Analytical Atomic Spectrometry, 63 (2016) 119-134. https://doi.org/10.1039/C5JA00310E

[41] I. Gaona, J. Serrano, J. Moros, J.J. Laserna, Evaluation of laser-induced breakdown spectroscopy analysis potential for addressing radiological threats from a distance, Spectrochimica Acta Part B: Atomic Spectroscopy, 96 (2014) 12-20. https://doi.org/10.1016/j.sab.2014.04.003

[42] C.M. Davies, H.H. Telle, A.W. Williams, Remote in situ analytical spectroscopy and its applications in the nuclear industry, Fresenius Journal of Analytical Chemistry, 355 (1996) 895-899. https://doi.org/10.1007/s0021663550895

[43] A.I. Whitehouse, J. Young, I.M. Botheroyd, S. Lawson, C.P. Evans, J. Wright, Remote material analysis of nuclear power station steam generator tubes by laser-

induced breakdown spectroscopy, Spectrochimica Acta Part B: Atomic Spectroscopy, 56 (2001) 821-830. https://doi.org/10.1016/S0584-8547(01)00232-4

[44] E.J. Judge, J.E. Barefield, J.M. Berg, S.M. Clegg, G.J. Havrilla, V.M. Montoya, L.A. Le, L.N. Lopez, Laser-induced breakdown spectroscopy measurements of uranium and thorium powders and uranium ore, Spectrochimica Acta Part B: Atomic Spectroscopy, 83-84 (2013) 28-36. https://doi.org/10.1016/j.sab.2013.03.002

[45] J.B. Sirven, A. Pailloux, Y.M. Baye, N. Coulon, T. Alpettaz, S. Gosse, Towards the determination of the geographical origin of yellow cake samples by laser-induced breakdown spectroscopy and chemometrics, Journal of Analytical Atomic Spectrometry, 24 (2009) 451-459. https://doi.org/10.1039/b821405k

[46] M.Z. Martin, S. Allman, D.J. Brice, R.C. Martin, N.O. Andre, Exploring laser-induced breakdown spectroscopy for nuclear materials analysis and in-situ applications, Spectrochimica Acta Part B: Atomic Spectroscopy, 74-75 (2012) 177-183. https://doi.org/10.1016/j.sab.2012.06.049

[47] G.F. Knoll, Radiation Detection and Measurement, 4 edn., (Wiley, 2010); b: Radiation Detection and Survey Devices.

[48] http://lem.ch.unito.it/didattica/infochimica/2008_Esplosivi/Classification.html and http://www.tracefireandsafety.com/VFRE-99/Recognition/High/high.htm

[49] J.P. Agrawal, High Energy Materials: Propellants, Explosives and Pyrotechnics (Wiley-VCH Verlag GmbH & Co, Weinheim, 2010). KGaA (ISBN:9783527326105) https://doi.org/10.1002/9783527628803

[50] V. Pal, M.K. Sharma, S.K. Sharma, A.K. Goel, Biological warfare agents and their detection and monitoring techniques. Defence science journal, 66 (2016) 445-457. https://doi.org/10.14429/dsj.66.10704

[51] R.J. Colton, J.N. Russell, Making the world a safer place Science, 299 (2003), 1324-1325. https://doi.org/10.1126/science.1080688

[52] F. Akhgari, H. Fattahi, Y. M. Oskoei, Recent advances in nanomaterial-based sensors for detection of trace nitroaromatic explosives, Sensors and Actuators B: Chemical B, 221 (2015) 867-878 https://doi.org/10.1016/j.snb.2015.06.146

[53] P.B. Chouke, A.K. Potbhare, N.P. Meshram, M. M. Rai, K.M. Dadure, K. Chaudhary, A.R. Rai, M. Desimone, R.G. Chaudhary, D.T. Masram, Bioinspired NiO nanospheres: Exploring in-vitro toxicity using Bm-17 and L. rohita liver cells, DNA degradation, docking and proposed vacuolization mechanism. ACS Omega, 7 (2022) 6869−6884. https://doi.org/10.1021/acsomega.1c06544

[54] M.S. Umekar, G.S. Bhusari, A.K. Potbhare, A. Mondal, B.P. Kapgate, M. Desimone R.G. Chaudhary, Bioinspired reduced graphene oxide based nanohybrids for photo-catalysis and antibacterial applications, Current Pharmaceutical Biotechnology, 22(13) (2021) 1759-1781. https://doi.org/10.2174/1389201022666201231115826

[55] G. Nichols, Applications of Nanotechnology in Military Medicine: From the Battlefield to the Hospital and Beyond, Homeland defence & security information analysis center, (2016), 3 (2).

[56] A. Dash, P.C. Mohapatra, Nano Robotics- A Review, AICTE National Conference on Modern Trends in Engineering Solutions (NCMTES), (2013).

[57] A. Krishnan, Enforced Transparency: A Solution to Autonomous Weapons as Potentially Uncontrollable Weapons Similar to Bioweapons. Lethal Autonomous Weapons: Re-Examining the Law and Ethics of Robotic Warfare, Oxford: Oxford University Press, (2021), 219-236. https://doi.org/10.1093/oso/9780197546048.003.0015

[58] R.A. Cavalcanti, Jr. Freitas, Nanorobotics Control Design: A Collective Behavior Approach for Medicine, IEEE Transactions on Nano-Bio-Science, 4 (2) (2005) 133-140. https://doi.org/10.1109/TNB.2005.850469

[59] L. Karthik, J. Nadar. Functional nanotube-based textiles: Pathway to next generation fabrics with enhanced sensing capabilities. Textile Research Journal, 75 (9) (2005) 670-681. https://doi.org/10.1177/0040517505059330

[60] S. Frank, S. Steven. Booklet on nanotechnology-innovation opportunities for tomorrow's defence. TNO Science and Industry, Future Technology Center, Netherlands, March (2006).

[61] G. Thilagavathi, A. Raja, T. Kannaian, Nanotechnology and Protective Clothing for Defence Personnel, Defence Science Journal, 58 (4) (2008) 451-459. https://doi.org/10.14429/dsj.58.1667

[62] A.B. Rashid, Md E. Hoque, Polymer nanocomposites for defence applications, Advanced Polymer Nanocomposites. (2022) 373-414. https://doi.org/10.1016/B978-0-12-824492-0.00015-5

Emerging Applications of Nanomaterials
Materials Research Foundations 141 (2023) 169-217

Materials Research Forum LLC
https://doi.org/10.21741/9781644902295-8

Chapter 8

Nanomaterials for Multifunctional Textiles

Kawser Parveen Chowdhury[1,2], Md. Abu Bin Hasan Susan[2], Saika Ahmed[2*]

[1]Department of Wet Process Engineering, Bangladesh University of Textiles (BUTEX), Dhaka-1208, Bangladesh

[2]Department of Chemistry, University of Dhaka, Dhaka-1000, Bangladesh

*saika@du.ac.bd

Abstract

The textile industry has been booming in recent decades due to new market explosion and consumer appeal for inventive apparel. Nanotechnology has played an exclusive role in such multifunctional high-performance textiles. Nanomaterials offer antimicrobial, flame retardance, self-cleaning, UV-protection, wrinkle-free, anti-static functionalities to textiles apart from their use for traditional aesthetic and decorative purposes. The nanoscale modified textiles offer new value-added functionalities while upgrading the existing aesthetic and physical properties. This chapter depicts an overall development on functional nanomaterials incorporated into textiles fabrics by different textile processes, featuring the aspects of nanotechnology to develop multifunctional textiles.

Keywords

Functional Nanomaterials, Textile Processes, Value-Added Functionalities, Upgradation of Physical Properties, Multifunctional Textiles

Contents

1.   Introduction

Textile industries have been experiencing a big revolution in recent decades, based on market and consumer demand for inventive multifunctional textiles [1,2]. In specific, nanotechnology played a pivotal role to develop high-performance value-added textiles, initiating new applications in various fields such as healthcare, sports, fashion, automotive, protection, etc. [3,4]. Such multifunctional textiles may provide further functionalities apart from traditional aesthetic and decorative purposes. Multifunctional textiles have truly grown into a great useful fashion trend for consumers based on their other applications accessible from a single product, which has generated enormous interest among the research community. The nanoscale modification of textile materials permits them to offer new value-added functionalities or upgrade the existing properties, while conserving not only their aesthetic performances like appearance, comfort, flexibility, hand feel, breathability, but also their functional performances like tensile and tear strength, pilling and abrasion resistance, and washing fastness [5,6]. Various functional nanomaterials such as nanoparticles, nanotubes, nanocomposites have been incorporated into textiles at various production stages like fiber spinning, yarn formation, finishing, or coating step [5,7].

This chapter provides an overall picture of the development in functional nanomaterials that have been incorporated in textile fabrics by different textile processes, featuring the aspects of nanotechnology to impart multifunctionality in textiles, and thus has been organized based on classified functionalities achieved in smart textiles, such as antimicrobial, self-cleaning, antistatic, flame retardance, crease recovery, etc., by strategic incorporation of different nanomaterials.

1.1   Functional processes

The routes used till now for incorporation of nanomaterials in textiles may be classified into two main categories: (i) application of nanomaterials into textile at the finishing or coating step (top-down approach), and (ii) application of nanomaterials on textile at the fiber production stage through spinning processes (bottom-up approach).

Multifunctional textiles can be produced by incorporating diffrent nanomaterials at finishing/coating stage into fabrics through various lamination or coating processes or by dyeing or printing processes [8]. These steps are more accepted by textile producers or manufacturers than impregnation at fiber producing stage or spinning process, because of the high cost-effectiveness, decreased production time, reduced infrastructure, and requirement of less machines required in these methods. On the contrary, fiber spinning processes provide huge extensibility for the textile industries as they can produce fine-tunable nanoengineered fibers. However, this method of choice is still at an evolving stage as it needs tuning before operation.

1.2   Innovative textiles based on nanotechnology

Textile materials have been growing into multifunctional for last few decades owing to the implementation of nanotechnology. Nanofinishing and nanocoating on textile substrates are being introduced in various application sectors. Nanocomposites have been applied on textiles by means of various laminating, coating, or finishing processes. Both nanocomposites and nanofibers have presented huge prospects in producing multifunctional nanotextiles (Fig. 1).

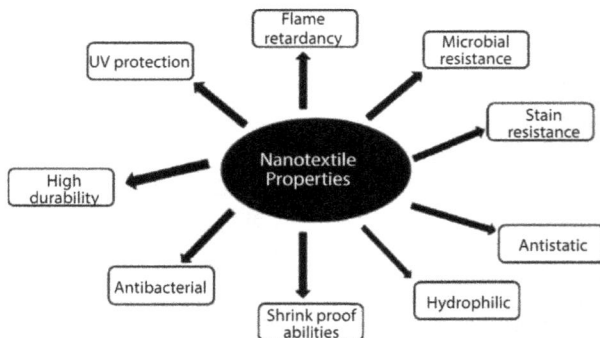

Figure 1. Various enhanced properties of multifunctional nanotextiles [9]. (Reprinted with permission from John Wiley & Sons (2000) [9]).

1.2.1 Nanomaterials

Nanomaterial, as defined by ISO (2015), is a material with any of the outward dimension of *ca.* 1 to 100 nm or with the surface or the internal structure of nano-dimension [10,11]. Nanomaterials have a huge potential associating different functional properties in textiles. Various nanomaterials that are most commonly used in different textile application fields can be classified as follows:

i) Nanomaterials based on carbon: Nanotubes [12-15] and nanofibers [16-18] of carbon and graphene [19,20]

ii) Inorganic nanoparticles (NPs): Nanoclay; e.g. saponite, mica, montmorillonite, laponite, attapulgite, vermiculite, etc. [21-25], metal; e.g. Au, Ag, Al, Pt, Cu, etc. [26-30], metal oxide; e.g. TiO_2, MgO, SiO_2, ZnO, CeO_2, etc. [31,32], and a gemstone, tourmaline

iii) Polymeric nanomaterials: e.g. chitosan [33-35]

iv) Hybrid nanomaterials: e.g. polyhedral oligomeric silsesquioxanes (POSS) [36-39]

v) Composite NPs: TiO_2/SiO_2 [40,41]

vi) Core-shell NPs: $(SiO_2)_{core}/(TiO_2)_{shell}$, Ag_{core}/Cu_{shell}, $(TiO_2)_{core}/Ag_{shell}$, $(ZnO)_{core}/Ag_{shell}$, etc. [42]

Nanomaterials, based on their shapes, can also be categorized into dimensional structures of 0D, 1D, 2D, and 3D [43]. The physical and chemical properties of NPs are quite different fundamentally from their bulk counterparts of an identical composition. Nanostructures classified based on dimensions are as follows (Fig. 2) [44]:

- 0D nanostructures: The three-dimensional free motion of electrons or any other particles is restricted in the structure of these materials. Examples are core-shell NPs, semiconductor quantum dots, hollow nanospheres, etc.

- 1D nanostructure: The free motion of electrons or other particles is restricted in two dimensions in such structures. Examples of this category are nanowires, nanotubes, nanorods, nanofilaments, etc.

- 2D nanostructure: One-dimensional electron movement gets restricted in these nanostructures, which include surface thin films, plates/nanodiscs or plates, and layered materials.

- 3D nanostructure: Nanophasic materials containing nanosize small grains fall in this group [44].

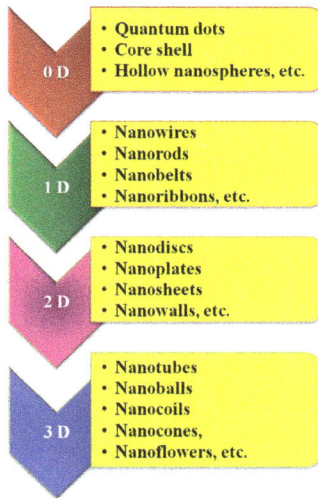

0 D
- Quantum dots
- Core shell
- Hollow nanospheres, etc.

1 D
- Nanowires
- Nanorods
- Nanobelts
- Nanoribbons, etc.

2 D
- Nanodiscs
- Nanoplates
- Nanosheets
- Nanowalls, etc.

3 D
- Nanotubes
- Nanoballs
- Nanocoils
- Nanocones,
- Nanoflowers, etc.

Figure 2. Types of nanostructures [44]. (Reprinted with permission from Elsevier (2021) [44]).

1.2.2 Nanofinishing

Nanofinishing is a practice where application of colloidal solutions of NPs or their ultrafine dispersions on a textile substrate results in upgradation of certain functionality. Nanofinishing is generally done by the method of pad-dry-cure. In the nanofinishing process, textile clothing materials are treated with nanosol or nanoemulsion, with an average droplet size of ca. 100–500 nm. For the padding mangle processes (both semicontinuous and continuous), the textile material is dipped into a dispersion of nanomaterials and further pressed by a roller pair with a view to imposing the nanomaterials to enter into the fabric and also to eliminating the excessive dispersion (Fig. 3) [45-47]. A padding mangle can control the rollers pressure and fabric speed passing within the padder. Then the solvent (mostly water) is removed during drying and curing the fabric, which ensures that the nanomaterials are fixed on the fabric. The total fabric is dipped into the nanomaterial liquor at a time, for this reason it is necessary to ensure homogeneous padding throughout the entire area of the fabric. The finishing route is treated as the most conventional one for its simplicity, fastness, and economic viability compared to the process of exhaustion [48].

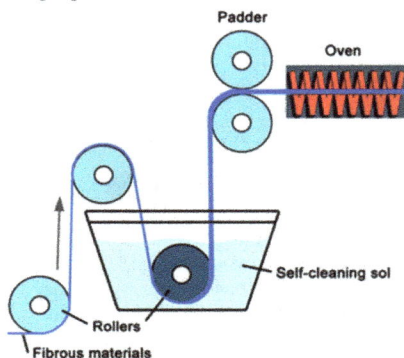

Figure 3. Common padding process of textile fabrics [49]. (Reprinted with permission from Royal Society of Chemistry (2011) [49]).

This method has showed various improvements than other conventional finishing processes as discussed below [48]:

- Nanofinishing usually needs smaller quantities of nanomaterials than bulk materials used in traditional finishing to obtain an equal effect.
- Nanofinishes are highly durable and do not alter the aesthetic property like hand feel, lustre of textile materials.
- Nanomaterials can be uniformly distributed on textile substrate by nanofinishing.
- Nanofinishing can achieve some functionalities that may appear challenging by conventional finishes.

Owing to these benefits, now a days, multifunctional finishes by using nanomaterials are becoming increasingly popular. Many textile industries have already commercialized their textile products where nanoemulsions or nanosols have been used for finishing and also employed innovative processes to introduce advantageous functionalities on their textile materials.

1.2.3 Nanocoatings

Nanocoating is a deposition process of applying a thin layer (1 to 100 nm) on a material for enhancement of desirable properties or to introduce novel functionalities. Coatings applied by traditional means exhibit a few limitations like (i) inadequate adhesion between the coating and the substrate, (ii) low flexibility, (iii) low strength, (iv) weak resistance to abrasion, and (v) short durability [46]. Nanocoating can effectively overcome these limitations of conventional coating techniques. Nanocoating techniques usually used are [5]: self-assembly, layer-by-layer (LbL), sol-gel, spin coating, dip coating, plasma or ion-beam assisted techniques, magnetron sputtering vapor deposition, electrochemical deposition, and pulsed laser deposition.

1.2.4 Nanofibers

Nanofibers within diameter of 50-500 nm are not achievable from regular melt-jet-dry-wet spinning as these produce fine microfibers with diameter 1–100 mm. Rather, methods which are in use to produce nanofibers include electrospinning, flash-spinning, melt blowing, force-spinning, gas jet technique, self-assembly, etc. [50,51]. Electrospinning is the ultimate beneficial technique because of its superior productivity, minimum cost, large porosity, and potentiality to manipulate size and morphology of nanofiber (for producing shapes like core-shell, beaded, ribbon, aligned, porous, etc.) [50,52,53]. Electrospun fibers have various outstanding features such as 3-4 times diameter, larger surface area with fine micropores, while such characteristics make them feasible to use in different applications such as energy harvesting, tissue engineering, nanosensor, filtration, protective clothing, wound healing, drug delivery, etc.

The electrospinning technique uses charged polymer melt or solution ejected via a spinneret with diameter in the micrometer range and lastly assembled on a collector. The plate and the spinnerets are subjected to the application of a large difference in potential (Fig. 4). The ejected fiber is forcibly stretched to lower the diameter into nano-level. The electrospinning process is dependent on several factors [5,54]:

- Properties of the polymer solution (molecular weight, concentration, elasticity, viscosity, surface tension, etc.),
- Conditions of processing (e.g. needle diameter, volume flow rate, applied voltage, and needle-collector distance),
- Atmospheric conditions (e.g. humidity, temperature, atmospheric pressure, etc.).

1.2.5 Nanocomposites

Nanocomposites, composed of at least one nanosize (<100 nm) reinforced material, are solid materials containing multiple phases. In polymer nanocomposites, the dispersed phases are nanomaterials, while polymeric substances are used as the matrices. The fibers and coatings of such polymer nanocomposites hold great possibility to manufacture multifunctional textiles with superior quality.

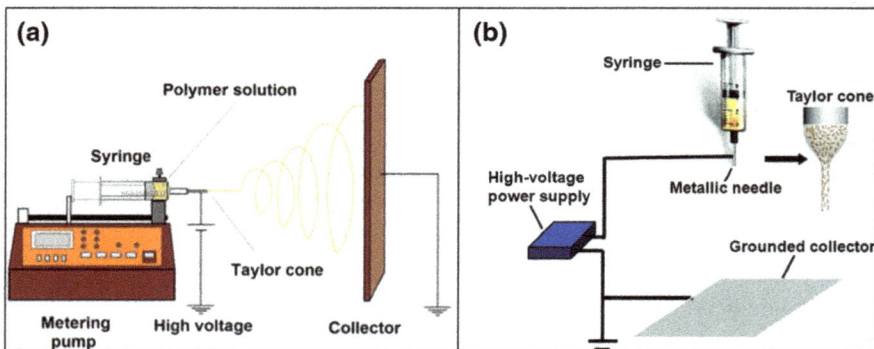

Figure 4. Instrumental set-up for electrospinning: (a) horizontal [55] (Reprinted with permission from Elsevier (2016) [55]), and (b) vertical [56] (Reprinted with permission from Royal Society of Chemistry (2014) [56]).

1.2.5.1 Fibers

Synthetic polymer nanocomposite fibers have gained special interest in recent times as they upgrade many of the features of synthetic fibers, described as follows:

- Nanocomposite fibers can be a charming option as synthetic fibers can be easily dyed after adding nanoclay [57].

- Polymer nanocomposite fibers exhibit great antimicrobial property after reinforcement with metal nanoparticles, like silver [27].

- These fibers reinforced with CNT or other nanoclay exhibit upgraded strength [58,59].

- Antistatic nanocomposite fibers can be manufactured after reinforcement with some nanoparticles like TiO_2, ZnO, and Ag [60-62].

- Polymer nanocomposite fibers reinforced with graphene or CNT show improved conductivity [13].

1.2.5.2 Coatings

The processes by which polymer nanocomposites are applied on textiles by coatings are called polymeric nanocomposite coatings. These are biggest rising path to produce multifunctional smart textiles with superior qualities [59], such as antimicrobial properties, protection against UV, self-cleaning, photoactivity, wrinkle prevention, etc.

2.   Technologies used for production of multifunctional textiles

This section summarizes some smart technologies employed to obtain multifunctional textiles holding huge potential for future.

2.1   Antimicrobial finish

Nanotechnology is the most interesting technology in the field of functionalized textiles for industrial applications as effective and long-lasting functionalization of textiles can be achieved through its use, with least sacrifice of desired properties like breathability and comfortability. NPs have got huge attention as antimicrobial agents due to their biological activity to fight microorganisms [63]. Antimicrobial property of NPs is measured by their efficacy against bacteria, protozoa, algae, fungi, virus, etc. Textiles materials like cotton, lyocell, viscose, linen, wool, and silk are prone to hosting many drug resistant pathogens because of their ideal condition like moisture, temperature, oxygen, and nutrients [64]. Currently, contamination by toxic microorganisms is of concern as it leads to some unpleasant properties in textiles, such as discoloration, unpleasant odors, allergic responses, spread of infectious diseases, and decline in the aesthetic value of textile materials [65]. Antimicrobial textiles are a new approach to resolve this existing problem in healthcare and hygiene sector. Moreover, research based on antimicrobial textiles to upgrade their performances are now flourishing because of the rising awareness about hygiene and health. Resistance against microbe thus has been a crucial prerequisite for all sorts of clothes [66]. There are different types of antimicrobial agents which have diverse structures like heavy metal ions, formaldehydes, organometallics, phenols, chitosan, quaternary ammonium salts, and organosilicones, and are generally used to convey antimicrobial properties to textiles via two different mechanisms (Fig. 5) [67].

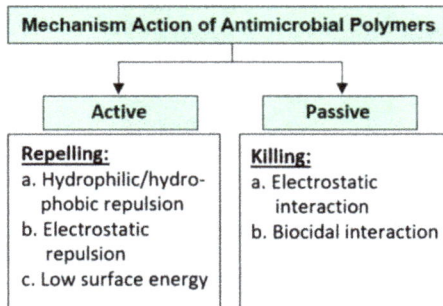

Figure 5. Classification of the mechanism of action of antimicrobial polymers [68].
(Reprinted with permission from MDPI (2016) [68]).

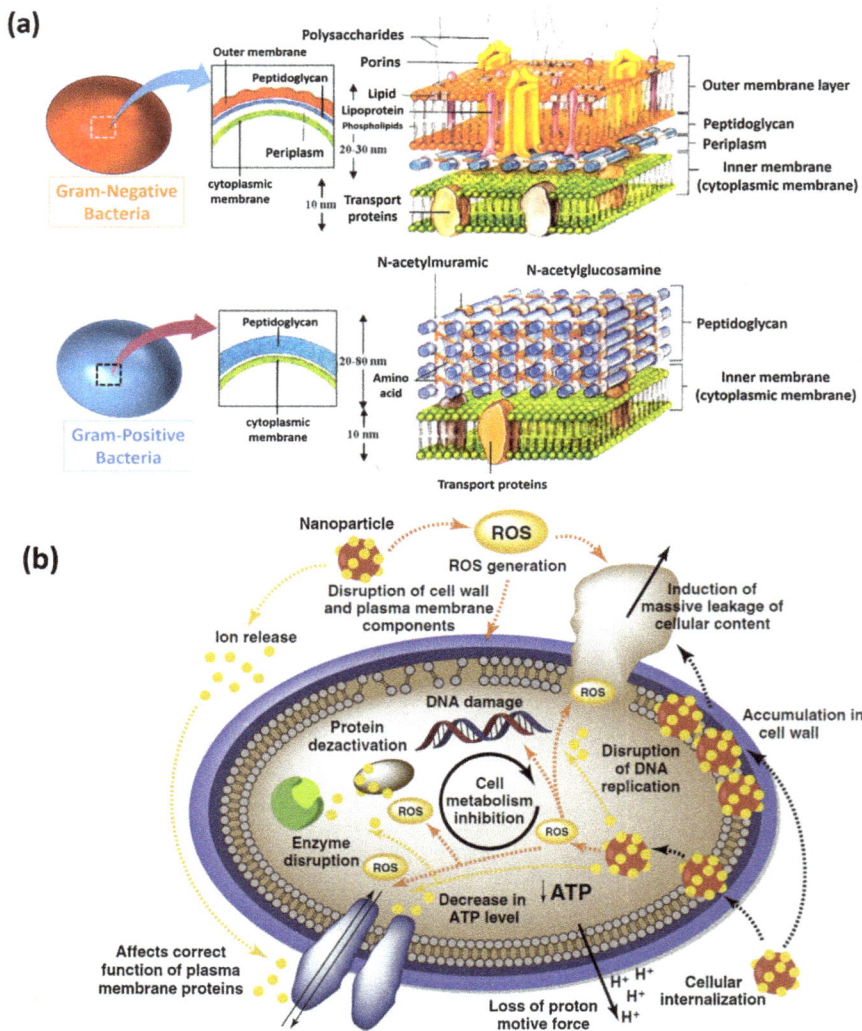

Figure 6. (a) Difference between the wall structures of the outer cells of gram-positive and negative bacteria [71] (Reprinted with permission from Royal Society of Chemistry (2008) [71]), and (b) schematic illustration showing the mechanism for antimicrobial action of metal ions [72] (Reprinted with permission from Elsevier (2016) [72]).

But these antimicrobial agents generally experience several flaws like action on non-target microorganisms, little persistence of antimicrobial finish, and toxicity to the environment [69]. NPs can be a good alternative to overcome these problems and there have been extensive research on the investigation of the probable impact of nanoscience and nanotechnology on textile technologies for functional finishing of different textiles [70]. NPs exhibit excellent antimicrobial behavior compared to traditional antimicrobial agents. Generally, NPs that can be incorporated with textile fabrics for antibacterial activity are classified as metal NPs, metal oxide NPs, and carbon-based NPs.

Bacteria can be either gram-positive or gram-negative, based on the constitute of the wall structure of the outer cell (Fig. 6) [71-72]. While the cell wall of a gram-positive bacteria has a broader layer of peptidoglycan, the outer membrane in gram-negative bacteria is made of lipopolysaccharide.

The additional periplasmic layer in gram-negative bacteria sandwiched between their plasma membrane and outer layer (Fig. 6(a)) allows the antimicrobial NPs to penetrate and finally destroy the cell [73]. Here, bactericidal mechanism of antimicrobial NPs and their effectiveness on textiles have been described thoroughly. An outline of bactericidal mechanism of metal ion NPs is expressed using schemes in Fig. 6(b). The possible mechanisms can be summarized as:

- Generation of Reactive Oxygen Species (ROS): Metal oxide NPs generate ROS like superoxide anions, hydroxyl radicals, hydrogen peroxide, etc. which can cause peroxidation of the polyunsaturated phospholipids as part of the structure of the cell, to cause destruction of DNA and ultimately cell death [74].

- Physical damage: Sharp edges of nanostructured material can injure bacterial cell wall membrane [75].

- Binding: Binding of nanomaterials on the cell wall of bacteria may damage integrity of the cell membrane and reduce efflux rate of the materials of the cytoplasm [76].

- Release of metal ions: NPs release metal ions to the medium that can prohibit DNA replication and ATP production and damage the cells [77,78].

In specific, Ag NPs show antimicrobial activity following a mechanism that includes disruption of cell wall and plasma membrane, decrease in ATP level and disintegration of the plasma membrane potential, introduction of pits and gaps in the wall of the cell, depriving ability of DNA replication and halting the cycle of the cell at the G2/M phase, interaction with thiol groups of protein, reduction in proton motive force, etc. CuO NPs exhibit antimicrobial activity via cellular internalization, introduction of leakage of intracellular substance, damaging the vital enzymes, disruption of DNA structure, decreasing permeability of cell wall, etc. The antimicrobial action of ZnO NPs proceeds through cellular internalization, generation of ROS, and disorganization of plasma membrane.

These NPs are gaining significant research attention as prospective antimicrobial agents for their unique antibacterial and antifungal activity, low cost, and toxicity [79]. Ag NPs have the ability to catalytically oxidize the bacteria structure by producing oxygen radicals via the following mechanism: [70]

$$H_2O + 1/2O_2 \rightarrow \text{Metal ion } H_2O_2 \rightarrow H_2O + [O] \tag{1}$$

The uses of nanocomposites prepared from antimicrobial agents have shown highly promising synergistic antimicrobial properties in textiles in comparison to single, monophasic nanomaterial. Ag NPs can be immobilized on silk fibers and nylon by following LbL deposition [80]. Colored, antimicrobial thin film could thus be prepared by sequentially dipping silk fibers and nylon in to a mixture of poly(diallyldimethylammonium chloride) (commonly known as PDADMAC) solution of low concentration and poly(methacrylic acid) (PMA) coated NPs. The LbL deposition was assessed from the K/S value of the coated fibers, measured with a reflectance spectrophotometer where S refers to the scattering and K is sorption coefficient of the fiber. Evaluation of the antimicrobial efficiency of these coated fibers against *Staphylococcus aureus* bacteria revealed that Ag was capped by PDADMAC/PMA and deposited as layers onto the fibers to lead to reduction of 80% in bacteria population for the silk fiber and 50% in case of nylon [80].

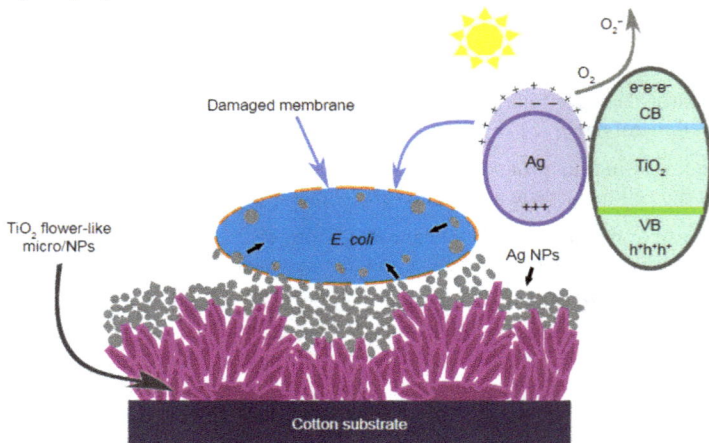

Figure 7. Mechanism of antibacterial action by TiO_2 and Ag NPs coated on cotton surface. CB: conduction band; VB: valence band; E. coli: Escherichia coli [81]. (Reprinted with permission from International Journal of Nanomedicine 2017:12 2593-2606; Originally published by and used with permission from Dove Medical Press Ltd [81]).

Li et al. reported a probable mechanism of antibacterial activity (Fig. 7) where the TiO_2 coated cotton displayed superior antibacterial performance than untreated cotton fabric [81]. The biggest impressive perception is that coating by TiO_2 may not serve any substance for bacteria, although the surface of cellulose cotton provides suitable condition in terms of bacterial growth and upholds fine respiration because of its textured structure.

The Ag NPs provide high sites of interaction with microorganisms. At first, Ag NPs attach to the cell wall of the bacteria and then penetrate through the cell membrane that kills bacteria by directly inhibiting its respiration process. Moreover, Ag NPs constrain uptake and release of phosphate, succinate, mannitol, glutamine, and proline in *E.coli* cells. Ag NPs not only destroy the membrane of the microorganism but also harm the respiration chain to affect the cell division process, resulting in cell death (Fig. 8) [82,83]. In addition, metallic Ag ionizes by reacting with moisture and such Ag^+ ions are highly active than metallic Ag. These ionized Ag produce many reactive species (photo-generated hole-electron pairs, ROS such as $\cdot O_2^-$, $\cdot OH$, etc.) under visible light on the cellulose cotton surface. Ionized Ag damages the structure of the cell wall and thereby kills the bacteria cell [84].

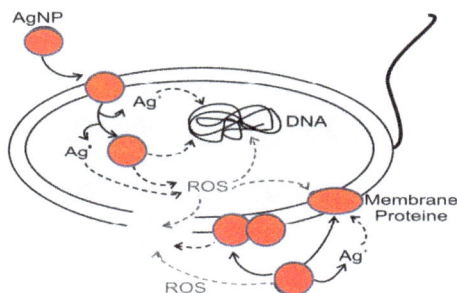

Figure 8. Probable interaction mechanism of bacterial cells with Ag containing compounds [85] (Reprinted with permission from American Chemical Society (2013) [85]).

Table 1 summarizes recent reports on textiles showing enhanced antibacterial activity due to incorporation of metal NPs.

Table 1: *Summarized data related to metal NPs used for achieving antibacterial performances.*

Types of NPs	Deposition process	Functionalized textile	Microbes	Reference
Ag	*In situ* synthesis	Cotton	*S. aureus* and *E. coli*	[86]
Ag	Dip-coating	Polyester, polyamide	*S. aureus, E. coli, C. albicans, P. aeruginosa* and *S. epidermidis*	[87-89]
Ag	Pad-dry-cure	Cotton	*S. aureus* and *E. coli*	[90]
Ag	Coating or deposition	Cotton	Gram-negative and Gram-positive bacteria	[92]
Ag-ammonia complex	Coating or deposition	Polyamide	*S. aureus* and *E. coli*	[93-95]
Ag	*In situ* synthesis	Polyester	*S. aureus* and *E. coli*	[96]
Cu NPs	*In situ* synthesis	Polyamide/nylon	*S. aureus*	[97]
Se	*In situ* synthesis	Wool	*S. aureus, E. coli* and *C. albicans*	[98]
CuO	*In situ* synthesis	Polyester	*S. aureus* and *E. coli*	[99]
TiO$_2$	*In situ* synthesis	Cotton	*S. aureus* and *E. coli*	[100]
TiO$_2$	Functionalization with 3-(trimethoxysilyl) propyl-*N,N,N*-dimethyloctadecyl am monium chloride and 3-glycidoxypropyl)trime thoxy-silane	Cotton	*S. aureus* and *E. coli*	[4]
Seaweed capped ZnO	Pad-dry-cure	Cotton	*S. aureus, E. coli, S. pyogenes,* and *K. aerogens*	[101]
SiO$_2$-Ag	Pad-dry-cure	Cotton/polyester blend (65/35)	*S. aureus*	[102]
CuO	*In situ* synthesis	Wool	*S. aureus* and *E. coli*	[99]
ZnO	Pad-dry-cure	Cotton	*S. aureus* and *E. coli*	[103]
ZnO	Pad-dry-cure	Cotton	*S. aureus* and *E. coli*	[104]
ZnO (30-60 nm)	Pad-dry-cure, Electroless deposition	Cotton/polyester	*M. luteus* and *E. coli*	[105]
Ag/TiO$_2$	Hydrothermal	Cotton	*S. aureus* and *E. coli*	[81]
Ag/ZnO, Ag/SiO$_3$	Pad-dry-cure	Cotton/polyester/p olyamide	*M. luteus* and *E. coli*	[106]
SiO$_2$ sols containing Ag, vinyl silica NPs	Sol-gel	polyamide /polyester	*S. aureus* and *E. coli*	[107]
Ag-doped silica complex NPs	Spin-coating	Different textile supports	*E. coli*	[108]

Chitosan Ag loaded chitosan NPs	Ionic gelation	Polyester	*S. aureus*	[109]
Colloidal Ag and TiO$_2$ NPs	Wet-chemical	Polyester	*S. aureus, E. coli* and *C. albicans*	[110]
Ag	Sol-gel	Polyamide	*S. aureus* and *E. coli*	[111]
Au, nano Ag colloids	Solvent swelling	Padded and nonpadded nonwoven polypropylene or polypropylene/poly ethylene	*S. aureus* and *E. coli*	[112]
Calcium phosphate-based Ag-doped powder	Wet chemical	Cotton and polyester	*E. coli*	[113]

2.2 UV protection

Ultraviolet (UV) protection is one of the major, yet desirable challenges in the textile industries. The radiation that comes to the earth from the sun consists of 50% visible, 45% infrared, and 5% UV radiations [114]. The UV radiation is categorized into UV-A, UV-B, and UV-C with the wavelength range of 400-320 nm, 320-280 nm, and 280-200 nm, respectively [98]. UV-A and UV-B radiations are the most harmful ones for both human skin and cloths [115]. Prolonged exposure to UV light increases the possibility of various toxic diseases such as sunburn, tanning, photodermatitis, erythema, and skin cancer [116]. In addition, UV radiation can cause several damages of the textiles such as reduction of chemical and mechanical properties and discoloration. Consequently, clothing and textiles should have a necessary property to protect against UV radiation [117]. Both refractive and dispersing UV radiations may get efficiently absorbed by certain metal oxide NPs like ZnO, TiO$_2$, SiO$_2$, Al$_2$O$_3$, and magnetite NPs, as these successfully block UV radiation owing to their suitable bandgap energies. These NPs guarantee both a sustainable and improved UV protection performance, in comparison with many organic UV absorbers [102,118,119]. Inorganic additives are generally preferred over the organic ones, due to their extraordinary properties like low toxicity and thermal and chemical stability against UV radiation. Chemical composition, particle size, shape, and crystallinity of ZnO and TiO$_2$ NPs possess a direct effect on their protective actions [120,121]. Fig. 9 shows the mechanism of UV protection by TiO$_2$ NPs on coated textiles.

Standard protocol for sun protection is to measure the efficiency of UV protection of fabrics in terms of the value of UV protection factor (UPF) [123]. The UPF of a fabric is determined as follows: [124]

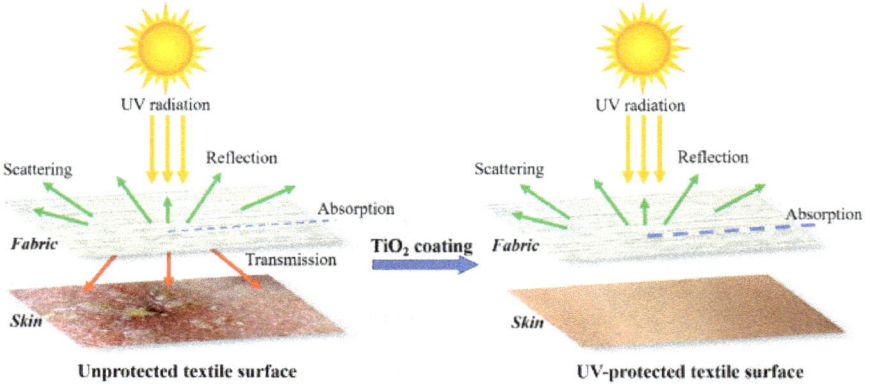

Figure 9. UV protection by TiO$_2$ NPs on coated textiles [122]. (Reprinted with permission from Elsevier (2021) [122]).

$$UPF = \frac{\sum_{\lambda=290}^{400} E(\lambda) \cdot \varepsilon(\lambda) \cdot \Delta(\lambda)}{\sum_{\lambda=290}^{400} E(\lambda) \cdot T(\lambda) \cdot \varepsilon(\lambda) \cdot \Delta(\lambda)}$$

(2)

where $E(\lambda)$ expressed in W/m^2/nm = solar irradiance, $T(\lambda)$ = spectral transmittance at the wavelength λ, $\varepsilon(\lambda)$ = erythema action spectrum, and $\Delta\lambda$) = wavelength interval. The UPF values of textiles represent their UV radiation blocking capabilities, as higher UPF values means higher protection capacity [125]. Table 2 categorizes the UPF and protection level.

Table 2: *Classification of UPF category measured from relative transmittance and protection level [123].*

UPF range	Protection category	UVBE$_{eryt}$ transmittance (%)
<15	Insufficient protection	>6.7
15–24	Good protection	6.7–4.2
25–39	Very good protection	4.1–2.6
40–50, 50+	Excellent protection	≤2.5

UV protection of textiles are decided by the following factors [126],

- Fiber composition (synthetic fibers transmit less UV than most natural fibers)
- Compactness of the weave (more tightly woven fabrics transmit less UV)
- Color (dark colors usually absorb more UV than light shades and therefore, possess higher UPFs)
- Stretchability (higher stretchable fabrics give reduced UPF)
- Moisture content (wet fabrics exhibit lower UV protection)
- Finishing (Presence of UV absorbing chemicals improve the UPF)

Washed cotton and poly cotton fabrics show higher UPF due to fabric shrinkage though worn and faded fabrics have lower UPF ratings [127]. Moreover, dyes are capable of enhanced UV blocking of fabrics and also UV protection is better provided by darker colors than lighter ones [129]. Different types of interactions of a dye with different fibers alter the physical performance of a fabric [129-136].

Untreated or control wool fabric allows significant transmittance of UV because of its porous nature (Fig. 10). On the other hand, Ag NPs treated wool fabrics show significant reduction of UV-A transmittance. Moreover, both the large sized (A- AgNPs) and non-spherical (D-AgNPs) NPs exhibit good UV protection capability than spherical shape and smaller sized Ag NPs (T-AgNPs) [134]. Table 3 summarizes data relating the use of metal NPs for UV protection performance.

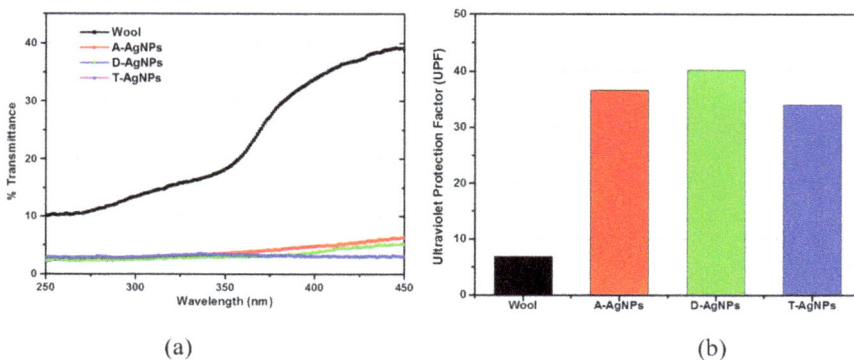

(a)   (b)

Figure 10. (a) UV transmittance of control and Ag NPs treated wool fabrics. (b) UPF of Ag NPs treated and control wool fabrics [134]. (Reprinted with permission from Springer Nature (2018) [134]).

Table 3: *Metal NPs for performance of UV protection*

Types of NPs	Deposition process	Functionalized textile	UV protection activity	Reference
Ag	*In situ* synthesis	Cotton	UPF >50	[86]
CuO	*In situ* synthesis	Polyester	64.24% compared with untreated sample	[99]
Se	*In situ* synthesis	Wool	61.82% compared with untreated sample	[98]
MgO and Al$_2$O$_3$	Electrospinning	polyacrylonitrile (PAN) nanofibrous mats	UPF >20-25	[133]
TiO$_2$	*In situ* synthesis	Cotton	Very good after 15 washing	[100]
TiO$_2$	Functionalization with 3-(trimethoxysilyl) propyl-*N,N,N*-dimethyloctadecy ammonium chloride and 3-glycidoxypropyl)trimethox y-silane	Cotton	UPF >150	[4]
FeTiO$_3$	Coating with polyurethane matrix	Cotton	UPF>50	[132]
Seaweed capped ZnO	Pad-dry-cure	Cotton	UPF>42	[101]
Ag	*In situ* synthesis	Wool	UPF>36	[134]
SiO$_2$-Ag	Pad-dry-cure	Cotton/polyester blend (65/35)	UPF>86	[102]
ZnO	Low-pressure oxygen plasma	Cotton	UPF> 65	[135]
ZnO	Electrospun ZnO nanocomposite fiber webs	Polypropylene nonwoven	UPF >40	[136]
ZnO	Pad-dry-cure	Cotton	UPF>60	[103]
ZnO	Pad-dry-cure	Cotton	UVA 76.3% and UVB 85.4% compared to untreated one	[105]
TiO$_2$, ZnO	Pad-dry-cure	Cotton	UPF>35 for ZnO, UPF>147 for TiO$_2$	[104]
CuO	*In situ* synthesis	Wool	Higher UV blocking	[99]

2.3   Self-cleaning

The notion of 'self-cleaning' for textiles is an inspiration from the nature. Lotus leaves have very clean surfaces to make it a superior example of self-cleaning. Nature has evolved super repellent surfaces or self-cleaning surfaces, known as "lotus effect." Lotus plant leaves can self-clean by resisting water and dirt particles. This phenomenon is related to the low surface energy associated with dual-size roughness of the lotus leaves, which include: epidermal cells with micro-scale papillose and hair-like waxy crystals in the nanoscale on the surface of the leaves (Fig. 11) [137].

Figure 11. Photograph and scanning electron microscopic images of surface of a lotus leaf: (a) a fresh leaf, (b) micro-structure, (c) nano-structure, (d) micro-structure of annealed leaf, (e) nano-structure of annealed leaf, (f) a droplet on an untreated leaf, and (g) a droplet placed on an annealed leaf, followed by tilting to 90° [137]. (Reprinted with permission from Elsevier (2016) [137]).

The self-cleaning textiles have some commercial implications such as, reduction of maintenance for protection of environment, cut back of time, energy, material, and cost for production, rendering textiles long lasting through increasing wire times, and saving people from the excessive laundry bill and efforts for cleaning. Self-cleaning textiles have a bright prospect for developing novel products by using NPs. They have the promises to be a standard feature of textiles of the next generation. The technology can prevent hygiene and pathogenic infection, help to lower the consumption of detergents, dry-cleansing agents, water and energy. They reduce high cleaning costs and save money, time, and labor. They may be potentially used in hospital garments, military uniforms, sportswear, and outdoor fabrics.

Self-cleaning textiles can be divided into: superhydrophobic based self-cleaning textiles and photocatalysis based self-cleaning textiles.

2.3.1 Superhydrophobicity and self-cleaning

Nanotechnology can be used for establishing 'superhydrophobic' textiles that resemble naturally self-cleaning surfaces, termed in other words, "lotus effect" (Fig. 12). The actual criteria for obtaining superhydrophobic textiles are as follows:

- a water contact angle > 150°,
- dual-size roughness of the material surface reminiscent to a lotus leaf,
- ease of water rolling on the surface with a tilting angle of < 10°, and
- very low surface energy (<10 dynes/cm) [138].

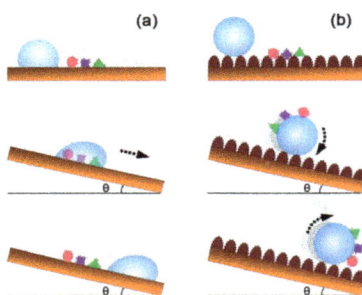

Figure 12. The effect of self-cleaning: (a) a normal smooth surface and (b) a tilted hydrophobic surface [139]. (Reprinted with permission from Royal Society of Chemistry (2012) [139]).

Superhydrophobicity based self-cleaning properties are reported in the literature. For instance, powder form of a dye, Rhodamine-B was used as a pollutant and was positioned on the surfaces of both untreated and treated fabrics. Upon pouring water over the surface, the untreated fabric underwent immediate wetting to cause coloring by the dye solution (Fig. 13a). Cotton fabric, with a large number of hydroxyl groups in the structure, is hydrophilic and exhibits high capacity of absorption. But treated fabric, due to hydrophobicity, showed self-cleaning behavior. The dye powders were instantly sorted and after dissolution were carried with water and left the fabric surface clean by the mechanism of "rolling water droplets" (Fig. 13b). The hydrophilic dye adhered weakly with the fabric surface and water droplets exerted capillary forces, which collectively assisted realizing the phenomenon [140]. The concept of superhydrophobicity may similarly be applied to other surfaces to protect them from microorganisms and aggressive environments.

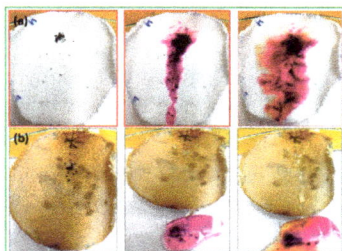

Figure 13. Photographs of the application of self-cleaning phenomenon to (a) untreated and (b) treated cotton fabric surface [141]. (Reprinted with permission from Elsevier (2022) [141]).

2.3.2 Photocatalysis and self-cleaning

To check photocatalytic activity, coffee stain of TiO_2 NPs treated cotton fabric was exposed under UV irradiation [142]. The photocatalytic degradation of coffee stain involved three steps; (a) adsorption of coffee stain, (b) adsorption of light by the used catalyst, and (c) charge transfer reactions to generate necessary radicals for stain degradation. Water-soluble and acidic colored components, polymers, and pigments in coffee primarily cause coffee stains on fabrics. The oligomeric or polymeric coloring agents in coffee are produced through condensation reaction and thermal degradation. They contain hydroxyl groups and conjugated double bonds, which attract fibers via van der Waals or dispersion forces and result in the affinity of coffee stains for cotton fabric [142]. The mechanism of degradation of coffee stain by photocatalysis is shown in Fig. 14.

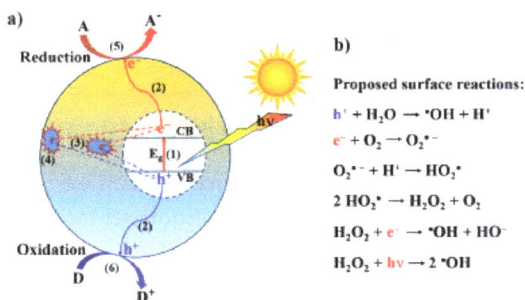

Figure 14. Mechanism of photocatalytic degradation of TiO_2: (a) steps involved are (1) photoexcitation, (2) separation followed by transfer of charge, (3) recombination of bulk charge, (4) recombination of surface charge, (5) reduction reactions at the surface, (6) oxidation reactions at the surface, (b) proposed surface reactions [122]. (Reprinted with permission from Elsevier (2021) [122]).

Here, e_{CB}^- is CB of electron and h_{VB}^+ is VB of hole. hv is the photon energy necessary to excite the electron of the semiconductor to travel from VB to the CB. Superoxide anion ($O_2^{\cdot-}$) is produced by oxygen, whereas the moisture (H_2O) present in the air converts hole (h^+) to hydroxyl radical ($OH^{\cdot}$). $O_2^{\cdot-}$ and H^+ are converted to perhydroxyl ion ($HO_2^{\cdot}$). All superoxide anion, hydroxyl radical and perhydroxyl ion are responsible for stain reduction.

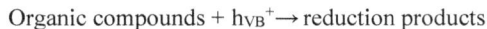

$$\text{Organic compounds} + OH^{\cdot} \rightarrow \text{degradation products}$$

$$\text{Organic compounds} + h_{VB}^+ \rightarrow \text{reduction products}$$

Nano-size of TiO_2 provides sufficient number of active sites to enhance coffee stain degradation. The ability of photocatalysis of TiO_2 coating is related to the phase, surface area, crystallinity, and density of surface $OH^{\cdot}$ groups and so on [143, 144]. Anatase phase of TiO_2 with high crystallinity exhibits superior activity for photocatalysis since it involves disruption in electronic band structure to a lesser extent [145].

Pakdel et al. applied TiO_2/Pt, TiO_2/Au, and TiO_2/Ag colloids to the fabrics surface to develop self-cleaning functionalized wool fabrics with photocatalytic activity. The incorporated metals (Pt, Au, Ag) synergistically improved the performance of self-cleaning and the efficiency depended on the nature and concentration of the dopants. The photo-induced self-cleaning property of the treated fabrics also depended on the light source and wavelength range of the radiation used [143].

2.3.3 Self-cleaning testing

When exposed to light, photocatalytic self-cleaning fabrics can decontaminate themselves. Different methods are employed to assess the activity of photocatalysis of textiles. By analyzing the degradation of a synthetic dye for instance, methylene blue, or natural colorants stain, which are frequently employed as model pollutants, photocatalytic efficiency is evaluated. Two different breakdown processes- solution discoloration and stain degradation- are used to address the photodegradation of the colorants [146-148]. Cut into bits, the fabrics after functionalization are immersed into a dye solution and are subjected to UV radiation. After a predetermined period of time, the dye solutions are systematically collected, and their concentration is determined using a UV-Vis spectrophotometer [50,148].

The fabric after functionalization is stained with the dye and exposed to UV radiation for a certain period in order to remove the stain. A spectrometer measures the color strength (K/S) values of the stained sample both for the exposed and unexposed areas. By comparing the K/S values of the same stain, the self-cleaning action can be determined [2,149,150].

$$\% \text{ Decrease in } \frac{K}{S} \text{ value} = \frac{(\frac{K}{S})_{unexposed} - (\frac{K}{S})_{exposed}}{(\frac{K}{S})_{unexposed}} \times 100 \qquad (3)$$

The concentration of the dye present on the surface of the fabric dictates K/S values. The reduction in the K/S indicates vanishing trend of the stains (Fig. 15) [41,148]. Furthermore, gas chromatography can be used to assess the self-cleaning fabric on the basis of the production of highly oxidative intermediates at the surface of cotton textiles which have been subjected to the stain. According to Meilert et al., wine, coffee, and perspiration stains that are left on the treated fabrics cause CO_2 to be released. These self-cleaning fabrics can thus be evaluated by monitoring the release of CO_2. Higher amount of CO_2 measured is indicative of enhanced self-cleaning properties [151].

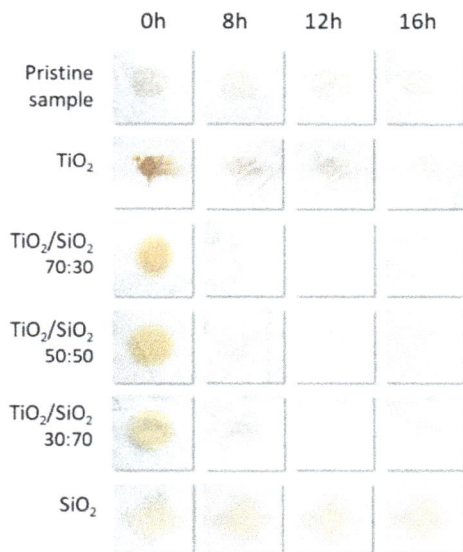

Figure 15. Photocatalytic stain removal on cotton fabrics [41]. (Reprinted with permission from Elsevier (2013) [41]).

2.4   Wrinkle resistance

Generally, resin treatment offers wrinkle resistance to fabric. However, this conventional method reduces the water absorbency, tensile strength, softness, abrasion resistance, breathability, and dyeability of the fabric. Nanomaterials can be used to prepare wrinkle resistant fabric for enhanced fabric strength and integrity and minimize the limitations associations associated with the use of resins. TiO_2 NPs improved the wrinkle resistance of cotton [152,153], SiO_2 NPs while served the same for silk fabrics [154]. The cross-linking reaction between the cellulose and the acid was catalyzed by TiO_2 NPs when they were irradiated with UV radiation. Similarly, SiO_2 NPs in presence of maleic anhydride served as a catalyst to upgrade wrinkle resistance of silk.

2.5   Flame retardant finish

Nanotechnology has drawn much appeal to both industries and academics for development of flame-retardant textiles by introducing nano-sized objects [155]. Most textiles burn readily when there is an ignition source. There are very a few textile materials that are non-combustible. Flame retardancy generally means that fiber or fabrics either do not ignite, or if ignited, do not propagate the fire once the source of ignition is removed. Some fibers are more flammable than others, e.g., cotton. Thermoplastic fibers like polyester and nylon although do not burn but melt and shrink, hence are not heat resistant. These fibers need chemical treatment to be flame retardant and can be used in protective clothing, uniforms (civil and military) and domestic upholstery fabric.

Flames are used to initiate combustion in the presence of oxygen from the atmosphere. The textile materials are thermally degraded before combustion and the degraded species may transform into combustible volatile products, which later ignite the flame when combined with oxygen (Fig. 16).

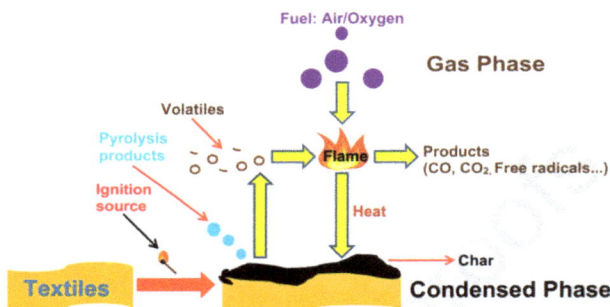

Fig. 16. The cycle of combustion for a representative textile material [156]. (Reprinted with permission from Elsevier (2020) [156]).

Flame retardant agents used to produce flame-retardant textiles slow down the combustion process. The process is affected by the chemicals at different stages of combustion such as during heating, decomposition, ignition, and flame spread to restrain overall combustion [157].

Three approaches reported so far have shown promising results for development of flame retardant textiles. These are: (i) deposition of coatings in nanodimension [158], (ii) introducing NPs in traditional back-coating [159,160], and (iii) nanostructuring of synthetic fibers [158]. Researchers, in particular, have made remarkable efforts regarding surface modifications which are able to implement and enhance different properties of textiles after treatments. The surface modification involves formation of coatings of micro- to nano-sizes. Novel coating can be applied on fibers of any kinds and may comprise a

Emerging Applications of Nanomaterials   Materials Research Forum LLC
Materials Research Foundations 141 (2023) 169-217   https://doi.org/10.21741/9781644902295-8

completely inorganic or a hybrid organic-inorganic material. This is achievable by different approaches such as, adsorption of NPs, LbL assembly or self-assembly of nanolayer films, silica-based sol-gel method, silica-based coatings and plasma treatments in other words, applications of plasma technology for surface grafting of nanofilms of polymers [158].

2.5.1 Adsorption of NPs

A certain degree of heat and fire protection may be achieved by applying coatings at the nano-level. Coated textiles are physically relatively thin films in contrast to bulk polymers.

Average flame retardant textiles, laminates, and coated fabrics are 3-5 mm thick and have a limited ability to generate a thick and surface-insulating char. The fibers deteriorate and may ignite when the temperature quickly approaches that of the ignition source (>500 °C). However, several researchers [158,161,162] reported that when applied to the surfaces of textile fibers, montmorillonite clays may form nanocoatings of inorganic species to generate thermal barriers typical for nanoceramics. To enhance the physicochemical interaction between fibers and particles, these studies are linked to concurrent plasma treatments. As reflecting metal films are deposited on the surfaces of the fabric to lower the impacts of thermal radiation from a fire source, the deposition of nanofilms and nanocoatings in the field of heat protective textiles may present necessary potential [162].

2.5.2 Layer-by-Layer assembly

Alongi et al. [158] and Horrocks [163] reviewed LbL surface treatments for nanocoatings. This process involved repeated deposition of NPs using different reagents at each adsorption phase to help many bilayers (BLs) adhere to the surfaces of the textiles and fabrics. To create a multilayer film, the substrate is alternatively dipped into a solution of an oppositely charged polyelectrolyte (or nanoparticle dispersion). As seen schematically in Fig. 17, this process explains how to prepare a structure of alternate layers of positively and negatively charged species stacked up on the surface of the substrate.

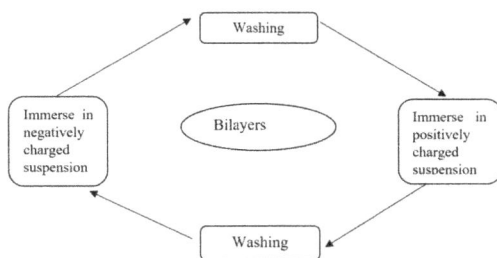

Figure 17. Cyclic L-b-L assembly process.

Two different LbL procedures that could result in thick coatings with high concentrations for efficient flame retardation have been clarified by Alongi et al. [158]. These are LbL nanocoatings either inorganic or hybrid or intumescent.

The flame retardancy characteristics of cotton are slightly improved by nanoparticulates like laponite, a negatively charged lamellar clay, combined with positively charged branched polyethylenimine, colloidal silica, and POSS. Additionally, melt dripping was significantly reduced when NPs such as Al_2O_3, SiO_2, or nanoplatelets of zirconium phosphate with various counterparts (for instance, poly(diallyldimethylammonium chloride)) were deposited using LbL [164]. However, vertical strip testing showed that fabrics treated with inorganic NPs using the LbL technique did not produce self-extinction.

To achieve flame retardancy, Laufer et al. used the LbL process to apply an intumescent multilayer nanocoating on cotton fabric using polyelectrolytes such as cationic chitosan (CH) and anionic phytic acid (PA). The flammability of multilayer CH-PA coatings on cotton fabrics was evaluated using a vertical flammability test at three different pH levels (4, 5, and 6). While uncoated cloth burnt entirely, multilayered coated fabrics displayed full self-extinguishing nature (Fig. 18a). All coated fabrics experienced a significant drop in heat release rate (HRR) compared to control fabrics. The most significant reduction in peak and total HRR values was seen in the fabric coated at pH 4. These values were reduced by 60% and 76%, respectively (Fig. 18b) [34].

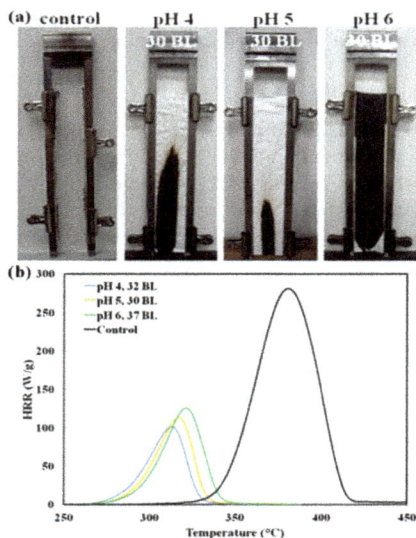

Figure 18. (a) Vertical flammability test for fabric without coating as the control and fabrics coated with 30 BLs of CH–PA; (b) Change of heat release rate with temperature for uncoated and 18 wt% CH–PA coated cotton fabrics [34]. (Reprinted with permission from American Chemical Society (2012) [34]).

Blends of cotton and polyester (70/30) with 5 BLs and 10 BLs of ammonium polyphosphate (APP) and CH and SiO_2 and APP have been reported for char forming or intumescent type properties [166]. Gaining vertical strip self-extinguishment with washing durability has been the main challenge. After 30 BLs of hybrid organic-inorganic deposition of a polycationic/polyanionic system, CH/poly(sodium phosphate) (PSP), self-extinguishing ability could be achieved effectively [166]. By branching poly(ethylelenimine) (PEI)/PSP polyelectrolyte system and coating at pH = 7 and then decreasing to a pH = 2, surfaces of cotton fabric can attain self-extinguishing properties in a single deposition step [167]. Another subsequent study on polyester-cotton fabric using a poly(allylamine)/PSP polyelectrolyte system made a comparative analysis of a 30 BL application to a "one pot" technique to achieve self-extinguish ability (17.5% add-on) and durability against five home washes at 30 °C (AATCC 135) and wash for 8 h [169]. The commercial viability of this procedure is now under investigation.

2.5.3 Sol-gel coating based on silica

On the surfaces of cotton, viscose, polyester, and cotton/polyester blends, two-step reactions of hydrolysis and condensation using alkyl silicates as the precursor can produce structures based on silica [158]. Stöber method was used to synthesize silica NPs on the denim fabric using a sodium silicate solution under an alkali pH of 12 using Seidlitzia rosmarinus (Keliab, a natural source of alkali) and also using ethanol/Keliab [169].

Higher heat resistance was one of the enhanced properties of the finished denim fabric. Self-extinguishing ability during vertical strip testing could hardly be achieved and a poor durability to water soaking was noted due to the formation of sol-gel through SiO_2-based coatings. Alongi et al. [158] reported that more addition of phosphorous-containing species is required to attain medium level of flame retardancy [158,163,170].

The basic sol-gel chemistry can be represented as given below [158]:

Hydrolysis:

$(RO)_4Si + H_2O \rightarrow (RO)_3SiOH + ROH$

$(RO)_3SiOH + H_2O \rightarrow (RO)_2Si(OH)_2 + ROH$

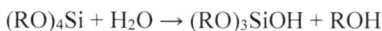

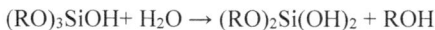

Aqueous Condensation:

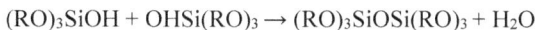

$(RO)_3SiOH + OHSi(RO)_3 \rightarrow (RO)_3SiOSi(RO)_3 + H_2O$

Alcoholic Condensation:

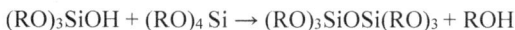

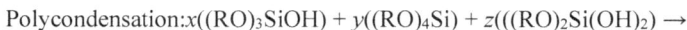

$(RO)_3SiOH + (RO)_4 Si \rightarrow (RO)_3SiOSi(RO)_3 + ROH$

Polycondensation:$x((RO)_3SiOH) + y((RO)_4Si) + z(((RO)_2Si(OH)_2) \rightarrow$

linear chains of $-Si(RO)_2O$ –with $-Si(RO)O$

Where $R = CH_3, C_2H_5$, etc.

Emerging Applications of Nanomaterials   Materials Research Forum LLC
Materials Research Foundations 141 (2023) 169-217   https://doi.org/10.21741/9781644902295-8

2.5.4 Plasma technique for surface grafting

The technology based on plasma is used to provide a novel nanocoating with required thermal and flame-retardant properties. In spite of limited success to develop flame retardant textiles, most plasma research works since 2005 are related to vacuum plasma. They are generally inappropriate for commercial textile processing system. Jama et al. reported that silicon-based films deposited by plasma can increase the flame-retardant property of the surfaces of polymers [171]. Both conventional and nanocomposite polyamide 6 films were activated by a cold nitrogen plasma and heated in a reactor for 20 min by the vapor of 1,1,3,3-tetramethyldisiloxane in an oxygen carrier gas. At about 800 °C, the coating of oxygenated polysiloxane changed into a silica-based structure, creating a thermal barrier effect that increased the flame retardancy of conventional polyamide 6 films by a medium amount and of nanocomposite polyamide 6 films by a large amount (Limiting oxygen index, LOI 45 vol%). According to Quede et al. (2005), a low-pressure plasma can be used to create coated samples that are wider and more consistent, yielding LOI values of 48 vol%. The transformation of the coated nanocomposite film into a silica-like structure produces the thermal shielding effect, which is responsible for these high LOI values [172].

However, to meet the demands of continuous processing, textile sector requires atmospheric plasma treatment [173]. The flash fire resistance of a conventional flame retardant fabric was improved by surface treatment using a nanoparticulate, Cloisite Na^+ functionalized with vinyl triphenyl phosphonium bromide (functionalized montmorillonite clay), and a silicon-based monomer, hexamethyldisiloxane following a process patented by Horrocks et al. in 2009 and 2011 [161,174].

The argon plasma alone, the plasma with a silicon-containing monomer, the plasma with nanoclay alone, and the plasma with a mixture of silicon-containing monomer and nanoclay were all applied to the meta-aramid woven fabric (GSM, 200 g/m^2). The fabric alone though gives out to ignite against 50 kW/m^2 heat fluxes but ignite against heat fluxes of 60 kW/m^2. The results show that heat release over time, including time-to-peak (TTP) and time-to-ignite (TTI), was observed. The advantage of this method is that it may be used to increase the heat and fire resistance of any textile material. According to Tata et al., polyester textiles are first etched by cold plasma before being submerged in aqueous solutions of hydrotalcite, nanometric TiO_2, and SiO_2 [175]. After a single wash at 30 °C for 30 minutes, only hydrotalcite treatment demonstrated increased fire resistance with higher TTI and durability. In a subsequent study, thermal stability was boosted by plasma surface activation and nanomontmorillonite coating [164]. When compared to untreated polyester, the treated polyester had a good enhancement of TTI of around 104% and a slight decrease in peak heat release rate (PHRR; ca.10%).

Following grafting of sodium silicate layers by atmospheric pressure plasma, cotton flannel and viscose materials had longer burning times during 45° testing [176]. The burning time increased due to crosslinking formed (OSiO) between both inter and intramolecular OH in cellulose chains and pendant $OSiO_2H$ groups. Multiplexed laser surface enhancement

(MLSE), which involves exposing pre-padded fabrics simultaneously to a UV laser (wavelength =308 nm) and treatment by atmospheric plasma in a variety of atmospheres, including argon, nitrogen, oxygen, CO_2, and mixtures, could also be used successfully to produce graft flame species on fabric surfaces [177]. This method demonstrated its potential to provide innovative and long-lasting flame resistance treatments on a variety of fabrics, including wool, cotton, polyester, and mixes, with no waste and little effluent. After receiving MLSE treatments, cotton twill furnishing fabrics pad-dried with a proprietary flame retardance gained improved flame resistance with durability [178,179].

The use of plasma for surface modification of fiber and fabric has a history of about 40 years, but the textile industry has been slow to adopt it because the majority of successful applications have been for low pressure plasma, which is less suited for textile processing than atmospheric plasma technologies [180]. Even though there were certain challenges, such as achieving a high level of surface deposition, atmospheric pressure plasma was successful in producing long-lasting flame retardancy [173].

2.6   Antistatic finish

Antistatic refers to the ability to reduce static electric charges on textiles while retaining just enough moisture to allow for electrical conduction. Synthetic fibers like nylon and polyester have low moisture regain and are vulnerable to the generation of static charge. On the other hand, cellulose fibers like cotton, lyocell linen, viscose, etc. show lower static charges due to higher moisture content. When treating materials, an antistatic agent is used to eliminate or reduce the development of static electricity on the surface of the material, which is often brought on by the triboelectric effect. In analogous to a surfactant, an antistatic agent consists of two parts: one is hydrophilic and the other is hydrophobic. The hydrophobic part interacts with the substance surface, while the hydrophilic part interacts with the moisture in the atmosphere by binding water molecules [181].

Nanotechnology has been widely used to enhance the antistatic qualities of synthetic textiles, as synthetic fibers often have poor antistatic capabilities. In fact, the development of antistatic fabrics through the use of conducting nanofillers has been the subject of numerous investigations [182]. Synthetic fibers can be made antistatic by adding TiO_2 and Ag NPs [183], ZnO whiskers NPs [60,184], Sb-doped SnO_2, and silane nanosol [61]. Since ZnO and TiO_2 NPs are electrically conductive, they can be utilized to remove static electricity from these fibers. Since some silane-based nanosols may absorb moisture through their surface hydroxyl groups, they have also been employed as antistatic agents. Multifunctional nanotextiles thus can be produced either by accumulating NPs within the fibers or chemical grafting of NPs onto the surface of the fibers or functional nanocoating of surfaces of textiles.

A mathematical representation of an electric force between charged objects is described by Coulomb's law, which states that objects with similar charges repel one another while those with opposing charges attract one another. The electric force between two charged particles is [185]:

$$F = \frac{k \times Q_1 \times Q_2}{r^2}$$

$$(4)$$

where Q_1 and Q_2 are the charges, r is the distance between the charged objects, and k is a proportionality constant.

Commercial antistatic membrane based on poly(tetrafluoroethylene) (PTFE) was produced with conductive NPs attached to the membrane [146]. Several techniques have so far been used to produce nanocoated textile materials, including LbL deposition, coating with nano-oxide dispersions, functionalization with inorganic sol-gel coating, and applying NPs to the surfaces of fibers using cross-linking agents like water-based polysiloxane emulsion and (poly)carboxylic acid. To accomplish antistatic properties, sol-gel coatings have been applied on the surfaces of the fiber [93]. After modification with hydrophilic substances or alkoxysilanes containing amino groups, different hydrophobic alkoxysilanes have been used for this purpose. Since they contain moisture inside the coatings but hydrophobicity on the outside, textiles coated with sol-gel exhibit antistatic properties. For instance, after finishing with fluorine hydrophobic substance and Ag NPs, polyester fabric exhibits good antistatic qualities [62].

By using coatings of ZnO NPs, Wasim et al. demonstrated development of antistatic property [187]. Ag NPs could minimize 60.4% of the static voltage of polyester fiber. On the other hand, the static voltage was reduced by 77.7% for a combination of Au and ZnO NPs. The use of Sb-doped SnO_2 also induced antistatic properties to polyacrylonitrile fibers [188]. Diffusion of the SnO_2 particles into the conductive channels generated in the fibers eventually led to antistatic property.

Memon et al. treated wool fabric with protease (3%, 6%, and 9%) and incorporated Ag NPs at various pH (i.e., pH = 4 and 7) as nanocoating and evaluated the antistatic effect [182]. The 3% and 6% protease treated wool fabrics showed better antistatic properties than the 9% protease treated fabrics. In addition, samples treated at pH 4 exhibited superior antistatic properties compared to samples treated at pH 7. The wool surface may be deteriorated and its antistatic properties weakened at higher pH levels. From 7000 V to 1400 V and even as low as 200 V, the electrostatic voltage dropped. Even after being washed in soap for 20 times, wool fiber can have good antistatic properties due to incorporating Ag NPs and scale stripping. Such a material effectively dissipates the static charge that has accumulated on the fabric [128]. Since silica gel particles on fiber collect moisture from air by amino and hydroxyl groups and bind water, silane nanosol enhances antistatic characteristics. The fibrils of Teflon membranes are permanently occupied by electrically conductive NPs, creating a network that prevents the development of isolated chargeable areas and voltage peaks, generally found in conventional anti-static materials. Due to the ease of removal by anti-static agents after a few washing cycles, this method can overcome the shortcomings of conventional methods [147]. Fig. 19 shows the mechanism of treatment of textiles by antistatic and fixing agents.

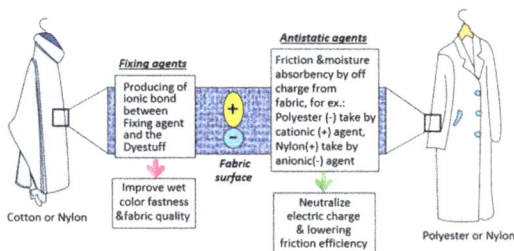

Figure 19. Mechanism of treatment of textiles using antistatic and fixing agents [186].
(Reprinted with permission from Springer (2017) [186]).

Textiles with antistatic finish can be used for manifold applications including upholstery fabrics, fabrics for operating rooms in hospitals, airbags for automobiles, filtration fabrics, conveyor belts, parachutes, carpets, and protective clothing for working with flammable liquids, powdered solids, and gases [62].

2.7 Hydrophilic finish textiles

Kundu et al. reported that TiO_2 and SiO_2 NPs along with PA and CS can be used to improve hydrophilic character of polyamide 66 (PA66) fabrics. PA 66 fabrics treated with SiO_2 NPs showed good hydrophilicity since the water contact angle showed an expected decrease from 112° to 29° for the treated fabrics from the untreated one [189].

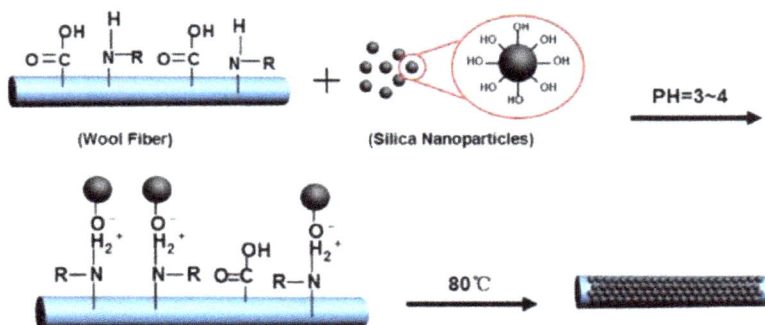

Figure 20. Formation of an ultrathin silica layer on a wool fiber exhibiting
superhydrophilicity [131]. (Reprinted with permission from American Chemical Society
(2010) [131]).

Capillary penetration causes water to penetrate through the hydrophilic fabric when it is placed on its surface. The original force directs the penetration of the wettability of the fiber. Surface energy and surface roughness have a significant impact on this wettability. According to Chen et al., the roughness of the surface was enhanced by a thin layer coating of nanoscale spherical protuberances made of silica on the wool surface (Fig. 20) [131]. Covalently bonded fatty acids are prevalent on the outer surface of wool fibers with predominating 18-methyleicosanoic acid [1,2].

Due to hydrophobicity of the exterior surface as determined by water contact angles, the surface energy of pristine wool fibers is low. But, by coating wool fiber with an ultrathin layer of silica, the surface wettability of the fiber can be enhanced. The presence of many Si-OH groups on the surface of the SiO_2 clusters gives the silica layer its hydrophilic character and increased wettability. The water contact angle on a planar SiO_2 surface is approximately 20° [35]. The wool fabric surface will exhibit a change in porosity when SiO_2 NPs of different sizes from 27 to 300 nm are deposited on the fibers. The spacing between the NPs increase as the diameter of SiO_2 NPs increases, which weakens capillary penetration of water droplets and results in less hydrophilicity. With respect to water absorption, wool fabric coated with 300 nm SiO_2 NPs had the lowest rate.

Furthermore, the coverage of SiO_2 coated wool fibers underwent noticeable shrinking as a result of reduction of the specific surface area by larger NPs to weaken the driving force of coalescence between the fiber and the NPs. Silica NPs with a smaller dimension is thus more suited for fabricating superhydrophilic wool textiles [131]. Superhydrophilic wool fabrics of this kind display good wash fastness after 20 cycles of dry cleaning in perchloroethylene, with a wetting time of less than 3 s.

3.   Current challenges and future scope

Replacement of the current marketed technologies and products, particularly in textile industries, has been a big challenge for nanotechnology. The application of nanotechnology or the usage of expensive nanomaterials may result in higher end product costs. However, due to the effectiveness of nanomaterials even at low concentrations to obtain equivalent or superior qualities to bulk materials, the cost of nano multifunctional textile goods can partially offset the expenses.

Despite providing several advancements and benefits for humanity, nanoengineered textile goods raise awareness around the world about their potential negative effects on the environment and health hazards [1,91,190,191]. A few recent studies have shown that textile and garment industries at present account for about 10% of global carbon emissions and that the dyeing and finishing chemicals they employ are responsible for 17–20% of industrial water pollution [1]. Furthermore, NPs even at high concentrations may be toxic to living beings and may impart negative impact on the environment.

The extensive usage of extremely small size NPs poses a greater risk to the health of humans and animals. In particular, the carbonaceous NPs may create problems related to inhalation and cause many other lethal diseases. In this regard, it is now important for textile manufacturers to ensure that the new textile products based on nanotechnology will not have a negative impact on either human health or the environment at any point during their lifespan, including during production, use, and end-of-life (Fig. 21). Safer nanotechnology-based methods are therefore crucial. Since NPs may have negative environmental impact, their toxicity and leaching will continue to remain as major issues [150].

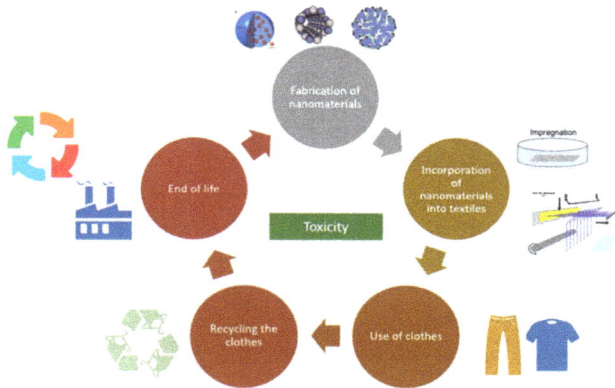

Figure 21. Life cycle of nanoengineered textiles [91]. (Reprinted with permission from Elsevier (2020) [91]).

Numerous studies and developments in nano multifunctional textiles till now are confined to the laboratory scale, which needs to be transformed to industrial scale. This is a big task for textile engineers and material scientists. Despite many challenges at the moment, scientists and researchers are looking forward to exploiting the promises of nanotechnology to develop multifunctional textiles.

Conclusions

Traditional textile treatments applied to textile substrates face many difficulties in terms of their durability and efficacy of diverse functional finishes. Producing multifunctional textiles with features like self-cleaning, repellency against oil, water, and dirt, UV protection, antimicrobial property, wear/tear resistance, flame retardancy, wrinkle resistance, antistatic characteristics, etc. is a significant opportunity for nanotechnology. Nanofinishing, nanocoating, and nanocomposite coating can significantly improve the inherent properties of textiles by adding to their functionality, comfort, and durability.

Since they offer unique and cutting-edge properties that make textile goods smart and intelligent, textiles treated with nanomaterials can be employed in a variety of technical sectors besides clothing. High-performance and intelligent textiles can be made using polymer nanocomposite fibers and nanofibers. In particular, because of their large ratio of surface area to volume, tunable diameter, better porosity, and improved functionalities, nanofibers are widely used in various technical fields, which inter alia include power generation, medical, filtration, and automotive use.

Various multifunctional textiles treated with a range of nanomaterials have recently been commercialized to compete with the present market offerings. Although there are a number of obstacles preventing the widespread use of multifunctional textiles treated with nanomaterials, the demand for fashionable clothing with appealing aesthetics and cutting-edge features has improved the market for multifunctional and intelligent textiles. However, the future of nano-enhanced textiles may be driven by the increasing demand for more advanced, intelligent, and efficient multifunctional textiles.

Acknowledgement

M.A.B.H. Susan and S. Ahmed acknowledge the Centennial Research Grant from the University of Dhaka, Bangladesh for supporting the themed research.

Competing financial interests

There are no competing financial interests to declare.

References

[1] A.K. Yetisen, H. Qu, A. Manbachi, H. Butt, M.R. Dokmeci, J.P. Hinestroza, M. Skorobogatiy, A. Khademhosseini, S.H. Yun, Nanotechnology in Textiles, ACS Nano. 10 (2016) 3042-3068. https://doi.org/10.1021/acsnano.5b08176

[2] A.P.S. Sawhney, B. Condon, K.V. Singh, S.S. Pang, G. Li, Modern applications of nanotechnology in textiles, Text. Res. J. 78 (2008) 731-739. https://doi.org/10.1177/0040517508091066

[3] R. Mishra, J. Militky, Nanotechnology in textiles: theory and application, 1st ed, Cambridge: Woodhead Publishing Ltd, Duxford, UK, 2019, pp. 102-245.

[4] S. Riaz, M. Ashraf, T. Hussain, M.T. Hussain, A. Rehman, A. Javid, K. Iqbal, A. Basit, H. Aziz, Functional finishing and coloration of textiles with nanomaterials, Color. Technol. 134 (2018) 327-346. https://doi.org/10.1111/cote.12344

[5] M. Joshi, A. Bhattacharyya, Nanotechnology - a new route to high-performance functional textiles, Text. Prog. 43 (2011) 155-233. https://doi.org/10.1080/00405167.2011.570027

[6] B. Mahltig, H. Haufe, H. Böttcher, Functionalisation of textiles by inorganic sol-gel coatings. J. Mater. Chem. 15 (2005) 4385-4398. https://doi.org/10.1039/b505177k

[7] W.D. Schindler, P.J. Hauser, (Eds.) Chemical finishing of textiles. 1st ed. Cambridge: Woodhead Publishing Ltd and CRC Press LLC, England, 2004, pp. 18-182.

[8] M. Syduzzaman, S.U. Patwary, Smart textiles and nano-technology: a general overview, J. Text. Sci. Eng. 05 (2015) 1-7. https://doi.org/10.4172/2165-8064.1000181

[9] M. Shabbir, S. Ahmed, J.N. Sheikh, (Eds.) Frontiers of Textile Materials: Polymers, Nanomaterials, Enzymes, and Advanced Modification Techniques, John Wiley & Sons and Scrivener Publishing LLC, USA, 2020, pp. 13-70. https://doi.org/10.1002/9781119620396

[10] International organization for standardization, 2008. Technical specification: Nanotechnologies - Terminology and definitions for nano-objects - Nanoparticle, nanofiber and nanoplate, 2008, ISO/TS 80004-2.

[11] International organization for standardization, 2010. Nanotechnologies-vocabulary - Part 1: Core terms, 2010, ISO/TS 80004-1.

[12] R. Haggenmueller, C. Guthy, J.R. Lukes, J.E. Fischer, K.I. Winey, Single wall carbon nanotube/polyethylene nanocomposites: Thermal and electrical conductivity, Macromolecules. 40 (2007) 2417-2421. https://doi.org/10.1021/ma0615046

[13] J. Ren, W. Bai, G. Guan, Y. Zhang, H. Peng, Flexible and Weaveable Capacitor Wire Based on a Carbon Nanocomposite Fiber, Adv. Mater. 25 (2013) 5965-5970. https://doi.org/10.1002/adma.201302498

[14] S. Pan, H. Lin, J. Deng, P. Chen, X. Chen, Z. Yang, H. Peng, Novel wearable energy devices based on aligned carbon nanotube fiber textiles, Adv. Energy Mater. 5 (2015) 1401438. https://doi.org/10.1002/aenm.201401438

[15] D. Zhang, M. Miao, H. Niu, Z. Wei, Core-spun carbon nanotube yarn supercapacitors for wearable electronic textiles, ACS Nano. 8 (2014) 4571-4579. https://doi.org/10.1021/nn5001386

[16] A.D. Lueking, L. Pan, D.L. Narayanan, C.E. Clifford, Effect of expanded graphite lattice in exfoliated graphite nanofibers on hydrogen storage, J. Phys. Chem. B 109 (2005) 12710-12717. https://doi.org/10.1021/jp0512199

[17] C. Kim, K.S. Yang, M. Kojima, K. Yoshida, Y. Kim, Y. Kim, M. Endo, Fabrication of electrospinning-derived carbon nanofiber webs for the anode material of lithium-ion secondary batteries, Adv. Funct. Mater. 16 (2006) 2393-2397. https://doi.org/10.1002/adfm.200500911

[18] Y. Yu, Q. Yang, D. Teng, X. Yang, S. Ryu, Reticular Sn nanoparticle-dispersed PAN-based carbon nanofibers for anode material in rechargeable lithium-ion batteries, Electrochem. Commun. 12 (2010) 1187-1190. https://doi.org/10.1016/j.elecom.2010.06.015

[19] X. Li, P. Sun, L. Fan, M. Zhu, K. Wang, M. Zhong, J. Wei, D. Wu, Y. Cheng, H. Zhu, Multifunctional graphene woven fabrics, Sci. Rep. 2 (2012) 395. https://doi.org/10.1038/srep00395

[20] M. Shateri-Khalilabad, M.E. Yazdanshenas, Fabricating electroconductive cotton textiles using graphene, Carbohydr. Polym. 96 (2013) 190-195. https://doi.org/10.1016/j.carbpol.2013.03.052

[21] M. Joshi, A. Bhattacharyya, N. Agarwal, S. Parmar, Nanostructured coatings for super hydrophobic textiles, Bull. Mater. Sci. 35 (2012) 933-938. https://doi.org/10.1007/s12034-012-0391-6

[22] P. Kiliaris, C.D. Papaspyrides, Polymer/layered silicate (clay) nanocomposites: An overview of flame retardancy, Prog. Polym. Sci. 35 (2010) 902-958. https://doi.org/10.1016/j.progpolymsci.2010.03.001

[23] G.V.R. Reddy, M. Joshi, B. Adak, B.L. Deopura, Studies on the dyeability and dyeing mechanism of polyurethane/clay nanocomposite filaments with acid, basic and reactive dyes, Color. Technol. 134 (2018) 117-125. https://doi.org/10.1111/cote.12332

[24] L. Tshniwal, Q. Fan, S.C. Ugbolue, Dyeable polypropylene fibers via nanotechnology, J. Appl. Polym. Sci. 106 (2007) 706-711. https://doi.org/10.1002/app.26719

[25] Y. Yu, X. Zhong, W. Gan, Conductive composites based on core-shell polyaniline nanoclay by latex blending, Colloid Polym. Sci. 287 (2009) 487-493. https://doi.org/10.1007/s00396-008-1989-5

[26] M. Esen, I. İlhan, M. Karaaslan, R. Esen, Investigation of electromagnetic and ultraviolet properties of nano-metal-coated textile surfaces, Appl. Nanosci. 10 (2020) 551-561. https://doi.org/10.1007/s13204-019-01122-1

[27] S.H. Jeong, S.Y. Yeo, S.C. Yi, The effect of filler particle size on the antibacterial properties of compounded polymer/silver fibers, J. Mater. Sci. 40 (2005) 5407-5411. https://doi.org/10.1007/s10853-005-4339-8

[28] H.J. Lee, S.Y. Yeo, S.H. Jeong, Antibacterial effect of nanosized silver colloidal solution on textile fabrics, J. Mater. Sci. 38 (2003) 2199-2204.

[29] H.Y. Ki, J.H. Kim, S.C. Kwon, S.H. Jeong, A study on multifunctional wool textiles treated with nano-sized silver, J. Mater. Sci. 42 (2007) 8020-8024. https://doi.org/10.1007/s10853-007-1572-3

[30] M.D. Teli, J. Sheikh, Modified bamboo rayon-copper nanoparticle composites as antibacterial textiles, Int. J. Biol. Macromol. 61 (2013) 302-307. https://doi.org/10.1016/j.ijbiomac.2013.07.015

[31] J.H. Xin, W.A. Daoud, Y.Y. Kong, A new approach to UV-blocking treatment for cotton fabrics, Text. Res. J. 74 (2004) 97-100. https://doi.org/10.1177/004051750407400202

[32] W. Duan, A. Xie, Y. Shen, X. Wang, F. Wang, Y. Zhang, J. Li, Fabrication of superhydrophobic cotton fabrics with UV protection based on CeO_2 particles, Ind. Eng. Chem. Res. 50 (2011) 4441-4445. https://doi.org/10.1021/ie101924v

[33] S.W. Ali, S. Rajendran, M. Joshi, Synthesis and characterization of chitosan and silver loaded chitosan nanoparticles for bioactive polyester, Carbohydr. Polym. 83 (2011) 438-446. https://doi.org/10.1016/j.carbpol.2010.08.004

[34] G. Laufer, C. Kirkland, A.B. Morgan, J.C. Grunlan, Intumescent multilayer nanocoating, made with renewable polyelectrolytes, for flame-retardant cotton, Biomacromolecules. 13 (2012) 2843-2848. https://doi.org/10.1021/bm300873b

[35] J.P. Chen, G.Y. Chang, J.K. Chen, Electrospun collagen/chitosan nanofibrous membrane as wound dressing, Colloids Surf. A: Physicochem. Eng. Asp. 313-314 (2008) 183-188. https://doi.org/10.1016/j.colsurfa.2007.04.129

[36] J. Alongi, G. Brancatelli, G. Rosace, Thermal properties and combustion behavior of POSS-and bohemite-finished cotton fabrics, J. Appl. Polym. Sci. 123 (2012) 426-436. https://doi.org/10.1002/app.34476

[37] S. Chen, X. Li, Y. Li, J. Sun, Intumescent flame-retardant and self-healing superhydrophobic coatings on cotton fabric, ACS Nano. 9 (2015) 4070-4076. https://doi.org/10.1021/acsnano.5b00121

[38] O. Baykuş, Ş.D. Dogan, U. Tayfun, A. Davulcu, M. Dogan, Improving the dyeability of poly (lactic acid) fiber using octa (aminophenyl) POSS nanoparticle during melt spinning, J. Text. Inst. 108 (2017) 569-578. https://doi.org/10.1080/00405000.2016.1175536

[39] A. Davulcu, M. Dogan, Production of dyeable polypropylene fiber using polyhedral oligomeric silsesquioxanes via melt spinning, Fibers Polym. 15 (2014) 2370-2375. https://doi.org/10.1007/s12221-014-2370-6

[40] K. Qi, X. Chen, Y. Liu, J.H. Xin, C.L. Mak, W.A. Daoud, Facile preparation of anatase/SiO$_2$ spherical nanocomposites and their application in self-cleaning textiles, J. Mater. Chem. 17 (2007) 3504-3508. https://doi.org/10.1039/b702887c

[41] E. Pakdel, W. Daoud, Self-cleaning cotton functionalized with TiO$_2$/SiO$_2$: focus on the role of silica, J. Colloid Interface Sci. 401 (2013) 1-7. https://doi.org/10.1016/j.jcis.2013.03.016

[42] M.J. Uddin, F. Cesano, D. Scarano, F. Bonino, G. Agostini, G. Spoto, S. Bordiga, A. Zecchina, Cotton textile fibres coated by Au/TiO$_2$ films: Synthesis, characterization and self cleaning properties, J. Photochem. Photobiol. A: Chem. 199 (2008) 64-72. https://doi.org/10.1016/j.jphotochem.2008.05.004

[43] L. Liu, L.H. Klausen, M. Dong, Two dimensional peptide based functional nanomaterials, Nano Today, 23 (2018) 40-58. https://doi.org/10.1016/j.nantod.2018.10.008

[44] M. Sathishkumar, S. Geethalakshmi, M. Saroja, M. Venkatachalam, P. Gowthaman, S.K. Verma, A.K. Das, Chapter Three - Antimicrobial activities of biosynthesized nanomaterials, Compr. Anal. Chem. 94 (2021) 81-172. https://doi.org/10.1016/bs.coac.2020.12.007

[45] A.D. Broadbent, An introduction to dyes and dyeing, Basic Principle of Textile Coloration, West Yorkshire: Society of Dyers and Colourists. 2001, pp. 174-180.

[46] I. Holme, Adhesion to textile fibres and fabrics, Int. J. Adhes. Adhes. 19 (1999) 455-463. https://doi.org/10.1016/S0143-7496(99)00025-1

[47] A.K.R. Choudhury, Introduction to finishing, in: Principles of Textile Finishing, Woodhead Publishing Series in Textiles. Woodhead Publishing Ltd, Duxford, 2017, pp. 1-19.

[48] M. Joshi, B.S. Butola, Application technologies for coating, lamination and finishing of technical textiles, in: M.L. Gulrajani, (Eds.), Advances in the Dyeing and Finishing of Technical Textiles, Woodhead Publishing Ltd, 2013, pp. 355-411. https://doi.org/10.1533/9780857097613.2.355

[49] W.S. Tung, W.A. Daoud, Self-cleaning fibers via nanotechnology: a virtual reality. J. Mater. Chem. 21 (2011) 7858-7869. https://doi.org/10.1039/c0jm03856c

[50] J.K. Patra, S. Gouda, Application of nano technology in textile engineering: An overview, J. Eng. Technol. Res. 5(5) (2013)104-111. https://doi.org/10.5897/JETR2013.0309

[51] R. Nayak, R. Padhye, I.L. Kyratzis, Y.B. Truong, L. Arnold, Recent advances in nanofibre fabrication techniques, Text. Res. J. 82 (2012) 129-147. https://doi.org/10.1177/0040517511424524

[52] S. Ramakrishna, K. Fujihara, W-E. Teo, T. Yong, Z. Ma, R. Ramaseshan, Electrospun nanofibers: Solving global issues, Mater. Today, 9 (2006) 40-50. https://doi.org/10.1016/S1369-7021(06)71389-X

[53] P.S. Kumar, J. Sundaramurthy, S. Sundarrajan, V.J. Babu, G. Singh, S.I. Allakhverdiev, S. Ramakrishna, Hierarchical electrospun nanofibers for energy harvesting, production and environmental remediation, Energy Environ. Sci. 7 (2014) 3192-3222. https://doi.org/10.1039/C4EE00612G

[54] B. Ghorani, N. Tucker, Fundamentals of electrospinning as a novel delivery vehicle for bioactive compounds in food nanotechnology, Food Hydrocoll. 51 (2015) 227-240. https://doi.org/10.1016/j.foodhyd.2015.05.024

[55] K. Riazi, J. Kübel, M. Abbasi, K. Bachtin, S. Indris, H. Ehrenberg, R. Kádár, M. Wilhelm, Polystyrene comb architectures as model systems for the optimized solution electrospinning of branched polymers. Polymer, 104 (2016) 240-250. https://doi.org/10.1016/j.polymer.2016.05.032

[56] C-L. Zhang, S-H. Yu, Nanoparticles meet electrospinning: recent advances and future prospects, Chem. Soc. Rev. 43 (2014) 4423-4448. https://doi.org/10.1039/c3cs60426h

[57] U. Tayfun, M. Dogan, Improving the dyeability of poly (lactic acid) fiber using organoclay during melt spinning. Polym. Bull. 73 (2016) 1581-1593. https://doi.org/10.1007/s00289-015-1564-4

[58] J.K.W. Sandler, S. Pegel, M. Cadek, F. Gojny, M. van Es, J. Lohmar, W.J.; Blau, K. Schulte, A.H. Windle, M.S.P. Shaffer, A comparative study of melt spun polyamide-12 fibres reinforced with carbon nanotubes and nanofibers, Polymer. 45 (2004) 2001-2015. https://doi.org/10.1016/j.polymer.2004.01.023

[59] J.W. Cho, D.R. Paul, Nylon 6 nanocomposites by melt compounding, Polymer. 42 (2001) 1083-1094. https://doi.org/10.1016/S0032-3861(00)00380-3

[60] Z. Zhou, L. Chu, W. Tang, L. Gu, Studies on the antistatic mechanism of tetrapod-shaped zinc oxide whisker, J Electrostat. 57 (2003) 347-354. https://doi.org/10.1016/S0304-3886(02)00171-7

[61] F. Zhang, J. Yang, Preparation of nano-ZnO and its application to the textile on antistatic finishing, Int. J Chem. 1 (2009) 18-22. https://doi.org/10.5539/ijc.v1n1p18

[62] T.W. Shyr, C.H. Lien, A.J. Lin, Coexisting antistatic and water-repellent properties of polyester fabric, Text. Res. J. 81 (2011) 254-263. https://doi.org/10.1177/0040517510380775

[63] M.K. Babu, K.B. Ravindra, Bioactive antimicrobial agents for finishing of textiles for health care products, J. Text. Inst. 106 (2014) 706-717. https://doi.org/10.1080/00405000.2014.936670

[64] R.B. Sadu, D.H. Chen, A.S. Kucknoor, Z. Guo, A.J. Gomes, Silver-doped TiO2/polyurethane nanocomposites for antibacterial textile coating, BioNanoSci. 4 (2014) 136-148. https://doi.org/10.1007/s12668-014-0125-x

[65] S. Islam, F. Mohammad, Natural colorants in the presence of anchors so-called mordants as promising coloring and antimicrobial agents for textile materials. ACS Sustain. Chem. Eng. 3 (2015) 2361-2375. https://doi.org/10.1021/acssuschemeng.5b00537

[66] Y. Gao, R. Cranston, Recent advances in antimicrobial treatments of textiles, Text. Res. J. 78 (2008) 60-72. https://doi.org/10.1177/0040517507082332

[67] B. Simoncic, B. Tomsic, Structures of novel antimicrobial agents for textiles - A review, Text. Res. J. 80 (2010) 1721-1737. https://doi.org/10.1177/0040517510363193

[68] K.S. Huang, C.-H. Yang, S.-L. Huang, C.-Y. Chen, Y.-Y. Lu, Y.-S. Lin, Recent advances in antimicrobial polymers, Int. J. Mol. Sci. 17 (2016) 1578. https://doi.org/10.3390/ijms17091578

[69] S. Islam, M. Shahid, F. Mohammad, Green chemistry approaches to develop antimicrobial textiles based on sustainable biopolymers-A review. Ind. Eng. Chem. Res. 57 (2013) 5245-5260. https://doi.org/10.1021/ie303627x

[70] R. Dastjerdi, M. Montazer, A review on the application of inorganic nano-structured materials in the modification of textiles: Focus on anti-microbial properties, Colloids Surf. B biointerfaces, 79 (2010) 5-18. https://doi.org/10.1016/j.colsurfb.2010.03.029

[71] S. Josset, N. Keller, M.-C. Lett, M.J. Ledoux, V. Keller, Numeration methods fortargeting photoactive materials in the UV-A photocatalytic removal of microorganisms, Chem. Soc. Rev. 37 (2008) 744-755. https://doi.org/10.1039/b711748p

[72] G. Wyszogrodzka, B. Marszałek, B. Gil, P. Dorozy nski, Metal-organic frameworks: mechanisms of antibacterial action and potential applications, Drug Discov. Today. 21 (2016) 1009-1018. https://doi.org/10.1016/j.drudis.2016.04.009

[73] E.D. Cavassin, L.F.P. de Figueiredo, J.P. Otoch, M.M. Seckler, R.A. de Oliveira, F.F. Franco, V.S. Marangoni, V. Zucolotto, A.S.S. Levin, S.F. Costa, Comparison of methods to detect the in vitro activity of silver nanoparticles (AgNP) against multidrug resistant bacteria, J. Nanobiotechnology, 13 (2015) 64. https://doi.org/10.1186/s12951-015-0120-6

[74] J.P. Ruparelia, A.K. Chatterjee, S.P. Duttagupta, S. Mukherji, Strain specificity in antimicrobial activity of silver and copper nanoparticles, Acta Biomaterialia. 4 (2008) 707-716. https://doi.org/10.1016/j.actbio.2007.11.006

[75] O. Akhavan, E. Ghaderi, Toxicity of graphene and graphene oxide nanowalls against bacteria, ACS Nano, 4 (2010) 5731-5736. https://doi.org/10.1021/nn101390x

[76] T. Parandhaman, A. Das, B. Ramalingam, D. Samanta, T.P. Sastry, A.B. Mandal, S.K. Das, Antimicrobial behavior of biosynthesized silica-silver nanocomposite for water disinfection: A mechanistic perspective, J. Hazard. Mater. 290 (2015) 117-126. https://doi.org/10.1016/j.jhazmat.2015.02.061

[77] C. Gunawan, W.Y. Teoh, C.P. Marquis, R. Amal, Cytotoxic origin of copper (II) oxide nanoparticles: Comparative studies with micron-sized particles, leachate, and metal salts, ACS Nano, 5 (2011) 7214-7225. https://doi.org/10.1021/nn2020248

[78] S.M. Dizaj, F. Lotfipour, M. Barzegar-Jalali, M.H. Zarrintan, K. Adibkia, Antimicrobial activity of the metals and metal oxide nanoparticles, Mater. Sci. Eng. C, 44 (2014) 278-284. https://doi.org/10.1016/j.msec.2014.08.031

[79] G. Zhao, S. E. Stevens Jr, Multiple parameters for the comprehensive evaluation of the susceptibility of Escherichia coli to the silver ion. Biometals. 11 (1998) 27-32. https://doi.org/10.1023/A:1009253223055

[80] S.T. Dubas, P. Kumlangdudsana, P. Potiyaraj, Layer-by-layer deposition of antimicrobial silver nanoparticles on textile fibers, Colloids Surf. A Physicochem. Eng. Asp. 289 (2006) 105-109. https://doi.org/10.1016/j.colsurfa.2006.04.012

[81] S. Li, T. Zhu, J. Huang, Q. Guo, G. Chen, Y. Lai, Durable antibacterial and UV-protective $Ag/TiO_2@$ fabrics for sustainable biomedical application, Int. J. Nanomedicine. 12 (2017) 2593-2606. https://doi.org/10.2147/IJN.S132035

[82] S. Zhang, Y.A. Tang, B. Vlaholic, A review on preparation and applications of silver-containing nanofibers, Nanoscale Res. Lett. 11 (2016) 80. https://doi.org/10.1186/s11671-016-1286-z

[83] M. Yu, Z. Wang, M. Lv, R. Hao, R. Zhao, L. Qi, S. Liu, C. Yu, B. Zhang, C. Fan, J. Li, Antisuperbug cotton fabric with excellent laundering durability, ACS Appl. Mater. Interfaces. 8 (2016) 19866-19871. https://doi.org/10.1021/acsami.6b07631

[84] M. Rana, B. Hao, L. Mu, L. Chen, P.C. Ma, Development of multi-functional cotton fabrics with Ag/AgBr-TiO₂, Compos Sci Technol. 122 (2016) 104-112. https://doi.org/10.1016/j.compscitech.2015.11.016

[85] S. Eckhardt, P.S. Brunetto, J. Gagnon, M. Priebe, B. Giese, K.M. Fromm, Nanobio silver: its interactions with peptides and bacteria, and its uses in medicine, Chem. Rev. 113 (2013) 4708-4754. https://doi.org/10.1021/cr300288v

[86] N. Čuk, M. Šala, M. Gorjanc, Development of antibacterial and UV protective cotton fabrics using plant food waste and alien invasive plant extracts as reducing agents for the in-situ synthesis of silver nanoparticles, Cellulose, 28 (2021) 3215-3233. https://doi.org/10.1007/s10570-021-03715-y

[87] V. Babaahmadi, M. Montazer, A new route to synthesis silver nanoparticles on polyamide fabric using stannous chloride, J. Text. Inst. 106 (2015) 970-977. https://doi.org/10.1080/00405000.2014.957468

[88] V. Ilic, Z. Saponjic, V. Vodnik, et al., Bactericidal efficiency of silver nanoparticles deposited onto radio frequency plasma pretreated polyester fabrics, Ind. Eng. Chem. Res. 49 (2010) 7287-7293. https://doi.org/10.1021/ie1001313

[89] J.C. Flores-Arriaga, R. García-Contreras, G. Villanueva-Sa'nchez, et al., Antimicrobial poly (methyl methacrylate) with silver nanoparticles for dentistry: a systematic review, Appl. Sci. 10 (2020) 4007. https://doi.org/10.3390/app10114007

[90] W. Yunping, Y. Yan, Z. Zhijie, W. Zhihua, Z. Yanbao, S. Lei, Fabrication of cotton fabrics with durable antibacterial activities finishing by Ag nanoparticles, Text. Res. J. 89 (2021) 867-880. https://doi.org/10.1177/0040517518758002

[91] C. Pereira, A.M. Pereira, C. Freire, T.V. Pinto, R.S. Costa, J.S. Teixeira, Nanoengineered textiles: from advanced functional nanomaterials to groundbreaking high-performance clothing, in: Handbook of functionalized nanomaterials for industrial applications, Elsevier, 2020, pp. 611-714. https://doi.org/10.1016/B978-0-12-816787-8.00021-1

[92] S.A. Jadhav, A.H. Patil, S.S. Thoravat, et al., A brief overview of antimicrobial nanotextiles prepared by in situ synthesis and deposition of silver nanoparticles on cotton, Nanotechnol. Russia. 16 (2021) 543-550. https://doi.org/10.1134/S2635167621040170

[93] T. Textor, M.M.G. Fouda, B. Mahltig, Deposition of durable thin silver layers onto polyamides employing a heterogeneous Tollens' reaction, Appl. Surf. Sci. 256 (2010) 2337-2342. https://doi.org/10.1016/j.apsusc.2009.10.063

[94] M. Montazer, A. Shamei, F. Alimohammadi, Synthesis of nanosilver on polyamide fabric using silver/ammonia complex, Mater. Sci. Eng. C 38 (2020) 170-176. https://doi.org/10.1016/j.msec.2014.01.044

[95] M. Montazer, A. Shamei, F. Alimohammadi, Synthesizing and stabilizing silver nanoparticles on polyamide fabric using silver-ammonia/PVP/UVC, Prog. Org. Coat. 75 (2012) 379-385. https://doi.org/10.1016/j.porgcoat.2012.07.011

[96] V. Allahyarzadeh, M. Montazer, N.H. Nejad, et al., In situ synthesis of nano silver on polyester using NaOH/Nano TiO$_2$, J. Appl. Polym. Sci. 129 (2013) 892-900. https://doi.org/10.1002/app.38907

[97] Z. Komeily-Nia, M. Montazer, M. Latifi, Synthesis of nano copper/nylon composite using ascorbic acid and CTAB, Colloids Surf. A Physicochem. Eng. Asp. 439 (2013) 167-175. https://doi.org/10.1016/j.colsurfa.2013.03.003

[98] M. Razmkhah, M. Montazer, A.B. Rezaie, M.M. Rad, Facile technique for wool coloration via locally forming of nano selenium photocatalyst imparting antibacterial and UV protection properties, J. Ind. Eng. Chem. 101 (2021) 153-164. https://doi.org/10.1016/j.jiec.2021.06.018

[99] A.B. Rezaie, M. Montazer, M.M. Rad, A cleaner route for nanocolouration of wool fabric via green assembling of cupric oxide nanoparticles along with antibacterial and UV protection properties, J. Clean. Prod. 166 (2017) 221-231. https://doi.org/10.1016/j.jclepro.2017.08.046

[100] M.E. El-Naggar, Th.I. Shaheen, S. Zaghloul, M.H. El-Rafie, A. Hebeish, Antibacterial activities and UV protection of the in situ synthesized titanium oxide nanoparticles on cotton fabrics, Ind. Eng. Chem. Res. 55 (2016) 2661-2668. https://doi.org/10.1021/acs.iecr.5b04315

[101] R. Pandimurugan, S. Thambidurai, UV protection and antibacterial properties of seaweed capped ZnO nanoparticles coated cotton fabrics, Int. J. Biol. Macromol. 105 (2017) 788-795. https://doi.org/10.1016/j.ijbiomac.2017.07.097

[102] N.F. Attia, M. Moussa, A.M.F. Sheta, R. Taha, H. Gamal, Synthesis of effective multifunctional textile based on silica nanoparticles, Prog. Org. Coat. 106 (2017) 41-49. https://doi.org/10.1016/j.porgcoat.2017.02.006

[103] B.A. Çakır, L. Leyla Budama, O. Topel, N. Hoda, Synthesis of ZnO nanoparticles using PS-b-PAA reverse micelle cores for UV protective, self-cleaning and antibacterial textile applications, Colloids Surf. A Physicochem. Eng. Asp. 414 (2012) 132-139. https://doi.org/10.1016/j.colsurfa.2012.08.015

[104] M.Z. Khan, V. Baheti, M. Ashraf, T. Hussain, A. Ali, A. Javid, A. Rahman, Development of UV Protective, Superhydrophobic and Antibacterial Textiles Using ZnO and TiO$_2$ Nanoparticles, Fibers Polym. 19 (2018) 1647-1654. https://doi.org/10.1007/s12221-018-7935-3

[105] A. Farouk, S. Moussa, M. Ulbricht, et al., ZnO-modified hybrid polymers as an antibacterial finish for textiles, Text. Res. J. 84 (2014) 40-51. https://doi.org/10.1177/0040517513485623

[106] M. Ibanescu, V. Musat, T. Textor, V. Badilita, B. Mahltig, Photocatalytic and antimicrobial Ag/ZnO nanocomposites for functionalization of textile fabrics, J. Alloys Compd. 610 (2014) 244-249. https://doi.org/10.1016/j.jallcom.2014.04.138

[107] L. Xu, Y. Shen, L. Wang, Y. Ding, Z. Cai, Preparation of vinyl silicabased organic/inorganic nanocomposites and superhydrophobic polyester surfaces from it,

Materials Research Forum LLC
https://doi.org/10.21741/9781644902295-8

Colloid Polym. Sci. 293 (2015) 2359-2371. https://doi.org/10.1007/s00396-015-3624-6

[108] Y. Shin, M. Park, H. Kim, F. Jin, S. Park, Synthesis of silver-doped silica complex nanoparticles for antibacterial materials. Bull. Korean Chem. Soc. 35 (2014) 2979-2984. https://doi.org/10.5012/bkcs.2014.35.10.2979

[109] S.W. Ali, M. Joshi, S. Rajendran, Novel, self-assembled antimicrobial textile coating containing chitosan nanoparticles, AATCC Rev. 11 (2011) 49-71.

[110] D. Mihailovic, Z. Saponjic, V. Vodnik, B. Pothonjak, P. Jovancic, J.M. Nedeljkovic, M. Radetic, Multifunctional PES fabrics modified with colloidal Ag and TiO_2 nanoparticles, Polym. Adv. Technol. 22 (2011) 2244-2249. https://doi.org/10.1002/pat.1752

[111] A.P. Periyasamy, M. Venkataraman, D. Kremenakova, J. Militkyet, Y. Zhou, Progress in sol-gel technology for the coatings of fabrics, Materials, 13 (2020) 1838. https://doi.org/10.3390/ma13081838

[112] N. Radic, B.M. Obradovic, M. Kostic, B. Dojcinovic, M. Hudcova, M.M. Kuraica, M. Cernak, Deposition of gold nanoparticles on polypropylene nonwoven pretreated by dielectric barrier discharge and diffuse coplanar surface barrier discharge, Plasma Chem. Plasma Process. 33 (2013) 201-218. https://doi.org/10.1007/s11090-012-9414-8

[113] M.E. Ureyen, A. Dogan, A.S. Koparal, Antibacterial functionalization of cotton and polyester fabrics with a finishing agent based on silver-doped calcium phosphate powders, Text. Res. J. 8 (2012) 1731-1742. https://doi.org/10.1177/0040517512445331

[114] U. Manzoor, M. Islam, L. Tabassam, S. U. Rahman, Quantum confinement effect in ZnO nanoparticles synthesized by co-precipitate method, Physica E Low Dimens. Syst. Nanostruct. 41 (2009) 1669-1672. https://doi.org/10.1016/j.physe.2009.05.016

[115] A. Rahdar, M.R. Hajinezhad, V.S. Sivasankarapillai, F. Askari, M. Noura, G. Z. Kyzas, Synthesis, characterization and intraperitoneal biochemical studies of zinc oxide nanoparticles in Rattus norvegicus, Appl. Phys. A, 126 (2020) 1-9. https://doi.org/10.1007/s00339-020-03535-0

[116] S.A. Noorian, N. Hemmatinejad, J.A. Navarro, Ligand modified cellulose fabrics as support of zinc oxide nanoparticles for UV protection and antimicrobial activities, Int. J. Biol. Macromol. 154 (2020) 1215-1226. https://doi.org/10.1016/j.ijbiomac.2019.10.276

[117] N.R. Dhineshbabu, S. Bose, UV resistant and fire retardant properties in fabrics coated with polymer based nanocomposites derived from sustainable and natural resources for protective clothing application, Compos. B. Eng. 172 (2019) 555-563. https://doi.org/10.1016/j.compositesb.2019.05.013

[118] A. Sedighi, M. Montazer, S. Mazinani, Fabrication of electrically conductive superparamagnetic fabric with microwave attenuation, antibacterial properties and UV

protection using PEDOT/magnetite nanoparticles, Mater. Des. 160 (2018) 34-47. https://doi.org/10.1016/j.matdes.2018.08.046

[119] A. Fouda, E.L. Saad, S.S. Salem, T.I. Shaheen, In-Vitro cytotoxicity, antibacterial, and UV protection properties of the biosynthesized Zinc oxide nanoparticles for medical textile applications. Microb. Pathog. 125 (2018) 252-261. https://doi.org/10.1016/j.micpath.2018.09.030

[120] A.H. Alrajhi, N.M. Ahmed, M. Al Shafouri, M.A. Almessiere, A.A.M. Ghamdi, Green synthesis of zinc oxide nanoparticles using salvia officials extract, Mater Sci Semicond Process., 125 (2021) 105641. https://doi.org/10.1016/j.mssp.2020.105641

[121] H.Y. Phin, Y.T. Ong, J.C. Sin, Effect of carbon nanotubes loading on the photocatalytic activity of zinc oxide/carbon nanotubes photocatalyst synthesized via a modified sol-gel method, J. Environ. Chem. Eng. 8 (2020) 103222. https://doi.org/10.1016/j.jece.2019.103222

[122] M.M. Rashid, B. Simončič, B. Tomšič, Recent advances in TiO_2-functionalized textile surfaces, Surf. Interfaces. 22 (2021) 100890. https://doi.org/10.1016/j.surfin.2020.100890

[123] D. Grifoni, L. Bacci, G. Zipoli, L. Albanese, F. Sabatini, The role of natural dyes in the UV protection of fabrics made of vegetable fibres. Dyes Pigm. 91 (2011) 279-285. https://doi.org/10.1016/j.dyepig.2011.04.006

[124] J. Campos Payá, P. Diáz-Garciá, I. Montava, P. Miró-Martínez, M. Bonet, A new development for determining the ultraviolet protection factor, J. Ind. Text. 45 (2015) 1571-1586. https://doi.org/10.1177/1528083714567238

[125] A. Awad, A.I. Abou-Kandil, I. Elsabbagh, M. Elfass, M. Gaafar, E. Mwafy, Polymer nanocomposites part 1: Structural characterization of zinc oxide nanoparticles synthesized via novel calcination method, J. Thermoplast. Compos. Mater. 28 (2015) 1343-1358. https://doi.org/10.1177/0892705714551241

[126] S. Shahidi, Novel method for ultraviolet protection and flame retardancy of cotton fabrics by low-temperature plasma, Cellulose. 21 (2014) 757-768. https://doi.org/10.1007/s10570-013-0127-9

[127] P.D. Dubrovski, D. Golob, Effects of woven fabric construction and color on ultraviolet protection, Text. Res. J. 79 (2009) 351-359. https://doi.org/10.1177/0040517508090490

[128] W. Wong, J.K. Lam, C. Kan, R. Postle, Influence of knitted fabric construction on the ultraviolet protection factor of greige and bleached cotton fabrics, Text. Res. J. 83 (2013) 683-699. https://doi.org/10.1177/0040517512467078

[129] C.A. Wilson, N.K. Bevin, R.M. Laing, B.E. Niven, Solar protection- effect of selected fabric and use characteristics on ultraviolet transmission, Text. Res. J. 78 (2008) 95-104. https://doi.org/10.1177/0040517508089660

[130] B. Bulcha, J. Leta Tesfaye, D. Anatol, R. Shanmugam, L.P. Dwarampudi, N. Nagaprasad, V.L.N. Bhargavi, R. Krishnaraj, Synthesis of zinc oxide nanoparticles by

hydrothermal methods and spectroscopic investigation of ultraviolet radiation protective properties, J. Nanomater. (2021) 8617290. https://doi.org/10.1155/2021/8617290

[131] D. Chen, L. Tan, H. Liu, J. Hu, Y. Li, F. Tang, Fabricating superhydrophilic wool fabrics, Langmuir, 26 (2010) 4675-4679. https://doi.org/10.1021/la903562h

[132] N.R. Dhineshbabu, S. Bose, Smart textiles coated with eco-friendly UV-blocking nanoparticles derived from natural resources, ACS Omega, 3 (2018) 7454-7465. https://doi.org/10.1021/acsomega.8b00822

[133] S. Dadvar, H. Tavanai, M. Morshed, UV-protection properties of electrospun polyacrylonitrile nanofibrous mats embedded with MgO and Al_2O_3 nanoparticles, J. Nanoparticle Res. 13 (2011) 5163. https://doi.org/10.1007/s11051-011-0499-4

[134] M. Shabbir, F. Mohammad, Multifunctional AgNPs@Wool: colored, UV-protective and antioxidant functional textiles, Appl. Nanosci. 8 (2018) 545-555. https://doi.org/10.1007/s13204-018-0668-1

[135] K. Jazbec, M. Sala, M. Mozetic, V. Alenka, G. Marija, Functionalization of cellulose fibres with oxygen plasma and ZnO nanoparticles for achieving UV protective properties, J. Nanomater. 16 (2015) 346739. https://doi.org/10.1155/2015/346739

[136] S. Lee, Developing UV-protective textiles based on electrospun zinc oxide nanocomposite fibers, Fibers Polym. 10 (2009) 295-301. https://doi.org/10.1007/s12221-009-0295-2

[137] M. Zhang, S. Feng, L. Wang, Y. Zheng, Lotus effect in wetting and self-cleaning, Biotribology, 5 (2016) 31-43. https://doi.org/10.1016/j.biotri.2015.08.002

[138] B. Adak, S. Mukhopadhyay, All-cellulose composite laminates with low moisture and water sensitivity, Polymer. 141 (2018) 79-85. https://doi.org/10.1016/j.polymer.2018.02.065

[139] Y.L. Zhang, H. Xia, E. Kim, H.B. Sun, Recent developments in superhydrophobic surfaces with unique structural and functional properties. Soft Matter. 8 (2012) 11217-11231. https://doi.org/10.1039/c2sm26517f

[140] J. Yuan, J. Wang, K. Zhang, W. Hu, Fabrication and properties of a superhydrophobic film on an electroless plated magnesium alloy, RSC Adv. 7 (2017) 28909-28917. https://doi.org/10.1039/C7RA04387B

[141] S. Pal, S. Mondal, P. Pal, A. Das, J. Maity, Fabrication of Ag NPs/silane coated mechanical and washing durable hydrophobic cotton textile for self-cleaning and oil-water separation application, J. Indian Chem. Soc. 99 (2022) 100283. https://doi.org/10.1016/j.jics.2021.100283

[142] B. Ohtani, Y. Ogawa, S. Nishimoto, Photocatalytic activity of amorphous - anatase mixture of titanium (IV) oxide particles suspended in aqueous solutions, J. Phys. Chem. B. 101 (1997) 3746-3752. https://doi.org/10.1021/jp962702+

[143] E. Pakdel, W.A. Daoud, R.J. Varley, X. Wang, Antibacterial textile and the effect of incident light wavelength on its photocatalytic self-cleaning activity, Mater. Lett. 318 (2022) 132223. https://doi.org/10.1016/j.matlet.2022.132223

[144] S. Lakshmi and R. Renganathan, S. Fujita, Study on TiO2-mediated photocatalytic degradation of methylene blue, J. Photochem. Photobiol. A Chem. 88 (1995) 163-167. https://doi.org/10.1016/1010-6030(94)04030-6

[145] A. Houas, H. Lachheb, M. Ksibi, M. Elaloui, C. Guillard, and I. Herrmann, Photocatalytic degradation pathway of methylene blue in water, Appl. Catal B: Envir, 31 (2001) 145-157. https://doi.org/10.1016/S0926-3373(00)00276-9

[146] R. Shishoo, Recent developments in materials for use in protective clothing, Int. J. Cloth. Sci. Technol. 14 (2002) 201-215. https://doi.org/10.1108/09556220210437167

[147] J.K. Patra, S. Gouda, Application of nano technology in textile engineering: An overview, J. Eng. Technol. Res. 5 (2013) 104-111. https://doi.org/10.5897/JETR2013.0309

[148] M. Ashraf, P. Champagne, A. Perwuelz, C. Campagne, A. Leriche, Photocatalytic solution discoloration and self-cleaning by polyester fabric functionalized with ZnO nanorods, J. Ind. Text. 44 (2015) 884-898. https://doi.org/10.1177/1528083713519662

[149] H. Yaghoubi, N. Taghavinia, E.K. Alamdari, Self-cleaning TiO_2 coating on polycarbonate: surface treatment, photocatalytic and nanomechanical properties, Surf. Coat. Technol. 204 (2010) 1562-1568. https://doi.org/10.1016/j.surfcoat.2009.09.085

[150] B.S. Kumar, Self-cleaning finish on cotton textile using sol-gel derived TiO_2 nano finish, IOSR-JPTE, 2 (2015) 2348-0181.

[151] K. Meilert, D. Laun, J. Kiwi, Photocatalytic self-cleaning of modified cotton textiles by TiO_2 clusters attached by chemical spacers, J. Mol. Catal. A Chem. 237 (2005) 101-108. https://doi.org/10.1016/j.molcata.2005.03.040

[152] H.Y. Chien, H.W. Chen, C.C. Wang, The study of non-formaldehyde crease-resist finishing fabrics treated with the compound catalyst of nanometer grade TiO_2 under UV light and different polycarboxylic acid, J. Hwa Gang Text. 10 (2003) 104-114.

[153] C.C. Wang, C.C. Chen, Physical properties of crosslinked cellulose catalyzed with nano titanium dioxide, J. Appl. Polym. Sci. 97 (2005) 2450-2456. https://doi.org/10.1002/app.22018

[154] X.Q. Song, A. Liu, C.T. Ji, H.T. Li, The effect of nano-particle concentration and heating time in the anti-crinkle treatment of silk. J. Jilin Instit. Technol. 22 (2001) 24-27.

[155] M.L. Gulrajani, D. Gupta, Emerging techniques for functional finishing textiles, Indian J. Fibre Text. Res. 36 (2011) 388-397.

[156] C.K. Kundu, Z. Li, L. Song, Y. Hu, An overview of fire retardant treatments for synthetic textiles: From traditional approaches to recent applications, Eur. Polym. J. 137 (2020) 109911. https://doi.org/10.1016/j.eurpolymj.2020.109911

[157] P. Pandit, K. Singha, V. Kumar, S. Maity, Advanced flame retardant agents for protective textiles and clothing, in: S. ul-Islam, B.S. Butola (Eds), Advances in Functional and Protective Textiles, Woodhead Publishing, 2020, pp. 397-414. https://doi.org/10.1016/B978-0-12-820257-9.00016-3

[158] J. Alongi, A. Frache, G. Malucelli, G. Camino, in: Kilinc-Balci FS (Eds), Handbook of fire resistant textiles. Cambridge: Woodhead Publishing, UK, 2013, pp. 63-70.

[159] S. Bourbigot, M. Le Bras, X. Flambard, M. Rochery, E. Devaux, J.D. Lichtenhan, in: Le Bras M, Wilkie CA, Bourbigot S, Duquesne S, Jama C (Eds), Fire retardancy of polymers: new applications of mineral fillers, Royal Society of Chemistry, London, 2005, pp. 189-195.

[160] S. Bourbigot, E. Devaux, X. Flambard, Flammability of polyamide-6/clay hybrid nanocomposite textiles, Polym. Degrad. Stab. 75 (2002) 397-402. https://doi.org/10.1016/S0141-3910(01)00245-2

[161] A.R. Horrocks, B.K. Kandola, S. Nazare', D. Price, Surface modification of fabrics for improved flash-fire resistance using atmospheric pressure plasma in the presence of a functionalized clay and polysiloxane, Polym. Adv. Technol. 22 (2011) 22-29. https://doi.org/10.1002/pat.1707

[162] A.R. Horrocks, Thermal (heat and fire) protection. in: R. Scott (Ed.), Textiles for protection, Cambridge, Woodhead Publishing Ltd, UK, 2005, pp. 398-440. https://doi.org/10.1533/9781845690977.2.398

[163] Horrocks, A. R, Recent developments in flame retardant textile finishes. In: K. L. Mittal, and T. Bahners (Eds.), Textile finishing, recent developments and future trends, Wiley-Scrivener. 2017, pp. 69-126. https://doi.org/10.1002/9781119426790.ch2

[164] F. Carosio, J. Alongi, A. Frache, Influence of surface activation by plasma and nanoparticle adsorption on the morphology, thermal stability and combustion behavior of PET fabrics, Eur. Polym. J. 47 (2011) 893-902. https://doi.org/10.1016/j.eurpolymj.2011.01.009

[165] F. Carosio, J. Alongi, Few durable layers suppress cotton combustion due to joint combination of layer by layer assembly and UV curing. RSC Adv. 5 (2015) 71482-71490. https://doi.org/10.1039/C5RA11856E

[166] A.J. Mateos, A.A. Cain, J.C. Grunlan, Deposition of flame retardant and conductive nanocoatings on fabric, Ind. Eng. Chem. Res. 53 (2014) 6409-6416. https://doi.org/10.1021/ie500122u

[167] A.A. Cain, S. Murray, K.M. Holder, C.R. Nolen, J.C. Grunlan, Intumescent nanocoating extinguishes flame on fabric using aqueous polyelectrolyte complex deposited in a single step, Macromol. Mater. Eng. 299 (2014) 1180-1187. https://doi.org/10.1002/mame.201400022

[168] M. Haile, M. Leistner, O. Sarwar, C.M. Toler, R. Henderson, J.C. Grunlan, A wash-durable polyelectrolyte complex that extinguishes flames on polyester-cotton fabric, RSC Adv. 6 (2016) 33998-34004. https://doi.org/10.1039/C6RA03637F

[169] S. Talebi, M. Montazer, Denim Fabric with Flame retardant, hydrophilic and self-cleaning properties conferring by in-situ synthesis of silica nanoparticles, Cellulose, 27 (2020) 6643-6661. https://doi.org/10.1007/s10570-020-03195-6

[170] K.A. Salmeia, S. Gaan, G. Malucelli, Recent advances for flame retardancy of textiles based on phosphorus chemistry, Polymers 8 (2016) 319. https://doi.org/10.3390/polym8090319

[171] A. Quédé, C. Jama, P. Supiot, M. Le Bras, R. Delobel, O. Dessaux, P. Goudmand, Elaboration of fire retardant coatings on polyamide-6 using a cold plasma polymerization process, Surf. Coat. Technol. 151-152 (2002) 424-428. https://doi.org/10.1016/S0257-8972(01)01646-2

[172] A. Que'de', B. Mutel, P. Supiot, O. Dessaux, C. Jama, M. Le Bras, R. Delobel, Plasma-assisted process for fire properties improvement of polyamide and clay nanocomposite-reinforced polyamide: A scale-up study. In: M. Le Bras, C. A. Wilkie, S. Bourbigot, S. Duquesne, C. Jama (Eds.), Fire retardancy of polymers. New applications of mineral fillers, Royal Society of Chemistry, London, 2005, pp. 276-290.

[173] T. Herbert, Atmospheric-pressure cold plasma processing technology. In R. Shishoo (Ed.), Plasma technologies for textiles, Cambridge: Woodhead Publishing Ltd, UK, 2007, pp. 79-128. https://doi.org/10.1533/9781845692575.1.79

[174] A.R. Horrocks, B. Kandola, S. Nazare', D. Price, Flash fire resistant fabric, 2009, UK Patent Application 0900069.6, January 5.

[175] J. Tata, J. Alongi, A. Frache, Optimization of the procedure to burn textile fabrics by cone calorimeter: Part II. Results on nanoparticle-finished polyester, Fire Mater. 36 (2012) 527-537. https://doi.org/10.1002/fam.1105

[176] V. Totolin, M. Sarmadi, S.O. Manolache, F.S. Denes, Environmentally friendly flame-retardant materials produced by atmospheric pressure plasma modifications, J. Appl. Polym. Sci. 124 (2012) 116-122. https://doi.org/10.1002/app.35087

[177] P. Mistry, (assigned to MTIX Ltd., UK). Treating materials with combined energy sources, 2017, US Patent No. 9,605,376 B2, May 28.

[178] A.R. Horrocks, S. Eivazi, M. Ayesh, B. Kandola, Environmentally sustainable flame retardant surface treatments for textiles: The potential of a novel atmospheric plasma/UV laser technology, Fibers, 6 (2018) 31-44. https://doi.org/10.3390/fib6020031

[179] M.Y. Wang, A.R. Horrocks, S. Horrocks, M.E. Hall, J.S. Pearson, S. Clegg, Flame retardant textile back-coatings. Part 1: Antimony-halogen system interactions and the effects of replacement by phosphorus-containing agents, J. Fire Sci. 18 (2000) 243-323. https://doi.org/10.1106/16WQ-NMXN-EM6Q-GDW9

[180] R. Shishoo, (Ed.), Plasma technologies for textiles, Cambridge: Woodhead Publishing Ltd, Wood-head Textiles Series No. 62., UK, 2007 pp. 50-100. https://doi.org/10.1533/9781845692575

[181] W.G. Dong, G. Huang, Research on properties of nano polypropylene/TiO2 composite fiber. J. Textile Res. 23 (2002) 22-23.

[182] H. Memon, H. Wang, S. Yasin, A. Halepoto, Influence of incorporating silver nanoparticles in protease treatment on fiber friction, antistatic, and antibacterial properties of wool fibers, J. Chem, (2018) 4845687. https://doi.org/10.1155/2018/4845687

[183] M.M. Hassan, Enhanced colour, hydrophobicity, UV radiation absorption and antistatic properties of wool fabric multi-functionalised with silver nanoparticles, Colloids Surf. A Physicochem. Eng. Asp. 581 (2019) 123819. https://doi.org/10.1016/j.colsurfa.2019.123819

[184] A. Yadav, V. Prasad, A.A. Kathe, S. Raj, D. Yadav, C. Sundaramoorthy, N. Vigneshwaran, Functional finishing in cotton fabrics using zinc oxide nanoparticles, Bull. Mater. Sci. 29 (2006) 641-645. https://doi.org/10.1007/s12034-006-0017-y

[185] Z. Li, Physics essay: the nature of charge, principle of charge interaction and coulomb's law. Appl. Phy. Res. 7 (2015) 52-55. https://doi.org/10.5539/apr.v7n6p52

[186] Z. Abdullaeva, Nanomaterials for Clothing and Textile Products, In: Nanomaterials in Daily Life, Springer. 2017, pp. 111-132. https://doi.org/10.1007/978-3-319-57216-1_6

[187] M. Wasim, M.R. Khan, M. Mushtaq, A. Naeem, M. Han, Q. Wei, Surface modification of bacterial cellulose by copper and zinc oxide sputter coating for UV-resistance/antistatic/antibacterial characteristics, Coatings. 10 (2020) 364. https://doi.org/10.3390/coatings10040364

[188] D. Wang, Y. Lin, Y. Zhao, L. Gu, Polyacrylonitrile fibers modified by nano-antimony-doped tin oxide particles, Text. Res. J. 74 (2004) 1060-1065. https://doi.org/10.1177/004051750407401206

[189] C.K. Kundu, L. Song, Y. Hu, Nanoparticles based coatings for multifunctional Polyamide 66 textiles with improved flame retardancy and hydrophilicity, J. Taiwan Inst. Chem. Eng. 112 (2020) 15-19. https://doi.org/10.1016/j.jtice.2020.07.013

[190] J. Rovira, J.L. Domingo, Human health risks due to exposure to inorganic and organic chemicals from textiles: a review, Environ. Res. 168 (2019) 62-69. https://doi.org/10.1016/j.envres.2018.09.027

[191] C.M. Wood, C. Hogstrand, F. Galvez, R.S. Munger, The physiology of waterborne silver toxicity in freshwater rainbow trout (Oncorhynchus mykiss) 1. The effects of ionic Ag+, Aquat. Toxicol. 35 (1996) 93-109. https://doi.org/10.1016/0166-445X(96)00003-3

Emerging Applications of Nanomaterials
Materials Research Foundations 141 (2023) 218-245

Materials Research Forum LLC
https://doi.org/10.21741/9781644902295-9

Chapter 9

Nanomaterials in Pharmaceuticals

Tahmina Foyez[1,2*], and Abu Bin Imran[3]

[1]Department of Hematology, University of North Carolina at Chapel Hill, Chapel Hill, NC 27599, USA

[2]Department of Pharmaceutical Sciences, School of Health and Life Sciences, North South University, Dhaka 1229, Bangladesh

[3]Department of Chemistry, Bangladesh University of Engineering and Technology, Dhaka-1000, Bangladesh

* tfoyez@unc.edu, tahminafoyez@gmail.com

Abstract

Nanotechnology is a relatively new but fast-expanding field that uses nanoparticles as analytical tools or as a controlled delivery system for pharmaceuticals. The unique properties of nanomaterials, including thermodynamic, electrical, structural capabilities, and optical make them intriguing options for pharmaceuticals, which substantially impact drug quality. Pharmaceutical nanotechnology, with its nano-engineered tools, is now well-established for drug delivery, diagnostics, prognosis, and disorder therapy. It has the potential to improve materials and medical devices as well as help to introduce new technologies in areas where traditional technologies have reached their limits. The present review focuses on nanomaterials and their applications in pharmaceuticals.

Keywords

Pharmaceutical, Nanomaterials, Polymer, Nanotechnology, Drug Delivery, Diagnosis, Nanomedicine

Contents

Materials Research Forum LLC

https://doi.org/10.21741/9781644902295-9

1. Introduction

Nanotechnology is a fast developing science that produces and uses nanosized particles measured in nanometers. It has revolutionized the distribution of drugs, diagnosis, and manufacture of biomaterials for medical applications, which led to new therapeutic approaches. The outstanding characteristics and specifications of nanoparticles make them more versatile and are expected to have a broad spectrum impact on existing technologies such as drug delivery, health sciences, therapies, and pharmaceuticals. Novel approaches to drug delivery systems are required to ensure drug efficacy and reduce adverse effects by delivering them to the targeted sites in the body. As a result, the pharmaceutical industry will develop its technology to meet the demands of patients. Pharmaceutical nanotechnology is one of the most comprehensive technologies where nanomaterials that are between 1 and 100 nanometers in size ranges are used [1, 2]. The Food and Drug Administration (FDA) advises that while evaluating the efficacy, public health impact, safety, or regulatory status of nanotechnology finished goods, any distinguishing characteristics that the use of nanotechnology may impart be properly considered [3]. The presence of a nanomaterial is characterized by three key aspects: size of particles, surface area of particles, and particle size distribution (PSD) [4, 5]. Pharmaceutical nanomaterials are manufactured using both top-down and bottom-up approaches. The top-down method includes the mechanical or chemical breakdown of a large materials into smaller pieces. In contrast, the bottom-up technique allows the precursor particles to increase in size by chemical reaction starting from atomic or molecular species [6, 7]. These two manufacturing processes result in various primary particles, aggregates, and agglomerates

(Fig. 1) [8]. The drug nanoparticles improve solubility and bioavailability, as well as the potential to cross the blood-brain barrier, enter the pulmonary system, and be absorbed via the tight connections of skin endothelial cells owing to their small particle size and large surface area [9]. Nanotechnology can be used in two ways: to incorporate drugs into nanocarriers and fabricate drug-contained nanoparticles from various materials with a huge surface area and consistent porous interior [10]. Nanoparticles utilized in pharmaceutical nanotechnology include polymeric nanoparticles, dendrimers, polymeric micelles, liposomes, polymer and drug conjugates, antibodies, and drug conjugates. System types commonly employed include stimuli sensitive, sustained and regulated delivery of drugs, bioactive, multifunctional and site-specific targeted delivery, biosensory and diagnostic, multifunctional delivery systems, and so on [11-19].

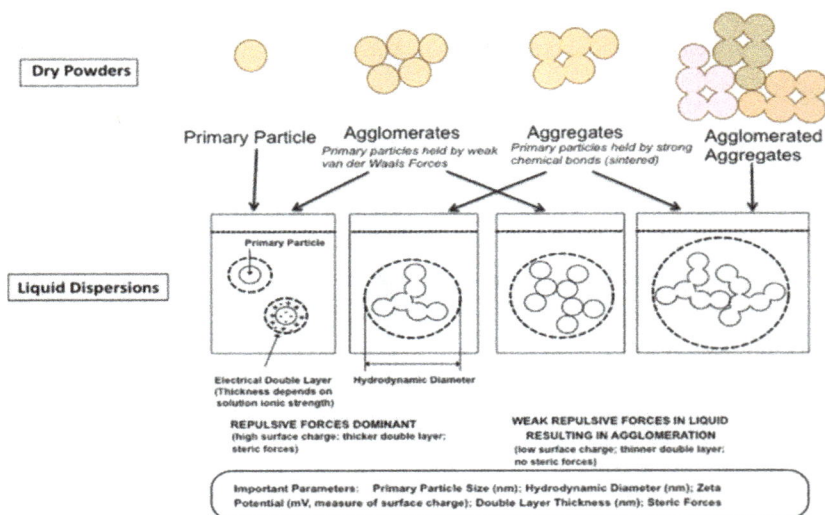

Figure 1. Schematic illustration of particle types. The figure has been reproduced with permission from [6]

The recent breakthroughs in nanomedicine, commercialization of pharmacological nano-tools, and worldwide interest shown by academia and governmental organizations all suggest that nano-based drug carriers have immense potential and scope (Fig. 2).

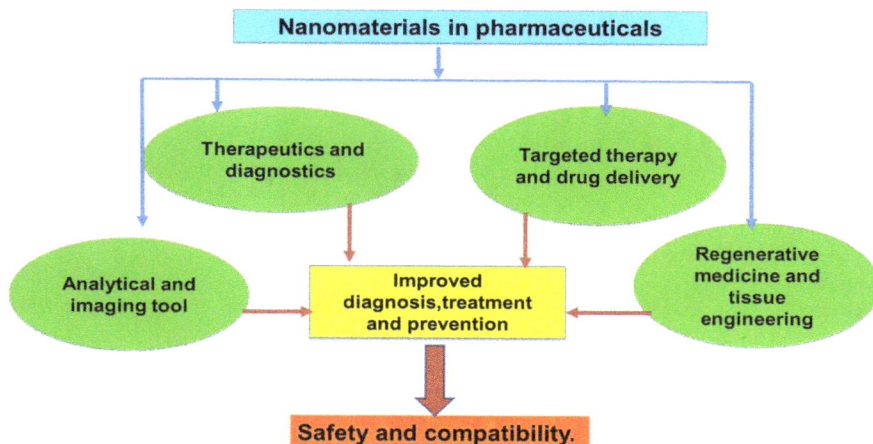

Figure 2. Nanomaterials for pharmaceutical applications

This chapter discusses the applications of nanomaterials in pharmaceutical research and industry to cure diseases with minimal side effects. The problems and prospects for the future development of nanomaterials for the pharmaceutical industry are also discussed

2. Application of nanomaterials in pharmaceutical

Adverse and therapeutic effects should be addressed when designing novel medicines using nanomaterials. The use of nanotechnology in medicine, diagnosis, image analysis, and biosensing could greatly affect people's health [20]. Three factors, characterization, safety, and environmental impact, are crucial components of nanoparticles in pharmaceuticals that must be regulated. Solid lipid nanoparticles, different metallic nanoparticles, antibody-based products, and polymeric drug/protein conjugates are all lipid-based nanoparticles approved by the FDA [21]. Nanomaterials impact pulmonology, ophthalmology, immunology, oncology, cardiology, endocrinology, targeting CNS, targeting cancerous cells, and vaccine formulations [20, 22]. Nanoparticles may increase pharmacokinetics and bioavailability by inhibiting phagocytic cell uptake. They recognize biomacromolecules on their surfaces as artificial receptors, which could be used to control cellular and extracellular processes in a variety of biological applications such as diagnostics, drug administration, therapy, and biosensing (Fig. 2). However, there are some drawbacks to use nanotechnology in pharmaceuticals, such as increased aggregation in a living system due to low solubility, biocompatibility, and high surface energy. In addition, low half-life due to rapid scavenging by the host immune system, site-specificity, and safety issues are not expected [10, 20].

2.1 Application of nanomaterials in the delivery of drugs

Low solubility, low selectivity, quick elimination, and degradation rate hinder drug delivery to cells. Nanomaterials in drug delivery bring up new frontiers because of their ability to pass through microvascular capillaries, high surface area, and high loading capacity [23]. They have increased stability, biodegradability, long-circulating, and effectiveness without higher doses [24]. Drugs are incorporated into nanoparticle delivery systems in two ways: during nanoparticle formation and by incubating the carrier with a drug solution [23, 25]. Drug solubility, absorbed drug clearance, drug dispersion, delivery route, and degradation mechanisms affect drug release patterns [25, 26]. The size of nanoparticles, drug solubility, surface charge, stability, permeability, biocompatibility, biodegradability, and toxicity must be addressed when choosing matrix materials. Liposomes, polymers, carbon based nanomaterials, nanogels, dendrimers, metal nanoparticles etc. increase the pharmacological and therapeutic effects of traditional medicines [10, 23, 24].

2.1.1 Lipid nanomaterials in the delivery of drugs

Lipid-based nanoparticles may be an alternative to traditional carriers in targeted drug delivery. They may protect their contents from deteriorating physiological conditions by using lipid nanoparticles. They may be used to deliver medications that aren't soluble in water. Lipid bilayer-encased vesicles, known as liposomes, have an aqueous core and an outer lipids coating. The liposome is used as a delivery method because it has biocompatibility and encapsulating ability [27, 28]. Phospholipids such as phosphatidylglycerol, phosphatidylserine, phosphatidylcholine, and phosphatidylethanolamine are fabricated into liposomes, which are tiny vesicles. Food, cosmetics, and pharmaceutical industries have employed these phospholipids [29-31]. Generally, three varieties of liposomes are utilized by their size and number of bilayers: small unilamellar vesicle (SUV) (20 to 100 nm), large unilamellar vesicle (LUV) (100 to 800 nm), and multilamellar vesicle (MLV) [32]. Liposomes can be designed to treat various medical conditions, including cancer and skin disease, through multiple routes of administration. Depending on lipid bilayers, liposomes may include one or more therapeutic compartments, with lipophilic compounds incorporated into the lipid membrane and hydrophilic compounds integrated into the aqueous core. Liposomes, solid lipid nanoparticles (SLNs), lipid nanoparticles (LNPs), lipid nanoemulsions (LNEs), and nanostructured lipid carriers (NLCs) are the different types of lipid-based nanoparticles (Fig. 3) [33]. Because liposomal lipids resemble biological membranes, they are biocompatible and biodegradable [34].

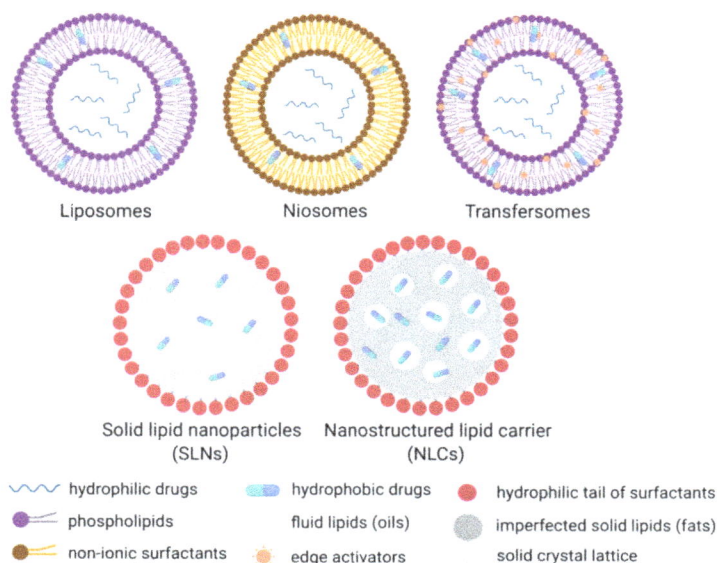

Figure 3. Different classes of lipid-based nanoparticles. The figure has been reproduced with permission from [35]

The advantages of liposomal formulations include improved cellular penetration, pharmacokinetics, drug release, decreased side effects, and toxicity [32]. Liposomes are used in vaccine adjuvants, medical diagnostics, and analytical biochemistry. The usage of organic solvents, instability, sluggish payload release after administration and expensive cost of liposomes motivated researchers to create SLNs and NLCs [36-38]. SLNs are solid-at-room-temperature lipid nanocarriers. They may accept drugs or other molecules between fatty acid chains. They have considerable promise in pharmaceutical nanotechnology because of their particle size after drug incorporation, biocompatibility, biodegradability, and enhanced efficiency. High initial release, coupled with homogeneity and cheap cost/ease of scaling up, increases the effect of drugs [39]. Since SLNs have limitations such as drug leakage through the matrix and reduced loading efficiency, a new generation of NLCs have been designed [33, 40]. In these NLCs formulations, mixing solid lipids with small amounts of liquid lipids to induce rearrangements of the matrix structure increased their qualities while preserving the original benefits of SLNs [28, 41]. SLNs have European Medicines Agency (EMA) and FDA approval and are commercially available for intravenous, dermal, and oral delivery [34]. Lipid-based nanoparticles drug delivery systems can carry drug molecules to target cells with a regulated release, making them ideal for various pharmaceutical companies. Almurshedi and colleagues explored pH-sensitive afatinib (AFT) liposomes for lung cancer. They discovered AFT-loaded liposomes induced cell death with a massive inhibitory effect on cancer cells. They also investigated that pH-sensitive liposomes release AFT quickly at pH 5.5, while cationic liposomes release AFT slowly at pH 7.5 [42]. The researchers also discovered that Turmeric-loaded NLC exhibited more excellent antioxidant activity than turmeric extract alone. NLC-turmeric has a higher antibacterial effect than free turmeric extract [43]. Lipid nanoemulsions (LNEs) are aqueous solutions containing submicron lipid droplets stabilized by surfactants. Therapeutic LNEs are plant-based lipid droplets 500nm in size maintained by phospholipids. LNEs are also employed to transport anesthetics, cancer treatments, and vitamins [44]. Whereas LNPs are similar to liposomes, except they enclose RNA and DNA. Pfizer and Moderna COVID-19 vaccines employ LNPs as mRNA carriers, making them the most frequently used non-viral gene delivery technology. LNPs lack the outer phospholipid bilayer structure of liposomes and instead contain genetic material in a micelle-like structure of cationic lipids. Surface functionalization of LNPs allows exact binding to target cells and boosts therapeutic effectiveness [45].

2.1.2 Carbon-based nanomaterials (CBNs) in drug delivery

CBNs are cylindrical structures produced from inorganic elements, including carbon, titanium, and metals like gold and silver (Fig. 4). CBN-based drug delivery systems provide higher efficiency, decreased toxicity, greater biodistribution, and enhanced patient compliance. Its drug delivery efficacy is impacted by shape, number of layers, size, and catalyst removal after synthesis [46, 47]. Examples of CBNs that have been extensively studied for drug delivery applications include CNTs, nanodiamonds, carbon nanohorns, carbon nanodots, and graphenes. It is possible to use SWNTs, MWNTs, and C60 fullerenes for drug administration because of their small size, geometry, and surface characteristics [46, 48]. CNTs have a huge surface area and unique surface chemistry, making them a good option for drug loading. CNTs-based drug delivery offers various benefits, including their tiny size (10 to 40 nm), ability to build a rod-like scaffold, and increased capacity to transport pharmaceuticals and transfer medications to the nucleus [46, 49, 50]. Liu et al. created chemically functionalized SWNTs and conjugated paclitaxel (PTX) to the PEG chain on the surface of the SWNTs. They then examined the suppression of cell proliferation in a murine breast cancer model. PTX had a tenfold increase in uptake due to the longer circulation time. They concluded that employing nanotubes as a drug delivery vehicle is effective for therapy with few adverse effects and low drug doses [51]. Treatment becomes more challenging in the central nervous system because drugs capable of crossing the blood-brain barrier are exceedingly limited. However, it has been demonstrated that certain functionalized CNT formulations can improve a compound's capacity to pass the blood-brain barrier [52]. Different carbon nanomaterials are shown in Fig.4 [53].

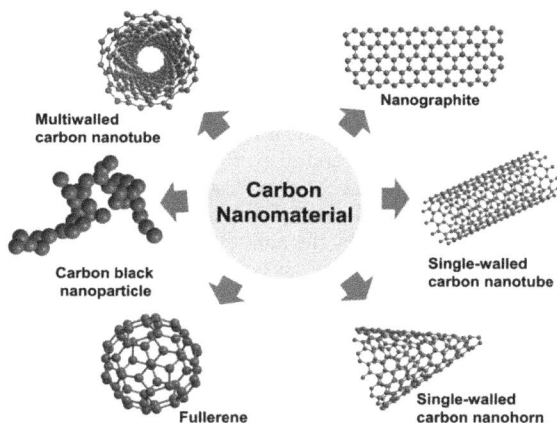

Figure 4. Schematic illustration of Carbon-based nanomaterials (CBNs) as an effective drug delivery system. The figure has been reproduced with permission from [53]

2.1.3 Polymeric nanomaterials in drug delivery

Polymeric nanoparticles provide substantial benefits over conventional drug delivery, including improved solubility in water, less antigenic activity, lower deactivation potential, and lowered toxicity [54]. It has been shown that stimuli-responsive, antibacterial, antioxidant, anticancer, and anti inflammatory activities are the most significant features of cationic polymeric nanosystems [55, 56]. In nature, biopolymers derived from living organisms can be found in abundance. They have the following benefits over traditional nanomaterials for drug delivery: they are nontoxic, biodegradable, biocompatible, and inexpensive. Albumin, gelatin, chitosan, dextran, cyclodextrin, HA, and starch are all examples of biodegradable polymeric nanoparticles [57]. Polysaccharides, a family of natural polymers, are particularly important. In terms of structure and pharmacokinetics, polysaccharides have a diverse molecular weight, polydispersity, electric charges, and chemical components. They have excellent biocompatibility, nontoxicity, biodegradability, nonimmunogenicity, and biological activity [58]. When it comes to hydrophilic drug delivery systems, chitosan is a good choice. It has garnered substantial interest as a carrier in innovative bioadhesive drug delivery nanosystems because of its reactive capabilities, ease of breakdown by enzymes, polycationic nature, and nontoxic degradation properties [55]. Drugs are chemically conjugated with chitosan derivatives such as *N*-Succinyl chitosan, glycol chitosan, and carboxymethyl chitosan. Cyclodextrins are polysaccharides that are used as nanocarriers [58-60]. The main benefits of cyclodextrins are sites where cationic or cell-targeting moieties can be added [55, 61]. HA nanosystems are used to make gels that deliver drugs to the eye. This is done so that insulin from eye drops can be absorbed through the cornea when HA is present. So, the gel keeps the medicine from being washed away by tears and gives it a long time to work at the site of action [62]. Peptides and polypeptides are the two main types of biopolymer type nanomaterials that have emerged as a vehicle for therapeutic molecular delivery to fight against cardiovascular disease, aging, and diabetes, as well as various chronic metabolic syndromes, cancer, and many degenerative disorders [63]. Albumin is a protein nanoparticle that disintegrates naturally and is not toxic or immunogenic. It is a great way to give a drug as it is easily absorbed by the tumor tissue and dissolves in water, making it easy to inject [57]. In drug delivery systems, protein nanoparticles have many disadvantages compared to nanoparticles made of inorganic materials or synthetic polymers. They are usually made from microorganisms, and making and purifying protein nanoparticles takes a long time [58, 64]. Using various synthetic polymers, polymeric prodrugs have been developed. Poly(L-glutamic acid) (PGA), poly(L-aspartic acid) (PAA), Polyethylene glycol, PEI, dendrimer, and some amphiphilic block copolymers like PEG-b-PAA, PEG-b-PGA, and N-(2-Hydroxypropyl) methacrylamide (HPMA) are all examples of polymeric prodrugs [58, 62]. PEI can be changed chemically into linear and branched forms with different molecular weights, which gives it beneficial physical and chemical properties [55, 65, 66]. Poly-L-lysine (PLL) has a lot of amines on its surface, and it can interact with negatively charged molecules [55].

2.1.4 Drug delivery based on nanogel

Polymer gels, which are 3D networks of synthetic or natural polymers, have attracted much interest as polymeric nanocarriers [67, 68]. These nanosized cross-linked structures have properties like controlling their size, having a large surface area for bioconjugation, being low in toxicity, being able to swell and de-swell quickly, responding to multiple stimuli, and having a large mesh size [69-72]. To make nanogels, there are two main routes: making nanogels from polymer precursors or making nanogels through heterogeneous polymerization of monomers. Physical and chemical cross-linking is often used to connect polymer precursors [73, 74]. The physical interaction between polymers in nanogel makes it possible to encapsulate hydrophobic drugs [75]. Nanogels solve some problems with drug delivery systems, such as being able to deliver the right amount of drug to the right place, keeping the drug in the blood longer and allowing the target cell to accept it in, breaking down slowly, being biocompatible, being able to be made into targeted drug delivery nanosystems, and having low cytotoxicity [69, 76]. Because of their hydrophobic/hydrophilic repeating units change depending on the pH, nanogels are considered a great way to carry drugs. They take positively charged drugs through electrostatic interactions at alkaline pH. When the pH is acidic, the drugs are released. Also, when the pH is acidic, the low polarity makes drug molecules go into the hydrophobic core of the nanogel and come out when the pH is neutral. In general, nanogels transport drug molecules through physical entrapment, covalent, and noncovalent interactions [69, 77].

Emerging Applications of Nanomaterials   Materials Research Forum LLC
Materials Research Foundations 141 (2023) 218-245   https://doi.org/10.21741/9781644902295-9

Drug release from nanogels can be controlled by diffusion, swelling, or active chemicals [76, 78].

2.1.5 Drug delivery based on metal nanomaterials

Scientists have focused on metal nanoparticles with unique physical and chemical properties in biomedical research. Metal-based nanoparticles have a high drug load and overcome the first-pass metabolism, photothermal behavioral potential, electrostatic charge, and surface chemistry of the molecules. Nanoparticles of silver, nickel, gold, iron, platinum, zinc, silica, gadolinium, and titanium dioxide (TiO_2) are often used in drug delivery systems [79-81]. Silicon nanoparticles work well as nanocarriers for a wide range of drug molecules. Mesoporous silica nanoparticles (MSN) with large surface areas resulted in high entrapment of either hydrophobic or hydrophilic drugs [81]. Noncrystalline drug entrapment in the mesoporous, high dispersibility with a large surface area, and an increase in the hydrophilic surface of MSN improve the drug's dissolution in water and bioavailability [82, 83]. Gold nanoparticles can also be considered promising pharmaceutical vehicles because of their chemical and physical properties. For example, they are easy to make in a wide range of sizes and shapes, from up to 100 nm, including spheres, hollows, nanoshells, rods, prims, and diamonds. They also absorb and scatter light strongly, have a high photothermal conversion rate, are biocompatible, and easily coat with different molecules. Lee et al. showed that gold nanoparticles could be used to deliver drugs for cancer therapy [84, 85]. TiO_2 is also used in the biomedical field because it is stable, biocompatible, and isn't toxic. It can destroy pathogenic substances like bacteria, viruses, fungi, and cancerous cells [86, 87].

2.1.6 Drug delivery based on dendrimer

Drug delivery is one of the most promising applications of dendrimmers, which are monodisperse, hyperbranched macromolecules with 3D structures [88, 89]. While classical nanomaterials and polymers may have higher cellular absorption, monodispersity, synergistic or multivalence effects, and good cellular uptake; the globular shape and well-defined surface functionalities of dendrimers make them superior for drug delivery systems [90, 91].

2.2 Nanomaterials in gene delivery

In order to influence the expression of a specific gene and the biosynthesis of related proteins implicated in the onset of disease, nucleic acids are delivered to the target cell or tissue and used as a gene therapy technique. Positive charges can be added to carbon nanomaterials by either covalently or non-covalently changing the carbon surface, allowing the negatively charged DNA and siRNA to be delivered to the cells [46]. To make electrostatic complexes with the negatively charged siRNA, tiny groups with distal ammonium functionalities were covalently incorporated into SWNTs. Even at very high concentrations, carbon nanoparticles used for gene therapy do not hurt cells as much as commercial gene transfection methods do. Using transporters made of carbon nanomaterials to deliver genes dramatically lowers the dose needed and improves cell absorption [46]. Dendrimers have received a lot of attention as gene delivery because their structures are flexible, their topologies have a lot of branches, and they are cationic, which allows binding to DNA at physiological pH [12]. A new way to get genes into the brain uses a serine-arginine-leucine (SRL) functionalized polyamidoamine (PAMAM) dendrimer [92]. Lipidic nanomaterials also seem to be a good choice for a gene delivery nanosystem because they are stable in storage, easy to make, and simple to sterilize, lyophilize, and pack [21]. Nanogel's cross-linked structure gives it better properties in gene delivery nanosystems. This made it possible to encase small fragments of nucleic acid like miRNA and siRNA in a stable way. Gene silencing has also been done successfully with polysaccharides like dextran and HA, and nanogels. Designing nanogels that can be disrupted by extracellular stimuli like light, ultrasound, temperature, or biological pH changes leads to fast disintegration and/or swelling [12, 77, 78, 93-104].

2.3 Nanomaterial in co-delivery systems

As nanomaterials can transport multiple therapeutic drugs, they can be used as an excellent way to make treatments more effective and have the best synergistic effect [105]. MSNs are metal-based nanomaterials used extensively as a co-delivery system [106]. Lipidic nanoparticles are better at encapsulating and incorporating drugs and genes. The formulations are an excellent way to deliver water-soluble or insoluble drugs while keeping side effects to a minimum and maximizing the delivery of active agents [21]. To treat Parkinson's disease, Cortesi and his team made NLCs with four types of levodopa. The results show that NLCs are attributed to a controlled drug release [107]. Using the appropriate nanosystem and dendrimer, researchers have been able to deliver genes and drugs to cells simultaneously, causing a synergistic effect at the target site of action [105, 108]. Hyperbranched structures and large holes in dendrimers enable the delivery of low molecular weight drugs via a co-delivery nanosystem, that is effective in delivering both drugs and nucleic acids to the same cell [109]. Co-treatment is considered an effective delivery system for polymeric nanoparticles because of their biocompatibility, stability, high efficiency in cellular uptake, and the ability to bind to genes. On the other hand, hydrophobic drugs can be encapsulated in hydrophobic polymer cores [106, 110]. SN38 anticancer medicines were successfully encapsulated in chitosan/CMD nanoparticles [111]. In nanotherapy, one of the most significant problems is the development of medication and gene delivery systems that selectively target specific cells without causing damage to normal healthy cells or tissues. Targeted therapy has the potential to improve efficacy while reducing adverse effects, which has led to the development of increasingly customized nanomedicines. Additionally, nanomedicine is promised in the local or targeted delivery of drug molecules, which results in the free dispersion of drug molecules throughout the body [112]. The steric hindrance of PEG allows it to be employed in the manufacture of stealth nanoparticles with limited absorption by the RES system and the creation of hydrophilic protective layers on the nanoparticles' surface [58, 77]. Targeting moieties such as aptamers, peptides, proteins carbohydrates folates, antibodies, vitamins, tiny molecules on nanomaterials may also boost treatment efficacy and selective accumulation in the targeted area [77, 113]. Nucleic acid, protein, and peptide-grafted CBNs have garnered interest too. Biomedical applications are constrained by the potential for functionalized CNTs to be turned into hazardous materials. Using CBNs as co-delivery systems for active targeting of overexpressed receptors on cells, specific targeting ligands may be chemically bonded to CBN surfaces [46]. Engineered nanotechnology has proven improved biocompatibility and multimodality to improve therapeutic results. Han and colleagues administered doxorubicin (DOX) and major vault protein (MVP) siRNA to human breast cancer MCF-7/ADR cells using a polyamidoamine (PAMAM) dendrimer linked with hyaluronic acid (HA) polysaccharide as an active target, lowering MVP expression and increasing DOX chemotherapeutic impact. Co-delivery of MVP siRNA and DOX significantly reduced MCF-7/ADR cell drug resistance [114]. Camptothecin (CPT) and 3,30-diindolylmethane (DIM) are two anticancer medicines that may be delivered via chitosan cross-linked with graphene oxide nanoparticles. Co-treatment of CPT and DIM boosted anticancer efficacy and prevented the harmful impact of CPT *in vivo* [115]. Imran et al. reported Ag nanoparticles incorporated hydrogel with antibacterial, biocompatibility, and wound healing capability [116].

3.   Approved pharmaceutical therapeutic nanosystems

Preclinical evaluation (clinical trial phases I, II, and III) and premarketing evaluation (phase IV) are required for FDA approval of nanoparticle-based delivery systems [117]. Approximately 70% of phase I evaluated drugs are considered for phase II, and the primary reason for phase I study termination is safety failure. In phase II, drugs are evaluated on a homogeneous patient population. Based on phase II results, drugs can proceed to the phase III study. About 33% of phase II drugs fail to progress to phase III due to inefficiency. Phase III is completed on a large patient population to ensure long-term safety. Only 25%–30% of phase III evaluated drugs are submitted to the FDA for approval [118]. Phase IV, or postmarketing, evaluation of the product occurs after it has been approved for sale. During this time, nanosystems drugs' efficacy and long-term safety are examined in populations that were not included in the phase III study. Phase IV of clinical trials is critical for nanoparticle delivery systems because of the efficiency and toxicity issues [119]. Many nanoparticles formulations have problems making it to the clinic because of how their studies were set up from biological, technological, and other perspectives. Biodistribution and controlling how nanoparticles move across the cell membrane are two examples of physical barriers. Increasing target site accumulation and reducing off-target activity or adverse effects is difficult in clinical application of the delivery system. Scaling up synthesis, improving performance, and anticipating performance are the major technical obstacles for nanoparticle-based drug carriers in clinical applications [118, 120, 121]. In the pharmaceutical industry, nanoparticle delivery systems already on the market or in clinical trials make up 15% of the market. The FDA-approved products can be given orally, topically, or systemically, depending on where they are meant to work and how the nanoparticle delivery system is being used [117]. Liposome nanocarriers, polymer drug conjugates, polymer protein conjugates, monoclonal antibody products, and some polymeric drugs are all FDA-approved nanomaterials (Table 1). Over 12 drug delivery systems that use liposomes are in different stages of clinical trial approval. Doxil is the first liposome-based drug delivery system that the FDA has approved. This product reduces the toxicity of DOX and makes it work better by putting the drug inside a unilamellar liposome that is linked to PEG to keep the drug stable in the bloodstream for a long time so that it can reach its target site [122]. The visudyne liposome is a drug that works when exposed to light. It can be used to treat people with neovascularization [121].

Table 1. FDA-approved nanomaterials-based pharmaceuticals [123-127]

Nanomaterial	Drug	Disease	Approval year
Liposome			
	Doxil	Multiple myeloma, cancer of the ovary	1995
	DaunoXome	Kaposi's cancer	1996
	AmBisome	Infections by fungus	1997
	Curosurf	Respiratory Distress Syndrome	1999
	Visudyne	Macular degeneration, myopia	2000
	Marqibo	Leukemia	2012
	Onivyde	Cancer of pancreas	2015
	Vyxeos	Myeloid leukemia	2017
	Onpattro	Amyloidosis	2018
	BNT162b2 vaccine	coronavirus disease	2020
	mRNA-1273 vaccine	Prevention of coronavirus disease	2020
Polymer Nanoparticles			
	Adagen	Immunodeficiency disease	1990
	Oncaspar	Lymphoblastic leukemia	1994
	Copaxone	Multiple sclerosis	1996
	Renagel	Chronic kidney disease	2000
	Peglntron	Hepatitis C infection	2001
	Eligard	Prostate cancer	2002
	Neulasta	Neutropenia, chemotherapy-induced	2002
	Somavert	Acromegaly	2003
	Macugen	Macular degeneration, neovascular age-related	2004
	Abraxane	Breast cancer, lung cancer, pancreatic cancer	2005
	Mircera	Anemia	2007
	Cimiza	Ankylosing spondylitis, Crohn's disease, psoriatic arthritis, rheumatoid arthritis,	2008

	Plegridy	Multiple sclerosis	2014
	Adynovate	Haemophilia	2015
Inorganic and metallic nanoparticles			
	INFeD	Anemia	1992
	DexFerrum	Iron deficiency anemia	1996
	Feridex	Imaging agent	1996
	Ferrlecit	Iron deficiency in chronic kidney disease	1999
	Venofer	Iron deficiency in chronic kidney disease	2000
	GastroMARK	Imaging agent	2001
	Feraheme	Iron deficiency in chronic kidney disease	2009
	Nanotherm	Glioblastoma	2010
	Injectafer	Iron-deficient anemia	2013
Nanocrystals			
	Megace ES	Anti-anorexic	2001
	Avinza	Psychostimulant	2002
	Focalin XR	Psychostimulant	2005
	EquivaBone	Bone substitute	2009
	Invega	Schizophrenia	2009
	Ryanodex	Malignant hypothermia	2014

4. Nanomaterial in vaccine technology

Nanoparticles outperform conventional vaccines and adjuvants. Solubilizing hydrophobic antigens with nanoparticles improve vaccine safety. They allow for a controlled release of antigens with less volume and dose [128, 129]. Modifying nanoparticles can make them more immunogenic with adjuvant, which helps to carry antigens for multiple pathogens at once securely. Researchers have also developed a spore-based vaccine that is effective against Bacillus subtilis and Clostridium tetani spores [130, 131]. Vaccines are the best way to prevent viruses like SARS-CoV-2. Imran et al., 2022, reviewed recent advances in COVID-19 vaccine development, highlighting the role of nanotechnology in vaccine production [132]. Peptide-based vaccines are the simplest to develop, validate, and prepare [133]. DNA vaccines can produce cellular immunity, including humoral immunity, and are currently the safest one. DNA vaccines encapsulated with specific nanoparticles prevent DNA degradation [134]. PSM-based therapeutic dendritic cell vaccine acts as both antigen peptide carrier and adjuvant. The shape of PSM objects is associated with absorption by circulating dendritic cells. Inguinal lymph nodes and spleens were heavily vaccinated. In contrast, popliteal lymph nodes respond better to intradermal vaccines [135]. Nanotechnology has a high potential for future vaccine development.

5. Application of nanomaterials in imaging

Imaging is being used more to diagnose and make treatment plans for diseases. Imaging data can now be used in clinical trials as objective, noninvasive ways to measure how well a therapy works. Nanotechnology can be used to find clinically significant markets with imaging techniques like magnetic resonance imaging (MRI), ultrasound, single photon emission computed tomography (SPECT), fluorescence microscopy, positron emission tomography (PET), and computed tomography (CT). Liposomes can be made magnetic so MRI can track their movement inside the body. With their nanometer-sized magnetite cores, these magnetic liposomes can be used as biocompatible MRI agents that could be used for both imaging and drug delivery [136]. Different contrast agents and radio-pharmaceuticals can be carried inside liposomes. Using positron emission tomography (PET), liposomes are also used to monitor real-time liposomal trafficking in tumors in mice [137]. Magnetic nanoparticles send drugs and genes to specific cells, separate cells, and label cells. Colloidal iron oxide and dextran [138] are utilized as MRI contrast agents. Iron oxide nanoparticles are widely used because they are biodegradable, safe for living things, superparamagnetic, and suitable for MRI applications. Insulin-coated iron oxide nanoparticles stuck to receptors on the surface of cells, blocking them from going inside the cells and lowering their toxicity [139].

6. Nanotechnology and safety issues

Nanotechnology is a new, promising technology that will significantly affect all industries. Therapeutics that use nanotechnology are becoming increasingly popular in the pharmaceutical industry. But the same new properties and traits that make it easier for drugs to get to where they need to go also pose new risks and toxicity. Some of these unique qualities are unknown or need to be investigated more. The limited information supporting either side makes it difficult to justify. Due to their low toxicity, nanoparticles made from natural or safe components, such as lipids, albumin, or nanoemulsions, appear to have the most potential. Other systems like CNTs show promise, but more research is needed on toxicity. Until then, CNT-based delivery systems for therapies are unlikely.

Conclusions

Nanomaterials can be used to deliver drugs in a controlled way and take images of specific areas. With its nanoengineering tools, pharmaceutical nanotechnology is expected to affect many aspects of diagnosing, predicting, and treating diseases. Many nanoparticles are made to carry substances that improve the pharmacological and therapeutic effects of drugs. Nanocarriers can better distribute active compounds in the body, protect them from disintegrating, and respond to biological barriers. The systems for delivering nanoparticles are being made to treat and cure many diseases. Gene therapy is hard to use clinically because it breaks down and gets rid of itself in the bloodstream, is taken up by cells that aren't the target, escapes from endosomes, and has toxic effects when the immune system is stimulated. Using nanocarriers to deliver a drug and a gene is better than just using

Emerging Applications of Nanomaterials   Materials Research Forum LLC
Materials Research Foundations 141 (2023) 218-245   https://doi.org/10.21741/9781644902295-9

chemotherapy alone. Co-delivery methods still have to deal with many problems, such as capacity, biocompatibility, stability, release kinetics, loading, and the effectiveness of tumor targeting. Many different nanoparticles can be used for bioactive delivery, and each has its benefits, which FDA has clinically approved. In addition to these efforts, many clinical trials are looking into new nanoparticle systems that are better than those already approved. The market for nanopharmaceuticals is growing, and that growth is expected to continue over the next few decades as new delivery systems based on nanoparticles are being made and new therapeutic approaches that need intelligent delivery systems to work.

Acknowledgments

T. Foyez gratefully acknowledges the administrative support received from North South University. A.B.Imran is thankful to the Committee for Advanced Studies and Research (CASR) at Bangladesh University of Engineering and Technology for funding.

References

[1] M.S. Arayne, N. Sultana, F. Qureshi, Review: nanoparticles in delivery of cardiovascular drugs, Pakistan Journal of Pharmaceutical Sciences, 20 (2007) 340-348.

[2] A.B. Imran, A.-N. Chowdhury, S. Joe, Innovations in Nanomaterials, 1st ed., Nova Science Publishers, Inc., NY, USA, 2015.

[3] Guidance for Industry, FDA, 2014 https://www.fda.gov/media/86377/download.

[4] E.A.J. Bleeker, W.H. de Jong, R.E. Geertsma, M. Groenewold, E.H.W. Heugens, M. Koers-Jacquemijns, D. van de Meent, J.R. Popma, A.G. Rietveld, S.W.P. Wijnhoven, F.R. Cassee, A.G. Oomen, Considerations on the EU definition of a nanomaterial: Science to support policy making, Regulatory Toxicology and Pharmacology, 65 (2013) 119-125. https://doi.org/10.1016/j.yrtph.2012.11.007

[5] D.R. Boverhof, C.M. Bramante, J.H. Butala, S.F. Clancy, M. Lafranconi, J. West, S.C. Gordon, Comparative assessment of nanomaterial definitions and safety evaluation considerations, Regulatory Toxicology and Pharmacology, 73 (2015) 137-150. https://doi.org/10.1016/j.yrtph.2015.06.001

[6] W. Luther, R. Nass, F. Schuster, M. Kallio, P. Lintunen, Industrial application of nanomaterials-changes and risks: Technology analysis, Future Technologies, 54 (2004) 19.

[7] G. Oberdörster, Safety assessment for nanotechnology and nanomedicine: concepts of nanotoxicology, Journal of Internal Medicine, 267 (2010) 89-105. https://doi.org/10.1111/j.1365-2796.2009.02187.x

[8] S. Soares, J. Sousa, A. Pais, C. Vitorino, Nanomedicine: Principles, Properties, and Regulatory Issues, Frontiers in Chemistry, 6 (2018) 360. https://doi.org/10.3389/fchem.2018.00360

[9] D.S. Kohane, Microparticles and nanoparticles for drug delivery, Biotechnology and Bioengineering, 96 (2007) 203-209. https://doi.org/10.1002/bit.21301

[10] R. Saini, S. Saini, S. Sharma, Nanotechnology: The future medicine, Journal of Cutaneous and Aesthetic surgery, 3 (2010) 32-33. https://doi.org/10.4103/0974-2077.63301

[11] S.T. Yerpude, A.K. Potbhare, P.R. Bhilkar, P. Thakur, P. Khiratkar, M. F Desimone, P.R. Dhongle, S.S. Sonawane, C. Goncalves, R. G. Chaudhary, Computational analysis of nanofluids-based drug delivery system: Preparation, current development and applications of nanofluids, Elsevier (2022) 335-364. https://doi.org/10.1016/B978-0-323-90564-0.00014-3

[12] P.B. Chouke, A.K. Potbhare, N.P. Meshram, M.M. Rai, K.M. Dadure, K Chaudhary, A.R. Rai, M. Desimone, R. G. Chaudhary D. T. Masram, Bioinspired NiO nanospheres: Exploring in-vitro toxicity using Bm-17 and L. rohita liver cells, DNA degradation, docking and proposed vacuolization mechanism, ACS Omega, 7 (2022) 6869−6884. https://doi.org/10.1021/acsomega.1c06544

[13] J.E.N. Dolatabadi, M.J.A.M. de la Guardia, Nanomaterial-based electrochemical immunosensors as advanced diagnostic tools, Analytical Methods, 6 (2014) 3891-3900. https://doi.org/10.1039/C3AY41749B

[14] M.-H. Qu, R.-F. Zeng, S. Fang, Q.-S. Dai, H.-P. Li, J.-T. Long, Liposome-based co-delivery of siRNA and docetaxel for the synergistic treatment of lung cancer, International Journal of Pharmaceutics, 474 (2014) 112-122. https://doi.org/10.1016/j.ijpharm.2014.08.019

[15] P. B. Chouke, K. M. Dadure, A. K. Potbhare, G. S. Bhusari, A. Mondal, K. Chaudhary, V. Singh, M. F. Desimone, R. G. Chaudhary, D. T. Masram, Biosynthesized δ-Bi2O3 nanoparticles from Crinum viviparum flower extract for photocatalytic dye degradation and molecular docking, ACS Omega, 7 (2022) 20983-20993. https://doi.org/10.1021/acsomega.2c01745

[16] H.B.M.Z. Islam, M. Susan, A.B. Hasan, A.B.J.I.P.J. Imran, High-strength potato starch/hectorite clay-based nanocomposite film: synthesis and characterization, Iranian Polymer Journal, 30 (2021) 513-521. https://doi.org/10.1007/s13726-021-00907-y

[17] N. Chowdhury, Solaiman, C.K. Roy, S.H. Firoz, T. Foyez, A.B. Imran, Role of Ionic Moieties in Hydrogel Networks to Remove Heavy Metal Ions from Water, ACS Omega, 6 (2020) 836-844. https://doi.org/10.1021/acsomega.0c05411

[18] M.R. Karim, M. Harun-Ur-Rashid, A.B. Imran, Highly Stretchable Hydrogel Using Vinyl Modified Narrow Dispersed Silica Particles as Cross-Linker, ChemistrySelect, 5 (2020) 10556-10561. https://doi.org/10.1002/slct.202003044

[19] H.B.M.Z. Islam, M.A.B.H. Susan, A.B. Imran, Effects of plasticizers and clays on the physical, chemical, mechanical, thermal, and morphological properties of potato starch-based nanocomposite films, ACS omega, 5 (2020) 17543-17552. https://doi.org/10.1021/acsomega.0c02012

[20] K.K. Jain, The role of nanobiotechnology in drug discovery, Drug Discovery Today, 10 (2005) 1435-1442. https://doi.org/10.1016/S1359-6446(05)03573-7

Materials Research Forum LLC
https://doi.org/10.21741/9781644902295-9

[21] Y.O. J.E.N Dolatabadi, Solid lipid-based nanocarriers as efficient targeted drug and gene delivery systems, TrAC Trends in Analytical Chemistry, 77 (2016) 100-108. https://doi.org/10.1016/j.trac.2015.12.016

[22] S.S. Mahmud, M. Moni, A.B. Imran, T. Foyez, Analysis of the suspected cancer-causing potassium bromate additive in bread samples available on the market in and around Dhaka City in Bangladesh, Food Science & Nutrition, 9 (2021) 3752-3757. https://doi.org/10.1002/fsn3.2338

[23] S. Tran, P.J. DeGiovanni, B. Piel, P. Rai, Cancer nanomedicine: a review of recent success in drug delivery, Clinical and Translational Medicine, 6 (2017) 44. https://doi.org/10.1186/s40169-017-0175-0

[24] L. Bregoli, D. Movia, J.D. Gavigan-Imedio, J. Lysaght, J. Reynolds, A. Prina-Mello, Nanomedicine applied to translational oncology: A future perspective on cancer treatment, Nanomedicine: Nanotechnology, Biology and Medicine, 12 (2016) 81-103. https://doi.org/10.1016/j.nano.2015.08.006

[25] B. Kumar, K. Jalodia, P. Kumar, H.K. Gautam, Recent advances in nanoparticle-mediated drug delivery, Journal of Drug Delivery Science and Technology, 41 (2017) 260-268. https://doi.org/10.1016/j.jddst.2017.07.019

[26] N.P. Truong, M.R. Whittaker, C.W. Mak, T.P. Davis, The importance of nanoparticle shape in cancer drug delivery, Expert Opinion on Drug Delivery, 12 (2015) 129-142. https://doi.org/10.1517/17425247.2014.950564

[27] J.D. Kingsley, H. Dou, J. Morehead, B. Rabinow, H.E. Gendelman, C.J. Destache, Nanotechnology: a focus on nanoparticles as a drug delivery system, Journal of Neuroimmune Pharmacology, 1 (2006) 340-350. https://doi.org/10.1007/s11481-006-9032-4

[28] J. Buse, A. El-Aneed, Properties, engineering and applications of lipid-based nanoparticle drug-delivery systems: current research and advances, Nanomedicine, 5 (2010) 1237-1260. https://doi.org/10.2217/nnm.10.107

[29] A. Alonso, F.M. Goñi, J.T. Buckley, Lipids favoring inverted phase enhance the ability of aerolysin to permabilize liposome bilayers, Biochemistry, 39 (2000) 14019-14024. https://doi.org/10.1021/bi001739o

[30] R. Banerjee, Liposomes: applications in medicine, Journal of Biomaterials Applications, 16 (2001) 3-21. https://doi.org/10.1106/RA7U-1V9C-RV7C-8QXL

[31] N. Moussaoui, M. Cansell, A. Denizot, Marinosomes, marine lipid-based liposomes: physical characterization and potential application in cosmetics, International Journal of Pharmaceutics, 242 (2002) 361-365. https://doi.org/10.1016/S0378-5173(02)00217-X

[32] A. Puri, K. Loomis, B. Smith, J.H. Lee, A. Yavlovich, E. Heldman, R. Blumenthal, Lipid-based nanoparticles as pharmaceutical drug carriers: from concepts to clinic, Critical Reviews in Therapeutic Drug Carrier Systems, 26 (2009) 523-580. https://doi.org/10.1615/CritRevTherDrugCarrierSyst.v26.i6.10

[33] B.S. Pattni, V.V. Chupin, V.P. Torchilin, New Developments in Liposomal Drug Delivery, Chemical Reviews, 115 (2015) 10938-10966. https://doi.org/10.1021/acs.chemrev.5b00046

[34] S. Weber, A. Zimmer, J. Pardeike, Solid Lipid Nanoparticles (SLN) and Nanostructured Lipid Carriers (NLC) for pulmonary application: A review of the state of the art, European Journal of Pharmaceutics and Biopharmaceutics, 86 (2014) 7-22. https://doi.org/10.1016/j.ejpb.2013.08.013

[35] T.T.H. Thi, E.J.A. Suys, J.S. Lee, D.H. Nguyen, K.D. Park, N.P. Truong, Lipid-Based Nanoparticles in the Clinic and Clinical Trials: From Cancer Nanomedicine to COVID-19 Vaccines, Vaccines, 9 (2021) 359. https://doi.org/10.3390/vaccines9040359

[36] J.E.N. Dolatabadi, H. Hamishehkar, M. Eskandani, H. Valizadeh, Formulation, characterization and cytotoxicity studies of alendronate sodium-loaded solid lipid nanoparticles, Colloids and Surfaces B: Biointerfaces, 117 (2014) 21-28. https://doi.org/10.1016/j.colsurfb.2014.01.055

[37] M.C. Teixeira, C. Carbone, E.B. Souto, Beyond liposomes: Recent advances on lipid based nanostructures for poorly soluble/poorly permeable drug delivery, Progress in Lipid Research, 68 (2017) 1-11. https://doi.org/10.1016/j.plipres.2017.07.001

[38] A. Beloqui, M.Á. Solinís, A. Rodríguez-Gascón, A.J. Almeida, V. Préat, Nanostructured lipid carriers: Promising drug delivery systems for future clinics, Nanomedicine: Nanotechnology, Biology and Medicine, 12 (2016) 143-161. https://doi.org/10.1016/j.nano.2015.09.004

[39] E. Rostami, S. Kashanian, A.H. Azandaryani, H. Faramarzi, J.E.N. Dolatabadi, K. Omidfar, Drug targeting using solid lipid nanoparticles, Chemistry and Physics of Lipids, 181 (2014) 56-61. https://doi.org/10.1016/j.chemphyslip.2014.03.006

[40] H.R. Oh, H.Y. Jo, J.S. Park, D.E. Kim, J.Y. Cho, P.H. Kim, K.S. Kim, Galactosylated Liposomes for Targeted Co-Delivery of Doxorubicin/Vimentin siRNA to Hepatocellular Carcinoma, Nanomaterials, 6 (2016) 141. https://doi.org/10.3390/nano6080141

[41] J. Ezzati Nazhad Dolatabadi, A. Azami, A. Mohammadi, H. Hamishehkar, V. Panahi-Azar, Y. Rahbar Saadat, A.A. Saei, Formulation, characterization and cytotoxicity evaluation of ketotifen-loaded nanostructured lipid carriers, Journal of Drug Delivery Science and Technology, 46 (2018) 268-273. https://doi.org/10.1016/j.jddst.2018.05.017

[42] A.S. Almurshedi, M. Radwan, S. Omar, A.A. Alaiya, M.M. Badran, H. Elsaghire, I.Y. Saleem, G.A. Hutcheon, A novel pH-sensitive liposome to trigger delivery of afatinib to cancer cells: Impact on lung cancer therapy, Journal of Molecular Liquids, 259 (2018) 154-166. https://doi.org/10.1016/j.molliq.2018.03.024

[43] N. Karimi, B. Ghanbarzadeh, H. Hamishehkar, B. Mehramuz, H.S. Kafil, Antioxidant, Antimicrobial and Physicochemical Properties of Turmeric Extract-

Loaded Nanostructured Lipid Carrier (NLC), Colloid and Interface Science Communications, 22 (2018) 18-24. https://doi.org/10.1016/j.colcom.2017.11.006

[44] S. Anuchapreeda, Y. Fukumori, S. Okonogi, H. Ichikawa, Preparation of Lipid Nanoemulsions Incorporating Curcumin for Cancer Therapy, Journal of Nanotechnology, 2012 (2012) 270383. https://doi.org/10.1155/2012/270383

[45] P.B. Ganesan, D.J.S.C. Narayanasamy, Lipid nanoparticles: Different preparation techniques, characterization, hurdles, and strategies for the production of solid lipid nanoparticles and nanostructured lipid carriers for oral drug delivery, Sustainable Chemistry and Pharmacy, 6 (2017) 37-56. https://doi.org/10.1016/j.scp.2017.07.002

[46] C. Cha, S.R. Shin, N. Annabi, M.R. Dokmeci, A. Khademhosseini, Carbon-Based Nanomaterials: Multifunctional Materials for Biomedical Engineering, ACS Nano, 7 (2013) 2891-2897. https://doi.org/10.1021/nn401196a

[47] J. Ezzati Nazhad Dolatabadi, Y. Omidi, D. Losic, Carbon Nanotubes as an Advanced Drug and Gene Delivery Nanosystem, Current Nanoscience, 7 (2011) 297-314. https://doi.org/10.2174/157341311795542444

[48] C. Shao-Yu, K. Ji-Lie, Advance in research on carbon nanotubes as diagnostic and therapeutic agents for tumor, Chinese Journal of Analytical Chemistry, 37 (2009) 1240-1246. https://doi.org/10.1016/S1872-2040(08)60125-5

[49] H. Ali-Boucetta, K. Kostarelos, Pharmacology of carbon nanotubes: Toxicokinetics, excretion and tissue accumulation, Advanced Drug Delivery Reviews, 65 (2013) 2111-2119. https://doi.org/10.1016/j.addr.2013.10.004

[50] P.N. Gurjar, S. Chouksey, G. Patil, N. Naik, S. Agrawal, Carbon Nanotubes: Pharmaceutical Applications, Asian Journal of Biomedical and Pharmaceutical Sciences, 3 (2013) 8-13.

[51] Z. Liu, K. Chen, C. Davis, S. Sherlock, Q. Cao, X. Chen, H. Dai, Drug delivery with carbon nanotubes for in vivo cancer treatment, Cancer Research, 68 (2008) 6652-6660. https://doi.org/10.1158/0008-5472.CAN-08-1468

[52] P.M. Costa, J.T.-W. Wang, J.-F. Morfin, T. Khanum, W. To, J. Sosabowski, E. Tóth, K.T. Al-Jamal, Functionalised Carbon Nanotubes Enhance Brain Delivery of Amyloid-Targeting Pittsburgh Compound B (PiB)-Derived Ligands, Nanotheranostics, 2 (2018) 168-183. https://doi.org/10.7150/ntno.23125

[53] X. Yuan, X. Zhang, L. Sun, Y. Wei, X. Wei, Cellular Toxicity and Immunological Effects of Carbon-based Nanomaterials, Particle and Fibre Toxicology, 16 (2019) 18. https://doi.org/10.1186/s12989-019-0299-z

[54] Z. Tang, C. He, H. Tian, J. Ding, B.S. Hsiao, B. Chu, X. Chen, Polymeric nanostructured materials for biomedical applications, Progress in Polymer Science, 60 (2016) 86-128. https://doi.org/10.1016/j.progpolymsci.2016.05.005

[55] S.K. Samal, M. Dash, S.V. Vlierberghe, D.L. Kaplan, E. Chiellini, C.V. Blitterswijk, L. Moroni, P. Dubruel, Cationic polymers and their therapeutic potential, Chemical Society Reviews, 41 (2012) 7147-7194. https://doi.org/10.1039/c2cs35094g

[56] W. Sun, X. Chen, C. Xie, Y. Wang, L. Lin, K. Zhu, X. Shuai, Co-Delivery of Doxorubicin and Anti-BCL-2 siRNA by pH-Responsive Polymeric Vector to Overcome Drug Resistance in In Vitro and In Vivo HepG2 Hepatoma Model, Biomacromolecules, 19 (2018) 2248-2256. https://doi.org/10.1021/acs.biomac.8b00272

[57] A. Mokhtarzadeh, A. Alibakhshi, M. Hejazi, Y. Omidi, J. Ezzati Nazhad Dolatabadi, Bacterial-derived biopolymers: Advanced natural nanomaterials for drug delivery and tissue engineering, TrAC Trends in Analytical Chemistry, 82 (2016) 367-384. https://doi.org/10.1016/j.trac.2016.06.013

[58] D. Ren, Protein Nanoparticle as a Versatile Drug Delivery System in Nanotechnology, Journal of Nanomedicine Research, 4 (2016) 00077. https://doi.org/10.15406/jnmr.2016.04.00077

[59] P. Yadav, A. Bandyopadhyay, A. Chakraborty, K. Sarkar, Enhancement of anticancer activity and drug delivery of chitosan-curcumin nanoparticle via molecular docking and simulation analysis, Carbohydrate Polymers, 182 (2018) 188-198. https://doi.org/10.1016/j.carbpol.2017.10.102

[60] M.A. Razi, R. Wakabayashi, Y. Tahara, M. Goto, N. Kamiya, Genipin-stabilized caseinate-chitosan nanoparticles for enhanced stability and anticancer activity of curcumin, Colloids and Surfaces B: Biointerfaces, 164 (2018) 308-315. https://doi.org/10.1016/j.colsurfb.2018.01.041

[61] C.O. Mellet, J.M.G. Fernández, J.M. Benito, Cyclodextrin-based gene delivery systems, Chemical Society Reviews, 40 (2011) 1586-1608. https://doi.org/10.1039/C0CS00019A

[62] V.G. Kadajji, G.V. Betageri, Water Soluble Polymers for Pharmaceutical Applications, Polymers, 3 (2011) 1972-2009. https://doi.org/10.3390/polym3041972

[63] S. Jaiswal, P. Mishra, Co-delivery of curcumin and serratiopeptidase in HeLa and MCF-7 cells through nanoparticles show improved anticancer activity, Materials Science and Engineering: C, 92 (2018) 673-684. https://doi.org/10.1016/j.msec.2018.07.025

[64] M. Dalmau, S. Lim, S.-W. Wang, Design of a pH-Dependent Molecular Switch in a Caged Protein Platform, Nano Letters, 9 (2009) 160-166. https://doi.org/10.1021/nl8027069

[65] A. Bergstrand, G. Rahmani-Monfared, A. Ostlund, M. Nydén, K. Holmberg, Comparison of PEI-PEG and PLL-PEG copolymer coatings on the prevention of protein fouling, Journal of Biomedical Materials Research. Part A, 88 (2009) 608-615. https://doi.org/10.1002/jbm.a.31894

[66] P. Kumari, S.V.K. Rompicharla, O.S. Muddineti, B. Ghosh, S. Biswas, Transferrin-anchored poly(lactide) based micelles to improve anticancer activity of curcumin in hepatic and cervical cancer cell monolayers and 3D spheroids, International Journal of

Biological Macromolecules, 116 (2018) 1196-1213.
https://doi.org/10.1016/j.ijbiomac.2018.05.040

[67] A.B. Imran, Design, development, characterization and application of smart polymeric hydrogel, in: M.M. Arezki (Ed.) Manufacturing Systems: Recent Progress and Future Directions, Nova Science Publishers, Inc., NY, USA, 2020.

[68] M. Harun-Ur-Rashid, A.B. Imran, Superabsorbent Hydrogels from carboxymethyl cellulose, in: M.I.H. Mondal (Ed.) Carboxymethyl Cellulose. Volume I: Synthesis and Characterization, Nova Science Publishers, Inc., NY, USA, 2019, pp. 159-182.

[69] M.M. Yallapu, M.K. Reddy, V. Labhasetwar, Nanogels: Chemistry to Drug Delivery, Biomedical Applications of Nanotechnology, (2007) 131-171. https://doi.org/10.1002/9780470152928.ch6

[70] M.K. Riaz, M.A. Riaz, X. Zhang, C. Lin, K.H. Wong, X. Chen, G. Zhang, A. Lu, Z. Yang, Surface Functionalization and Targeting Strategies of Liposomes in Solid Tumor Therapy: A Review, International Journal of Molecular Sciences, 19 (2018) 195. https://doi.org/10.3390/ijms19010195

[71] N. Rahmanian, H. Hamishehkar, J.E.N. Dolatabadi, N. Arsalani, Nano graphene oxide: A novel carrier for oral delivery of flavonoids, Colloids and Surfaces B: Biointerfaces, 123 (2014) 331-338. https://doi.org/10.1016/j.colsurfb.2014.09.036

[72] J.A. Luckanagul, C. Pitakchatwong, P.R.N. Bhuket, C. Muangnoi, P. Rojsitthisak, S. Chirachanchai, Q. Wang, P. Rojsitthisak, Chitosan-based polymer hybrids for thermo-responsive nanogel delivery of curcumin, Journal of Carbohydrate Polymers, 181 (2018) 1119-1127. https://doi.org/10.1016/j.carbpol.2017.11.027

[73] R.T. Chacko, J. Ventura, J. Zhuang, S. Thayumanavan, Polymer nanogels: A versatile nanoscopic drug delivery platform, Advanced Drug Delivery Reviews, 64 (2012) 836-851. https://doi.org/10.1016/j.addr.2012.02.002

[74] A.E. Ekkelenkamp, M.R. Elzes, J.F.J. Engbersen, J.M.J. Paulusse, Responsive cross-linked polymer nanogels for imaging and therapeutics delivery, Journal of Materials Chemistry B, 6 (2018) 210-235. https://doi.org/10.1039/C7TB02239E

[75] H. Zhang, Y. Zhai, J. Wang, G. Zhai, New progress and prospects: The application of nanogel in drug delivery, Materials Science and Engineering: C, 60 (2016) 560-568. https://doi.org/10.1016/j.msec.2015.11.041

[76] L.E. Nita, A.P. Chiriac, A. Diaconu, N. Tudorachi, L. Mititelu-Tartau, Multifunctional nanogels with dual temperature and pH responsiveness, International Journal of Pharmaceutics, 515 (2016) 165-175. https://doi.org/10.1016/j.ijpharm.2016.10.017

[77] A. Pich, A. Tessier, V. Boyko, Y. Lu, H.-J.P. Adler, Synthesis and Characterization of Poly(vinylcaprolactam)-Based Microgels Exhibiting Temperature and pH-Sensitive Properties, Macromolecules, 39 (2006) 7701-7707. https://doi.org/10.1021/ma060985q

[78] D. Li, C.F. van Nostrum, E. Mastrobattista, T. Vermonden, W.E. Hennink, Nanogels for intracellular delivery of biotherapeutics, Journal of Controlled Release, 259 (2017) 16-28. https://doi.org/10.1016/j.jconrel.2016.12.020

[79] A. Goyal, G. Rath, Recent Advances in Metal Nanoparticles in Cancer Therapy, Journal of Drug Targeting, 26 (2017) 1-45. https://doi.org/10.1080/1061186X.2017.1400553

[80] M.R. Díaz, P.E. Vivas-Mejia, Nanoparticles as Drug Delivery Systems in Cancer Medicine: Emphasis on RNAi-Containing Nanoliposomes, Pharmaceuticals, 6 (2013) 1361-1380. https://doi.org/10.3390/ph6111361

[81] S. Nambiar, E. Osei, A. Fleck, J. Darko, A.J. Mutsaers, S. Wettig, Synthesis of curcumin-functionalized gold nanoparticles and cytotoxicity studies in human prostate cancer cell line, Applied Nanoscience, 8 (2018) 347-357. https://doi.org/10.1007/s13204-018-0728-6

[82] Y. Zhou, G. Quan, Q. Wu, X. Zhang, B. Niu, B. Wu, Y. Huang, X. Pan, C. Wu, Mesoporous silica nanoparticles for drug and gene delivery, Acta Pharmaceutica Sinica B, 8 (2018) 165-177. https://doi.org/10.1016/j.apsb.2018.01.007

[83] W. Jiang, B.Y.S. Kim, J.T. Rutka, W.C.W. Chan, Advances and challenges of nanotechnology-based drug delivery systems, Expert Opinion on Drug Delivery, 4 (2007) 621-633. https://doi.org/10.1517/17425247.4.6.621

[84] A. Carvalho, A.R. Fernandes, P.V. Baptista, Chapter 10 - Nanoparticles as Delivery Systems in Cancer Therapy: Focus on Gold Nanoparticles and Drugs, in: S.S. Mohapatra, S. Ranjan, N. Dasgupta, R.K. Mishra, S. Thomas (Eds.) Applications of Targeted Nano Drugs and Delivery Systems, Elsevier, 2019, pp. 257-295. https://doi.org/10.1016/B978-0-12-814029-1.00010-7

[85] D.M. Connor, A.-M. Broome, Gold nanoparticles for the delivery of cancer therapeutics, Advances in Cancer Research, 139 (2018) 163-184. https://doi.org/10.1016/bs.acr.2018.05.001

[86] J. Zhou, M.A. Frank, Y. Yang, A.R. Boccaccini, S. Virtanen, A novel local drug delivery system: Superhydrophobic titanium oxide nanotube arrays serve as the drug reservoir and ultrasonication functions as the drug release trigger, Materials Science and Engineering: C, 82 (2018) 277-283. https://doi.org/10.1016/j.msec.2017.08.066

[87] G.D. Venkatasubbu, S. Ramasamy, V. Ramakrishnan, B. Kumar, Folate targeted PEGylated titanium dioxide nanoparticles as a nanocarrier for targeted paclitaxel drug delivery, Journal of Advanced Powder Technology, 24 (2013) 947-954. https://doi.org/10.1016/j.apt.2013.01.008

[88] P. Kesharwani, A.K. Iyer, Recent advances in dendrimer-based nanovectors for tumor-targeted drug and gene delivery, Drug Discovery Today, 20 (2015) 536-547. https://doi.org/10.1016/j.drudis.2014.12.012

[89] M. Ghaffari, G. Dehghan, F. Abedi-Gaballu, S. Kashanian, B. Baradaran, J. Ezzati Nazhad Dolatabadi, D. Losic, Surface functionalized dendrimers as controlled-release

Emerging Applications of Nanomaterials   Materials Research Forum LLC
Materials Research Foundations 141 (2023) 218-245   https://doi.org/10.21741/9781644902295-9

delivery nanosystems for tumor targeting, European Journal of Pharmaceutical Sciences, 122 (2018) 311-330. https://doi.org/10.1016/j.ejps.2018.07.020

[90] F. Abedi-Gaballu, G. Dehghan, M. Ghaffari, R. Yekta, S. Abbaspour-Ravasjani, B. Baradaran, J. Ezzati Nazhad Dolatabadi, M.R. Hamblin, PAMAM dendrimers as efficient drug and gene delivery nanosystems for cancer therapy, Applied Materials Today, 12 (2018) 177-190. https://doi.org/10.1016/j.apmt.2018.05.002

[91] J. Li, H. Liang, J. Liu, Z. Wang, Poly (amidoamine) (PAMAM) dendrimer mediated delivery of drug and pDNA/siRNA for cancer therapy, International Journal of Pharmaceutics, 546 (2018) 215-225. https://doi.org/10.1016/j.ijpharm.2018.05.045

[92] A. Zarebkohan, F. Najafi, H.R. Moghimi, M. Hemmati, M.R. Deevband, B. Kazemi, Synthesis and characterization of a PAMAM dendrimer nanocarrier functionalized by SRL peptide for targeted gene delivery to the brain, European Journal of Pharmaceutical Sciences, 78 (2015) 19-30. https://doi.org/10.1016/j.ejps.2015.06.024

[93] H. Gotoh, C. Liu, A.B. Imran, M. Hara, T. Seki, K. Mayumi, K. Ito, Y. Takeoka, Optically transparent, high-toughness elastomer using a polyrotaxane cross-linker as a molecular pulley, Science Advances, 4 (2018) eaat7629. https://doi.org/10.1126/sciadv.aat7629

[94] A. Yasumoto, H. Gotoh, Y. Gotoh, A.B. Imran, M. Hara, T. Seki, Y. Sakai, K. Ito, Y. Takeoka, Highly Responsive Hydrogel Prepared Using Poly(N-isopropylacrylamide)-Grafted Polyrotaxane as a Building Block Designed by Reversible Deactivation Radical Polymerization and Click Chemistry, Macromolecules, 50 (2017) 364-374. https://doi.org/10.1021/acs.macromol.6b01955

[95] K. Ohmori, A.B. Imran, T. Seki, C. Liu, K. Mayumi, K. Ito, Y. Takeoka, Molecular weight dependency of polyrotaxane-cross-linked polymer gel extensibility, Chemical Communications, 52 (2016) 13757-13759. https://doi.org/10.1039/C6CC07641F

[96] A.B. Imran, K. Esaki, H. Gotoh, T. Seki, K. Ito, Y. Sakai, Y. Takeoka, Extremely stretchable thermosensitive hydrogels by introducing slide-ring polyrotaxane cross-linkers and ionic groups into the polymer network, Nature Communications, 5 (2014) 5124. https://doi.org/10.1038/ncomms6124

[97] A.B. Imran, T. Seki, Y. Takeoka, Recent advances in hydrogels in terms of fast stimuli responsiveness and superior mechanical performance, Polymer Journal, 42 (2010) 839-851. https://doi.org/10.1038/pj.2010.87

[98] A.B. Imran, T. Seki, K. Ito, Y. Takeoka, Facile synthesis of sliding poly (NIPA) gels using a vinyl modified polyrotaxane as a cross-linker, Transactions of the MRS of Japan, 35 (2010) 841-844. https://doi.org/10.14723/tmrsj.35.841

[99] A.B. Imran, T. Seki, K. Ito, Y. Takeoka, Hydrophobic and hydrophilic polyrotaxane based movable cross-linkers for thermo-sensitive poly (N-isopropylacrylamide) gels, J Transactions of the Materials Research Society of Japan, 35 (2010) 291-297. https://doi.org/10.14723/tmrsj.35.291

[100] M. Harun-Ur-Rashid, A.B. Imran, T. Seki, M. Ishii, H. Nakamura, Y. Takeoka, Angle-independent structural color in colloidal amorphous arrays, ChemPhysChem, 11 (2010) 579-583. https://doi.org/10.1002/cphc.200900869

[101] A.B. Imran, T. Seki, K. Ito, Y. Takeoka, Poly (N-isopropylacrylamide) gel prepared using a hydrophilic polyrotaxane-based movable cross-linker, Macromolecules, 43 (2010) 1975-1980. https://doi.org/10.1021/ma902349j

[102] M. Harun-Ur-Rashid, A.B. Imran, T. Seki, Y. Takeoka, M. Ishii, H. Nakamura, Template synthesis for stimuli-responsive angle independent structural colored smart materials, Transactions of the Materials Research Society of Japan, 34 (2009) 333-337. https://doi.org/10.14723/tmrsj.34.333

[103] H. Murayama, A.B. Imran, S. Nagano, T. Seki, M. Kidowaki, K. Ito, Y. Takeoka, Chromic slide-ring gel based on reflection from photonic bandgap, Macromolecules, 41 (2008) 1808-1814. https://doi.org/10.1021/ma0715627

[104] A.B. Imran, T. Seki, T. Kataoka, M. Kidowaki, K. Ito, Y. Takeoka, Fabrication of mechanically improved hydrogels using a movable cross-linker based on vinyl modified polyrotaxane, Chemical Communications, (2008) 5227-5229. https://doi.org/10.1039/b810290b

[105] B. Xiao, L. Ma, D. Merlin, Nanoparticle-mediated co-delivery of chemotherapeutic agent and siRNA for combination cancer therapy, Expert Opinion on Drug Delivery, 14 (2017) 65-73. https://doi.org/10.1080/17425247.2016.1205583

[106] S.S. Qi, J.H. Sun, H.H. Yu, S.Q. Yu, Co-delivery nanoparticles of anticancer drugs for improving chemotherapy efficacy, Drug Delivery, 24 (2017) 1909-1926. https://doi.org/10.1080/10717544.2017.1410256

[107] R. Cortesi, E. Esposito, M. Drechsler, G. Pavoni, I. Cacciatore, M. Sguizzato, A. Di Stefano, L-dopa co-drugs in nanostructured lipid carriers: A comparative study, Materials Science and Engineering: C, 72 (2017) 168-176. https://doi.org/10.1016/j.msec.2016.11.060

[108] N.S. Gandhi, R.K. Tekade, M.B. Chougule, Nanocarrier mediated delivery of siRNA/miRNA in combination with chemotherapeutic agents for cancer therapy: Current progress and advances, Journal of Controlled Release, 194 (2014) 238-256. https://doi.org/10.1016/j.jconrel.2014.09.001

[109] W. Yang, Y. Cheng, T. Xu, X. Wang, L.-p. Wen, Targeting cancer cells with biotin-dendrimer conjugates, European Journal of Medicinal Chemistry, 44 (2009) 862-868. https://doi.org/10.1016/j.ejmech.2008.04.021

[110] Y. Li, T. Thambi, D.S. Lee, Co-Delivery of Drugs and Genes Using Polymeric Nanoparticles for Synergistic Cancer Therapeutic Effects, Advanced Healthcare Materials, 7 (2018) 1700886. https://doi.org/10.1002/adhm.201700886

[111] A. Afkham, L. Aghebati-Maleki, H. Siahmansouri, S. Sadreddini, M. Ahmadi, S. Dolati, N.M. Afkham, P. Akbarzadeh, F. Jadidi-Niaragh, V. Younesi, M. Yousefi, Chitosan (CMD)-mediated co-delivery of SN38 and Snail-specific siRNA as a useful

anticancer approach against prostate cancer, Pharmacological Reports, 70 (2018) 418-425. https://doi.org/10.1016/j.pharep.2017.11.005

[112] D.R. Khan, M.N. Webb, T.H. Cadotte, M.N. Gavette, Use of Targeted Liposome-based Chemotherapeutics to Treat Breast Cancer, Breast Cancer: Basic and Clinical Research, 9 (2015) 29421. https://doi.org/10.4137/BCBCR.S29421

[113] S. Bamrungsap, Z. Zhao, T. Chen, L. Wang, C. Li, T. Fu, W. Tan, Nanotechnology in therapeutics: a focus on nanoparticles as a drug delivery system, Nanomedicine, 7 (2012) 1253-1271. https://doi.org/10.2217/nnm.12.87

[114] M. Han, Q. Lv, X.-J. Tang, Y.-L. Hu, D.-H. Xu, F.-Z. Li, W.-Q. Liang, J.-Q. Gao, Overcoming drug resistance of MCF-7/ADR cells by altering intracellular distribution of doxorubicin via MVP knockdown with a novel siRNA polyamidoamine-hyaluronic acid complex, Journal of Controlled Release, 163 (2012) 136-144. https://doi.org/10.1016/j.jconrel.2012.08.020

[115] A. Deb, N.G. Andrews, V. Raghavan, Natural polymer functionalized graphene oxide for co-delivery of anticancer drugs: In-vitro and in-vivo, International Journal of Biological Macromolecules, 113 (2018) 515-525. https://doi.org/10.1016/j.ijbiomac.2018.02.153

[116] A. Rahman, T. Foyez, M.A.B.H. Susan, A.B. Imran, Self-Healable and Conductive Double-Network Hydrogels with Bioactive Properties, Macromolecular Chemistry Physics, 221 (2020) 2000207. https://doi.org/10.1002/macp.202000207

[117] H. Ragelle, F. Danhier, V. Préat, R. Langer, D.G. Anderson, Nanoparticle-based drug delivery systems: a commercial and regulatory outlook as the field matures, Expert Opinion on Drug Delivery, 14 (2017) 851-864. https://doi.org/10.1080/17425247.2016.1244187

[118] R. Bawa, Regulating Nanomedicine - Can the FDA Handle It?, Current Drug Delivery, 8 (2011) 227-234. https://doi.org/10.2174/156720111795256156

[119] J. Arrowsmith, P. Miller, Phase II and Phase III attrition rates 2011-2012, Nature Reviews Drug Discovery, 12 (2013) 569-569. https://doi.org/10.1038/nrd4090

[120] Food and drug adminstration, Guidance for industry: ANDA submissions-content and format of Abbreviated New Drug Applications, 2014.

[121] G. Pillai, Nanomedicines for cancer therapy: An update of FDA approved and those under various stages of development, SOJ Pharmacy and Pharmaceutical Sciences, 1 (2014) 13. https://doi.org/10.15226/2374-6866/1/1/00109

[122] U. Bulbake, S. Doppalapudi, N. Kommineni, W. Khan, Liposomal Formulations in Clinical Use: An Updated Review, Pharmaceutics, 9 (2017) 12. https://doi.org/10.3390/pharmaceutics9020012

[123] J.M. Caster, A.N. Patel, T. Zhang, A. Wang, Investigational nanomedicines in 2016: a review of nanotherapeutics currently undergoing clinical trials, Wiley Interdisciplinary Reviews: Nanomedicine and Nanobiotechnology, 9 (2017) e1416. https://doi.org/10.1002/wnan.1416

[124] D. Bobo, K.J. Robinson, J. Islam, K.J. Thurecht, S.R. Corrie, Nanoparticle-Based Medicines: A Review of FDA-Approved Materials and Clinical Trials to Date, Pharmaceutical Research, 33 (2016) 2373-2387. https://doi.org/10.1007/s11095-016-1958-5

[125] A.C. Anselmo, S. Mitragotri, Nanoparticles in the clinic: An update, Bioengineering & Translational Medicine, 4 (2019) e10143. https://doi.org/10.1002/btm2.10143

[126] O.S. Fenton, K.N. Olafson, P.S. Pillai, M.J. Mitchell, R. Langer, Advances in Biomaterials for Drug Delivery, Advanced Materials, 30 (2018) 1705328. https://doi.org/10.1002/adma.201705328

[127] A. A. H. Abdellatif, A. F. Alsowinea, Approved and marketed nanoparticles for disease targeting and applications in COVID-19, Nanotechnology Reviews, 10 (2021) 1941-1977. https://doi.org/10.1515/ntrev-2021-0115

[128] P. Malik, G.K. Inwati, T.K. Mukherjee, S. Singh, M.J.J.o.M.L. Singh, Green silver nanoparticle and Tween-20 modulated pro-oxidant to antioxidant curcumin transformation in aqueous CTAB stabilized peanut oil emulsions, Journal of Molecular Liquids, 291 (2019) 111252. https://doi.org/10.1016/j.molliq.2019.111252

[129] T. Foyez, A.B. Imran, Nanotechnology in vaccine development and constraints, in: K. Pal (Ed.) Nanovaccinology outbreak as targeted therapeutics, Wiley-Scrivener 2022. https://doi.org/10.1002/9781119858041.ch1

[130] G. Gnanamoorthy, K. Ramar, A. Padmanaban, V.K. Yadav, K.S. Babu, V. Karthikeyan, V. Narayanan, Implementation of ZnSnO3 nanosheets and their RE (Er, Eu, and Pr) materials: Enhanced photocatalytic activity, Advanced Powder Technology, 31 (2020) 1209-1219. https://doi.org/10.1016/j.apt.2019.12.028

[131] V.K. Yadav, R. Suriyaprabha, S.H. Khan, B. Singh, G. Gnanamoorthy, N. Choudhary, A.K. Yadav, H. Kalasariya, A novel and efficient method for the synthesis of amorphous nanosilica from fly ash tiles, Materials Today: Proceedings, 26 (2020) 701-705. https://doi.org/10.1016/j.matpr.2020.01.013

[132] M. Harun-Ur-Rashid, T. Foyez, I. Jahan, K. Pal, A.B. Imran, Rapid diagnosis of COVID-19 via nano-biosensor-implemented biomedical utilization: a systematic review, RSC advances, 12 (2022) 9445-9465. https://doi.org/10.1039/D2RA01293F

[133] M. Khan, A.U. Khan, M.A. Hasan, K.K. Yadav, M.M.C. Pinto, N. Malik, V.K. Yadav, A.H. Khan, S. Islam, G.K. Sharma, Agro-Nanotechnology as an Emerging Field: A Novel Sustainable Approach for Improving Plant Growth by Reducing Biotic Stress, Applied Sciences, 11 (2021) 2282. https://doi.org/10.3390/app11052282

[134] G. Gnanamoorthy, D. Ali, V.K. Yadav, G. Dhinagaran, K. Venkatachalam, V. Narayanan, New construction of Fe3O4/rGO/ZnSnO3 nanocomposites enhanced photoelectro chemical properties, Optical Materials, 109 (2020) 110353. https://doi.org/10.1016/j.optmat.2020.110353

[135] G. Gnanamoorthy, P. Priya, D. Ali, M. Lakshmi, V.K. Yadav, R. Varghese, A new CuZr2S4/rGO and their reduced graphene oxide nanocomposities enhanced photocatalytic and antimicrobial activities, Chemical Physics Letters, 781 (2021) 139011. https://doi.org/10.1016/j.cplett.2021.139011

[136] S. Soenen, J. Cocquyt, L. Defour, P. Saveyn, P.V.D. Meeren, M.D.J.M. Cuyper, M. Processes, Design and development of magnetoliposome-based theranostics, Materials and Manufacturing Processes, 23 (2008) 611-614. https://doi.org/10.1080/10426910802160635

[137] N. Oku, Y. Tokudome, H. Tsukada, S. Okada, Real-time analysis of liposomal trafficking in tumor-bearing mice by use of positron emission tomography, Biochimica et Biophysica Acta (BBA)-Biomembranes, 1238 (1995) 86-90. https://doi.org/10.1016/0005-2736(95)00106-D

[138] Y. Gandon, J.-F. Heautot, F. Brunet, D. Guyader, Y. Deugnier, M. Carsin, Superparamagnetic iron oxide: clinical time-response study, European Journal of Radiology, 12 (1991) 195-200. https://doi.org/10.1016/0720-048X(91)90072-4

[139] A.K. Gupta, C. Berry, M. Gupta, A. Curtis, Receptor-mediated targeting of magnetic nanoparticles using insulin as a surface ligand to prevent endocytosis, IEEE Transactions on Nanobioscience, 2 (2003) 255-261. https://doi.org/10.1109/TNB.2003.820279

Emerging Applications of Nanomaterials
Materials Research Foundations 141 (2023) 246-269

Materials Research Forum LLC
https://doi.org/10.21741/9781644902295-10

Chapter 10

Nanotechnology/Nanosensors for the Detection of Pathogens

Raju Ahmed[1], Mahmuda Nargis[2,3], Abu Bin Ihsan[3,4]*

[1]College of Public Health, University of Iowa, Iowa City, IA 52242, United States

[2]Department of Pharmacy, Dhaka International University, Dhaka, Bangladesh

[3]Department of Pharmaceutical Engineering, Faculty of Engineering, Toyama Prefectural University, 5180 Kurokawa, Imizu, Toyama 939-0398, Japan

[4]Department of Pharmacy, ASA University Bangladesh, Dhaka, Bangladesh

*ilincon@gmail.com

Abstract

Early and accurate detection of pathogen is an important key to prevent and treat pathogen originated health issues. Conventional diagnostic methods are relatively laborious, time-consuming, costly, and require sophisticated instruments. To overcome the limitations of traditional instruments, nanosensors have emerged as a promising alternative. Due to their easy fabrication, high surface-to-volume ratio, and great biocompatibility, nanosensors have been shown to be an ideal technique for sensing. Herein, we provide an overall impression of the application of different types of nanosensors in disease diagnosis, and in the monitoring of water and food quality along with their probable future orientations.

Keywords

Aptamers, COVID-19, Food Safety, Infectious Disease, Microorganisms, Nanotechnology, Polymer, Polymeric Chain Reaction (PCR), SARS-CoV-2

Contents

Materials Research Forum LLC
https://doi.org/10.21741/9781644902295-10

Materials Research Forum LLC

https://doi.org/10.21741/9781644902295-10

1. Introduction

Infectious illnesses that harm people, animals, and plants are on the rise today [1]. A wide variety of viruses (e.g. SARS-CoV2; *Mycobacterium bovis*; Tobacco mosaic virus etc.) and bacteria (e.g. *Salmonella; Campylobacter; Bartonella species etc.*) infect humans, animals, and plants. Diagnosis of disease is an important step for the treatment and prevention of communicable diseases. The capacity to identify illness-related metabolites, proteins, nucleic acids, cells, and pathogens are critical not only for disease diagnosis in the medical context and health technology but also in industrial, agricultural, and environmental research development [2]. Controlling outbreaks and averting large-scale epidemics need rapid and accurate identification of pathogens and timely reporting of the outbreaks and events of communicable diseases. Microorganisms may be detected, reported and tracked with the use of microbial pathogen detection [3]. Since conventional detection procedures like biochemical and immunological recognition tests, Polymerase Chain Reactions (PCRs), and cell culture might take hours to days to complete a test, and clinicians are sometimes left to make speculations about the identity of the pathogen or toxin. There is a pressing need to create new sensors because of the limitations of contemporary detection techniques. Recently, nanomaterials have emerged as a particularly promising and useful method in this regard. Greater surface/volume proportions, advantageous physical and chemical characteristics, dispersibility, and exclusive interactions at the nanoscale level are all being used to develop nanomaterial-based biosensors that have started to outperform conventional pathogen detection approaches in terms of sensitivity and precision while using less sample volume, time, and assay cost [4]. Nanosensors have transformed to be noteworthy for the detection of analytes from explosives [5] to nucleic acids [6], proteins [7], viruses [8], bacteria [9], toxins [10], carcinogens [11], clinical examinations [12], food processing [13], ecological monitoring [14], and fighting the biological terrorism [15], over the past decades. Nanosensors are particularly useful in medical diagnostics, water and food quality monitoring, and other chemical sensing applications. However, in this chapter, we will mainly focus on nanosensor technology in virus detection and we will shed some light on its use in the detection of bacteria. Finally, we will wrap things off with a few words on the conclusion and future outlook.

2. Use of nanosensors in infectious disease diagnosis

Infectious diseases have appeared to be an important public health concern around the world. Controlling the spread of infectious diseases begins with the rapid and precise diagnosis of infected people. With the rapid advancement of nanoscience and nanotechnology, nanosensors have emerged as a new tool for diagnosing infectious diseases. Among a wide variety of devices, some special features like sensitivity, selectivity, firmness, and speed have made nanosensors competitive for detecting infectious pathogens (both bacteria and viruses) [16].

Emerging Applications of Nanomaterials
Materials Research Foundations 141 (2023) 246-269

Materials Research Forum LLC
https://doi.org/10.21741/9781644902295-10

2.1 The use of nanosensors in the detection of viruses

2.1.1 Nanosensors in SARS-CoV-2 detection

Severe Acute Respiratory Syndrome (SARS) is a viral disease that affects the respiratory system, is caused by a novel strain of corona virus known as Severe Acute Respiratory Syndrome Coronavirus-2 (SARS-CoV-2). Since the beginning of 2019, more than 4.5 million people have lost their lives across the world due to SARS. Traditional diagnosis methods including Reverse Transcriptase-Polymerase Chain Reaction (RT-PCR), laser desorption/ionization-mass spectrometry, antibody-profiling, Computed-Tomography (CT) scan, immunoassays, lateral flow biosensors along with some other methods are being used by the healthcare provider for diagnosing SARS-CoV-2. However, in addition to some advantages, these traditional approaches do have several disadvantages, including limited specificity, extended detection time, high cost, and the need for highly trained personnel to operate sophisticated equipment. Simultaneously, nanotechnology-enabled diagnostics methods have emerged which are classified based on the features of the nanomaterials [17-19]. Recently, for SARS-CoV-2 identification, nanosensors have attained substantial attention because of their durability, high specificity and selectivity compared to the traditional methods [20]. For the diagnosis of SARS-CoV-2, several types of nanosensors have been used. Among them, the aptameric nanosensors, Molecularly Imprinted Polymers (MIP) based nanosensors and magnetic nanosensors have been discussed here.

2.1.1.1 Aptameric nanosensors

It is possible to construct nanosensors that can perform real-time detection, do not need any onerous procedure to prepare the sample, and can detect many targets within a short time thanks to the use of oligonucleotides collectively known as aptamers that can fold into three-dimensional shapes. Sensors that use aptamers and nanostructures as signaling elements are called aptameric nanosensors. In certain circumstances, aptameric nanosensors may detect viruses or their components at concentrations from picomolar to even as low as a femtomolar level. The creation of aptameric nanosensors relies on the proper selection of aptamers. Using a 3D printing platform with improved capabilities, it is possible to generate nanostructures with a wide variation for the intelligent fabrication of aptamer-based nanosensors to detect SARS-CoV-2. Light scattering, colorimetry, fluorescence, Raman scattering, and electrochemical signals are just a few of the signal transducer concepts that may be used with aptameric nanosensors to recognize a wide variety of targets [21]. A comprehensive summary of the aptamer-based nanosensors that have been developed for the detection of SARS-CoV-2 is shown in *Table 1*.

Table 1: Different types of aptameric nanosensors and their working principles for the detection of SARS-CoV-2. Reprinted with permission from Materials Letters [22]. Copyright (2022) Elsevier Publisher.

S. No.	Sensor Principle	Aptamer	Nanomaterials	Recognition element (SARS-CoV-2)	Analytical Sensitivity	Limit of Detection	References
1.	Surface Enhanced Raman Scattering	Biotinylated DNA aptamer	Silver Nanoparticles	Inactivated whole virus	5.5×10^4 to 1.4×10^6 $TCID_{50}/$ mL	5.5×10^4 $TCID_{50}/$ mL	[23]
2.	SLIP- Surface Enhanced Raman Scattering	Thiolated DNA aptamer	Silver Nanoparticles	Spike protein	1 nM to 1 pM	1 fM	[24]
3.	Nanoparticle Surface Energy Transfer Spectroscopy	Rhodamine 6G linked DNA aptamer	Gold Nanostar	Spike protein and virus particles	10-500 virus/mL	130 fg/mL (spike protein), 8 virus/ mL	[25]
4.	Photo-electrochemical Signal	Amino terminal DNA aptamer	Chitosan/graphitic carbon nitride cadmium selenide quantum dots nanocomposite	Receptor binding domain	0.5 to 32 nM	0.12 nM	[26]
5.	Electrochemical Signal	Thiolated DNA aptamer	Gold coated platinum nanoparticles	Nucleocapsid protein	0.025 to 50 ng/ mL	8.33 pg/mL	[27]
6.	Electrophoretic Mobility Shift Assay and Surface Plasmon Resonance	DNA and RNA aptamer	Graphene-oxide	Nucleocapsid protein	1×10^{-9} M	6.25×10^{-19} M	[28]

nM-Nanomolar; pM-Picomolar; fM-Femtomolar; pg-Picogram; fg- Femtogram; M-Molar; mL-Millilitre; $TCID_{50}$-Median tissue culture infectious dose; DNA-Deoxy ribonucleic acid; RNA-Ribonucleic acid.

2.1.1.2 Electrochemical nanosensors based on molecularly imprinted polymers (MIPs)

Polymers have widespread applications not only in solution chemistry [29-35] but also for advanced materials [36-44]. Because of the most recent advances in polymer technology and material science, it is now possible to create highly selective molecularly imprinted polymers (MIPs). A plastic antibody is a MIP that structurally resembles an antibody [45,46]. To prepare MIPs, researchers combined functional monomers, initiators, and cross-linkers with the correct pattern in the appropriate solvent. After synthesis, when the pattern is removed from MIP, it leaves template-specific association spots, allowing the analyte to be selectively bound [47]. Sensors that integrate electrochemical methods with Molecularly Imprinted Technology (MIT) are considered to be a useful approach [48]. When used in conjunction with electrochemical techniques, MIT is thought of as a viable strategy for effective sensing [46].

2.1.1.3 Magnetic nanosensors

Smaller particles, such as mRNA, proteins, enzyme activity, and pathogens, maybe detected using magnetic nanosensors that operate in the range of femtomole (0.5–30 fmol) [49]. Pathogen purification and detection are made easier with the use of magnetic nanoparticles in complex matrices. Fluorescent [50], Raman spectra [51], electrochemical [52], Mass Spectra [53], microcantilever [54], quartz crystal microbalance [55], surface plasmon resonance [56], and many other sensing platforms can be used with magnetic nanoparticles. Researchers have developed a "chemical nose" for infection detection using magnetic nanoparticles. Membranes of bacteria block the q-MNP fluorescent polymer, which results in an array of fluorescence response properties. Different pathogens were quantified using response patterns and linear discriminant analysis [57]. Additionally, a dependable and sensitive quartz-crystal-microbalance biosensor for the identification of marine pathogens were developed based on the augmentation of the response of functionalized magnetic nanoparticles in an exterior magnetic field [58]. Nanocomposites based on QMNP–urease were devised for identifying the pathogen, and the sensor was constructed by of qMNP-Urease noncovalent conjugation. Differential affinity for bacterial cells and the ability to sense the competition for associating to the surfaces of the cell results in a quick dislodgment of urease from the qMNP surface by the nanocomposites.

2.1.2 Nanosensors for the detection of the human immunodeficiency virus (HIV)

One of the most common retroviruses is the human immunodeficiency virus (HIV), which is a lentivirus. After entering the circulation, HIV infects CD41 T cells and begins to multiply. AIDS is the disease that results as a consequence, and it is one of the most concerning public health issues. More than 35 million individuals have been affected, according to a WHO study [59]. HIV-1 and HIV-2 are two distinct classes of the virus, and HIV-1 is the most common type of disease triggering mediators. An electrochemiluminescence (also called electrogenerated chemiluminescence and abbreviated ECL) nanosensor was developed by Babamiri et al. to identify the HIV-1 gene. They observed a considerable upsurge in signal after the process called hybridization. With this sensor they were able to identify a range of 3.0 fM to 0.3 nM. When compared to the non-complementary sequences, they found that their nanosensor had a higher level of selectivity [60]. To identify HIV-1 related glycoprotein 41(Gp41), Lu et al. developed a nanosensor. A synthetic peptide containing 579-613 Gp41 residues was used to modify the surface of the piezoelectric nanosensor. A nanosensor surface was shown to have a high affinity for the target peptide and to be capable of binding only to Gp41. According to their findings, the detection limit is 2 ng/mL [61]. An optical nanosensor for HIV-1 detection was displayed by Shafiee et al. They stated that the target, after being adsorbed, had an impact on the shift with 10 pM resolution and inspected with a range of 10^4 to 10^8 copies/mL [62].

2.1.3 Nanosensors for the detection of Hepatitis B virus

For decades, scientists have known about the hepatitis virus, which may cause both transitory and chronic hepatitis [63]. It is estimated that almost a million people die from complications related to cirrhosis and malignant liver development caused by Hepatitis B virus (HBV) infection every year. Moreover, 15% to 40% of affected individuals will develop liver cirrhosis, hepatic failure, or liver cancer, and sadly 15% to 25% will die [64]. Research published by Hassen et al. investigated the use of electrochemical impedance spectroscopy to identify the presence of the HBV. The first step was to encapsulate magnetic nanoparticles inside a gold electrode, then used to modify DNA probes. After the characterizations, they demonstrated excellent DNA hybridization and immobilization with various amounts of complementary DNA. Furthermore, they showed that this nanosensor could detect HBV DNA at 50 pmol levels and attained a maximum saturation level of 12.65 nM [65]. Besides these, in the human serum, Uzun et al. found a surface antibody of HBV utilizing an optical nanosensor. In the kinetic investigations the scientists used human serum samples that were HBV surface antibody-positive. They computed the value of the nanosensor's detection limit and also disclosed that this nanosensor followed the Langmuir adsorption isotherm model [66]. An electrochemical nanosensor developed by Istek et al. for detecting DNA hybridization that was sequence-selective and linked to the detection of HBV has been described. They carried out the nanosensor's selectivity test in the presence of both the target DNA along with the presence of other DNA sequences. They also determined a 0.86 g/mL detection limit value [67].

2.1.4 Nanosensors for the detection of human papillomavirus (HPV)

Scientists are examining whether there is any linkage of human papillomavirus (HPV) with cervical cancer, the second most common type of malignancy among women. Treating the disease in the primary stage is vital for a successful outcome of cancer treatment. For this reason, early diagnosis is a critical step as it allows the affected person to be taken care of at an initial phase. Therefore, a positive or negative result may be useful, even if it is a binary result. The common screening tool, i.e., the Pap smear, has limited sensitivity and specificity [68,69]. A microfluidic nanosensor for HPV-16 E7 antibody identification was developed by Inan et al. were able to identify a level of 2.87 ng/mL in the blood. They used serum samples to test the nanosensor's accuracy and got far better results than the control samples. Using this nanosensor, they demonstrated that it might be used as a screening tool for HPV-associated malignancies [70]. According to the researchers, a colorimetric nanosensor developed by Teengam et al. can detect synthetic HPV. A DNA-dependent multiplex paper-based nanosensor was developed by the team. Silver nanoparticles with a detection limit of 1.03 nM were used to determine the target concentrations. In contrast to non-complementary DNA targets and mismatches of one or two bases, the nanosensor was highly selective for complementary oligonucleotides [71]. To detect HPV, Peng et al. developed a nanosensor based on 2D nanosheets. These nanosheets reduced the fluorescence of dye-labeled significantly single-stranded DNA and showed distinct binding attractions for both double-stranded and single-stranded DNA. This nanosensor appeared to be highly specific in HPV-18 diagnosis with a detection limit as low as 100 pM [72].

2.1.5 Nanosensors for the detection of Ebola virus

Ebola virus infections are critically dangerous, causing a rapid development to a condition characterized by a shock-like state, multiple organ failure, and bleeding following an incubation period [73]. For quick, point-of-care detection of the virus, such enhanced superiorities (of detection method) would be ideal [74]. The Ebola virus may be detected using a digital nanosensor developed by Natesan et al. with a flow cell test and a smartphone fluorescence reader, their nanosensor collected particular antibodies with microarrayed recombinant antigens. A piece of hardware that is attached onto the backside of a smartphone and provides a user interface to perform the process, interact with the cloud facility, and gather test data was shown. Ebola virus may also be detected using an electrochemical nanosensor developed by Ilkhani et al. A low detection limit was obtained after all the processes were optimized, and the sample was first marked with a conjugate of streptavidin alkaline phosphatase. Finally, they were able to achieve specificity and repeatability [75]. A nanosensor developed by Yanik et al. can identify entire viruses. Group-specific antibodies and a light transmission impact were employed to build their nanosensors. They applied the nanosensor in a range that was spanning in between three orders of magnitude. Ebola glycoprotein-specific antibodies were altered, and the transmission spectra were recorded [76].

2.1.6 Zika virus detection with nanosensors

Studies on how to develop more straightforward diagnostic tests have been necessitated by the Zika virus outbreak in the Western Hemisphere [77]. The Zika virus may be detected using a graphene-based nanosensor developed by Afsahi et al. were able to quantify Zika viral antigen levels were as low as 450 pM. By identifying the antigen of the Zika virus in human serum, these nanosensors demonstrated a potentially suitable diagnostic technique. They are selective in nature which was confirmed by using Japanese encephalitis NS1 [78].

The Zika virus electrochemical nanosensor was developed by Kaushik et al. It was discovered that by using electrochemical impedance spectroscopy, the nanosensor can detect Zika virus at concentrations as low as 10 pM and as high as 1 nM, and the identification limit is less than 10 pM [79]. To detect the Zika virus a method was developed by Song et al. which is based on isothermal amplification mediated by a reverse transcription loop. This nanosensor, in an actual sample, can detect the Zika virus with a sensitivity of 5 pfu, according to the researchers [80].

2.1.7 Nanosensors for the detection of influenza virus

Influenza is a transmittable disease which is considered as a leading cause of several clinical problems and significant economic disruptions [81]. For in-field detection of influenza virus, performances of traditional approaches are inadequate as they are difficult and time consuming. For that reason, researchers are required to develop effective alternative methods that are more efficient and less time-consuming [82]. In order to detect influenza, Tam et al. presented a paper describing the immobilization of DNA using carbon nanotubes. They used a nanosensor surface and altered a DNA probe on it and characterized the basic interaction. A 0.5 nM detection limit was established for the hybridization of the DNA probe with the target DNA [83]. In order to identify influenza viruses, Vollmer et al. investigated an optical technique. Single virions were bound to a whispering-gallery mode by changing the resonance frequency of a single viron. They used virus-sized nanoparticles to assist the sensing mechanism [84]. A portable optical nanosensor based on an aptamer was developed by Bai et al. to detect the H5N1 strain of avian influenza virus. An increase in the refraction index may be attributed to the immobilized aptamers captured by AIV H5N1. Refraction index values were shown to be directly related to AIV concentration after the optimization of experimental conditions [85]. Piezoelectric and optical nanosensors were developed by Emir Diltemiz et al. to detect influenza virus hemagglutinin, the virus primary protein. In order to bind the sugar, they used 4-aminophenylboronic acid. They altered the surfaces of the nanosensor with thiol groups and subsequently immobilized 4-aminophenylboronic acid and sialic acid. Also, the detection limits for piezoelectric and optical nanosensors were found to be 4.7×10^{22} and 1.28×10^{21} µM, respectively [86].

2.1.8 Nanosensors for the detection of viruses other than those mentioned in this section

An optical nanosensor and microfluidic systems were used in research by Jin et al. to develop a viral diagnostic tool for the detection of human adenovirus. It took only 30 minutes for the researchers to detect 10 copies of adenovirus. The medical usefulness of this nanosensor has been validated, and it has been deemed to have a sensitive and fast detection with cheap price, short detection time, and little complexity [87]. Guerreiro et al. showed a genetically encoded switch-on fluorescent nanosensor based on a cyclized green fluorescent protein (cVisensor) and a site cleavable with adenoviral protease. Nanosensor optimization and viral detection in mammalian cells were then used to monitor adenovirus infection in living cells. cVisensor cells were used in a flow cytometry-based experiment 48 hours after infection, which revealed an infectious particle concentration of less than 10^5 infectious particles/mL [88]. For the detection of the Middle East respiratory syndrome (MERS) coronavirus, Kim et al. developed a colorimetric nanosensor based on different dsDNA-shielded gold nanoparticles. Using their nanosensor, they were able to confirm the presence of viral molecules and the color changes of gold nanoparticles in the UV-vis wavelength range. A 30 bp Middle East respiratory syndrome coronavirus, was shown to be distinguishable using a colorimetric nanosensor with an accuracy of 1 pmol/µL [89]. Optic nanosensors have been studied by Trzaskowska et al. to detect tuberculosis (TB) in sputum. After designing a portable surface plasmon resonance apparatus, they measured the reaction to TB secretory protein to compare it to a typical desktop surface plasmon resonance platform. In addition to detecting *M. tuberculosis* secretory protein, they claimed that their nanosensor was able to distinguish between cultures of TB bacteria at a concentration of 1×10^4 CFU/mL without interference from either of the other two bacterial species [90]. For the detection of infective murine norovirus, Weerathunge et al. developed a colorimetric nanosensor. Incorporating enzyme-mimic catalytic gold nanoparticles with a murine viral aptamer led to discovery of a blue hue in the norovirus. [91].

2.2 The use of nanosensors in the detection of bacteria

When it comes to public health, successful pathogen identification is essential since antibiotic-resistant bacteria are posing a greater danger to public health than ever before. There is a growing need for improved detection techniques that are easier to use, quicker to perform, and more reliable so that patients can get the most effective treatment. Different types of nanosensor-based detection methodology have been developed for the detection of bacteria. Some of them are discussed below.

2.2.1 Sensors based on nanoparticles for bacterial disease detection

2.2.1.1 Application of nanosensors for the detection of multiple anti-microbial resistant pathogens (SERS detection)

Multiple Anti-microbial Resistant Pathogens can be detected using nanosensor-based Surface-enhanced Raman scattering (**SERS**). Due to its high sensitivity, selectivity, and multiplexing capabilities, surface-enhanced **SERS** has become a popular analytical method for biosensing. For the isolation and detection of various bacterial infections via magnetic separation and **SERS**, a novel bionanosensor has been designed. *E. coli, Salmonella typhimurium*, and methicillin-resistant *Staphylococcus aureus* was effectively isolated and identified, with the lowest concentration for each of the organisms discovered at 10^1 colony-forming units per mL (CFU/mL). Bionanosensors may be used to identify bacterial infections both individually and in multiplex systems, which might lead to new point-of-care diagnostic devices, as well as breakthroughs in biomedical applications [92].

2.2.1.2 Multiplexed nanosensors for remote near-infrared (NIR) detection of bacteria

A set of NIR fluorescence nanosensors have been developed and utilized for distant fingerprinting of therapeutically significant microorganisms by researchers. These nanosensors are based on single-walled carbon nanotubes (SWCNTs) and provide ultra-low background and excellent tissue penetration in the NIR optical tissue transparency window. Functional hydrogel arrays with 9 distinct sensors are chemically tuned to detect the release of metabolic and particular virulence factors like siderophores, lipopolysaccharides, proteases, and DNAses. 6 significant clinical isolates of bacteria (*Staphylococcus aureus, Escherichia coli...*) are treated with these hydrogels and distant (≥25 cm) NIR imaging enables bacterium identification and distinction. It is possible to distinguish between the two primary pathogens, P. aeruginosa as well as S. aureus by using sensors encoded with spectral information (900 nm, 1000 nm, and 1250 nm). By using NIR fluorescent nanosensors in this form of multiplexing, remote detection and differentiation of significant infections are possible [93].

2.2.1.3 Nanosensors for the detection of protease and engineered phage-infected bacteria

To detect tobacco etch virus (TEV) protease and bacteria infected with a modified bacteriophage, and researchers built a peptide-graphene oxide nanosensor. They employed graphene oxide (GO) as an absorbent while simultaneously suppressing fluorescence from a peptide with the following sequence: RKRFRENLYFQSCP. TEV protease has a detection limit of 51 ng/μL based on fluorescence intensity. In addition, the investigators have been able to identify germs using the sensor system. Target bacteria (*E. coli*) were infected using TEV protease genes carrying engineered bacteriophages, which resulted in its (TEV protease genes) translation and release. A concentration of 10^4 CFU/mL of bacteria may be estimated using this method. There is the possibility for this method to be used as a multiplex detection platform for numerous bacterial species [94].

2.2.1.4 Use of plasmonic nanosensors for the identification of urease positive bacteria

The microorganism responsible for an illness must be identified before anti-microbial therapy may be tailored to the individual patient. Protease, oxidoreductase and lipase enzymes, as well as other bacterial enzymes, maybe detected at the point of care (POC). Santopolo et al. described a novel approach for promptly and accurately identifying urease-producing bacteria at ultra-low concentrations. Long bacterial growth processes are replaced by a 10-minute magnetic bead capture operation in this technology. An enzyme-substrate is added to the nanoparticle suspensions, and the presence of urease-positive bacteria is detected using a plasmonic signal generating the process that generates blue-or red-colored mixtures. The NH3-dependent the assemblage of gold nanoparticles in the presence of bovine serum albumin (BSA) generates these visually discernible colorimetric signals. Even in the presence of a significant number of urease-negative bacteria, the suggested approach can identify *Proteus mirabilis* with a limit of detection of 10^1 cells/mL and an assay duration of 40 minutes. The *P. mirabilis* can be detected at clinically significant levels within minutes, making it is suited for point-of-care testing of urease-positive pathogens [95].

2.2.1.5 Use of nanosensors for the detection of *Escherichia coli* O157:H7

Enterohemorrhagic *Escherichia coli* O157:H7 is one of the major causes of numerous food- and waterborne diseases and thus puts a severe threat to public health and sanitation. Several traditional approaches for identifying bacteria, like PCR, ELISA, and fluorescent immunoassays. However, while these methods are very sensitive and specific, they are time-consuming and need expensive equipment. It is also common for these procedures to require substantial sample preparation and long readout durations. Banerjee et al. presented a diagnostic method based on multiparametric magneto-fluorescent nanosensors (MFnS) which are not only more straightforward but also more sensitive. The nanosensors can detect bacterial contamination at concentrations as low as 1 colony-forming unit (CFU) using a combination of magnetic relaxation and fluorescence measurements. For sensitive screening at low target CFUs, these MFnS may be augmented with fluorescence measurements of larger CFU samples. Combining all of these characteristics allows the identification and measurement of broad-spectrum contaminants, from aquatic reservoirs to commercially manufactured foods [96].

2.2.2 Nanosensors for detection of pathogenic bacteria in water

It is extensively known that the drinking water contains a wide variety of harmful microorganisms. It is possible to reduce the risk of drinking water poisoning by learning more about monitoring and detection systems. Recently pathogen detection in water has received more attention. A variety of approaches exist to identify harmful microorganisms in water with their own merits and downsides. These include both culture-based and biochemical approaches. However, all these procedures share similar drawbacks: they are labor-intensive and time-consuming. Because they are usually qualitative, they can't be used to make treatment decisions (Li and Gu 2011). Enzyme-Linked Immunosorbent Assay, as well as PCR (Navas et al., 2006) and other immunological procedures, can only be performed in a laboratory setting (Crowther 1995). The use of nanomaterials to detect DNA is an exciting alternative to traditional methods. The use of metallic nanoparticles conjugated DNA probes instead of fluorescently labeled DNA probes is promising since it reduces the need for expensive and complex apparatus. Nanomaterials have a wide range of physical and chemical properties that make them attractive probes. Because of their compact size, they have a large surface area to volume ratio which facilitates them to attach with nucleic acids. When suspended in water, gold nanoparticles exhibit a red color due to surface plasmon resonance-induced optical absorption at 525 nm. A single-stranded DNA, which is complementary to the target DNA, is attached with the gold nanoparticles to generate the probe. The gold nanoparticle probes that have been manufactured are then allowed to hybridize with the target DNA sequence. A polymer network is formed as a result of hybridization. As a consequence of the constricting network, the conjugated gold nanoparticles formed aggregates and changed color from red to purple. Only tiny synthetic oligonucleotides (up to 20–30 bp) targets are being tracked in the aforementioned reports, which again do not reflect sequences of pathogenic DNA or PCR products. A variety of gold nanoparticle based lateral flow immunoassays for the detection of pathogens have been published in attempt to improve target detection performance, although most of those techniques are not used in reality. Aptamers have shown to be quite beneficial in the detecting process. When aptamers are attached to gold nanoparticles, they have a distinctive capability of detecting the entire target bacterium without separating its DNA [97].

2.2.3 Food pathogen detection with nanosensors

Pathogenic bacteria that cause food poisoning are considered to be a major threat to public health because of their capability to survive on food, rapidly increase in number, and the risk of cross-contamination. Microbiological contamination maybe it happened at any point in the food supply chain, from primary production to processing to distribution to retail to the consumer's storage and preparation of meals. Therefore, for food processing industry the identification of pathogens in food processing facilities or food material, nutrient measurement, and food safety alerts is essential. In food microbiology for the detection of harmful pathogens, nanosensors or nanobiosensors are being used successfully. Nanosensors have been adapted for food, taste, water, and healthcare applications, among other uses (Pathakoti et al., 2017; Singh et al., 2017). In nanobiosensing technology, a low-priced nano bioluminescent spray is used to detect pathogens. The pathogens react with the bioluminescent spray and produce visible light, indicating contamination. Nano cantilevers can detect pathogens, and a newly designed fast biosensor can detect the development of Escherichia coli bacteria (Pathakoti et al., 2017). Nanoparticles for active packagings, such as nanosilver, nano-gold, titanium-oxide, and zinc-oxide, are thought to release into the food and react with bacteria (Garca et al. 2010; Singh et al. 2017; Thakur and Ragavan 2013). However, it is not clear yet how these nanoparticles interact with food [98].

2.2.4 Monitoring of food quality and safety using nanosensors

Advances in nanotechnology, particularly in the food business, have the potential to have a significant impact on society. The emerging field of developing intelligent food packing systems that will focus on food safety and security is expected to grow exponentially soon. Nanosensor advancements, nanosensor integration into food containers, and intelligent packaging breakthroughs are all critical to the future of food security. This novel packaging method may aid detecting, tracking, monitoring, recording, and the collaboration throughout the supply chain. As a result of the connectivity of nanosensors, wireless nanosensor networks will have a considerable influence on practically every aspect of our everyday lives and will revolutionize the world around us [99].

Emerging Applications of Nanomaterials   Materials Research Forum LLC
Materials Research Foundations 141 (2023) 246-269   https://doi.org/10.21741/9781644902295-10

2.2.5 Nanosensors for the detection of plant pathogens

Plant the disease is one of the key contributors to restraining crop productivity. To control a plant disease, both in a greenhouse or field, the crucial first step is the precise detection of the pathogen. Current methods, for instance, the Quantitative Polymerase Chain Reaction, require a substantial amount of target tissue and a series of tests to effectively detect plant pathogens. Conventional diagnostic procedures have a common drawback: they are not highly sensitive and take a longer time. As a result, strategies for improving the accuracy and swiftness of plant-pathogen diagnosis at a lower cost are required. Fast and precise detection of a specific biological marker is now possible because of the emergence of nanotechnology, nanoparticles, and quantum dots (QDs). Pathogen detection may be improved by using biosensors, nanoimaging, nanostructured platforms, QDs, and nanopore DNA sequencing techniques to improve sensitivity, specificity, speed, and cost-effectiveness. To further aid farmers in preventing epidemic diseases, nanodiagnostic kit technology can swiftly and inexpensively identify possible dangerous plant pathogens. It is possible to construct a suitable integrated disease management system based on this technology, which might alter crop environments to control crop pathogens. Additionally, microfluidic devices in combination with nanotechnology may be successfully used to identify particular pathogens and poisons in molecular plant pathology. For instance, micro-PCR can run 40 PCR cycles in less than six minutes. Smart agricultural systems might be built using nanoscale devices with unique features soon. One possible use of these nanodevices is to detect plant disease before the farmer is even aware of it, which would allow for early intervention. Devices of this kind may be able to react to certain events, detect the issue, and take a suitable disease management measure. Thus, nanosmart technologies would function both as a defensive and a means of early warning. Nanodevices that can do thousands of measurements quickly and affordably will be accessible in the next decade. The downsizing of biochip technology to the nanoscale level is expected to continue in plant disease diagnostics in the future [100].

Conclusion and future outlook

In the last decade, there have been a lot of significant advances in nanotechnology. Reduced need for sample augmentation and culture time has been achieved by the use of methods like magnetic resonance, fluorescence, and colorimetric analysis, amongst many others. They all aim to identify trace pathogen amounts promptly, no matter how different their processes and designs are [101]. Additionally, nanosensors have a variety of capabilities, including practicable maneuver, remarkable performance, quick reaction time, enhanced level of specificity and sensitivity, compact size for transportability and real-time analysis [102]. For signal amplification studies, researchers use nanocomposites including nanofilm [103], quantum dot [104], or gold nanoparticle [105] for the improvement of the specificity and sensitivity of the approaches. Many nanobiosensors have already been produced, but the objective of affordable, high speed, multiplexed medical diagnostic lab-on-a-chip tools has yet to be achieved. Despite the fact that, there is still more work to be done, pathogens and illnesses will soon be detected like never before, thanks to these sensors [106]. Eventually, sensor developers will be able to make nanosensors that are more intelligent, achieve significant improvements in sensing performance, and have real-world applications by solving all of these problems. Several new technologies, including Artificial Intelligence (AI), machine learning, and the Internet of Things (IoT) could be used to make sensors smarter. Large amounts of data may be collected with these techniques, which can be analyzed in real-time. Every area of sensor performance is going to get a boost. The processes of analyte-recognition component interactions will be extensively explained. Therefore, more highly specialized components (natural or synthetic) with suitable dimensions will be used for target identification. It is also possible that additional research into other kinds of nanomaterials and their uses in optical sensing might help the further enhancement of the sensitivity of the sensors. Last but not least, the new sensors will be far more practical in the field. Standard yet irreducible preparation methods will be investigated in order to reduce the effect of external factors on the device performance while still being user-friendly. The future of complex test specimens monitoring, particularly food supplies, will include multi-target sensors. By integrating indirect colorimetric sensing with the red-to-blue spectrum, sensor designers may be able to add more arrays of colors to the system. Fluorescence sensors may be able to identify many objects simultaneously by using the individual fluorophores' distinct hues. Robotic systems should be used to synthesize nanoparticles in terms of material production. Thus, nanosensor production will be more efficient, critical to achieving high practicability in the long term. The practical uses of nanosensors are still quite restricted, even after decades of research. For that reason, the actual application of these findings requires further efforts going forward [107].

References

[1] A.A. Malik, C. Nantasenamat, T. Piacham, Molecularly Imprinted Polymer for human viral pathogen detection, Mater. Sci. Eng. C. 77 (2017) 1341-1348. https://doi.org/10.1016/j.msec.2017.03.209

[2] I. Kamal, Prospects of Some Applications of Engineered Nanomaterials: A review, Open Access J. Biomed. Engin. Biosci. 2 (2018) 245. https://doi.org/10.32474/OAJBEB.2018.02.000149

[3] M. Owusu, B. Nkrumah, G. Acheampong, E.K. Mensah, A.A-K. Komei, F.K. Sroda, S. David, S. Emery, L.M. Robinson, K. Asante, D. Opare, Improved detection of microbiological pathogens: role of partner and non-governmental organizations. BMC Infect Dis 21 (2021) 303. (2021) https://doi.org/10.1186/s12879-021-05999-8

[4] C.E. Rowland, C.W. Brown III, J.B. Delehanty and I.L. Medintz, Nanomaterial-based sensors for the detection of biological threat agents, Mater. Today 19(8) (2016) 464-477. https://doi.org/10.1016/j.mattod.2016.02.018

[5] Q. Zhang, D. Zhang, Y. Lu, Y. Yao, S. Li, Q. Liu, Graphene oxide-based optical biosensor functionalized with peptides for explosive detection, Biosens. Bioelectron. 68 (2015) 494-499. https://doi.org/10.1016/j.bios.2015.01.040

[6] K. Bartold, A. Pietrzyk-Le, K. Golebiewska, W. Lisowski, S. Cauteruccio, E. Licandro, F. D'Souza, W. Kutner, Oligonucleotide determination via peptide nucleic acid macromolecular imprinting in an electropolymerized CG-rich artificial oligomer analogue, ACS Appl. Mater. Interfaces 10 (2018) 27562-27569. https://doi.org/10.1021/acsami.8b09296

[7] B. Osman, L. Uzun, N. Be̦ sirli, A. Denizli, Microcontact imprinted surface plasmon resonance sensor for myoglobin detection, Mater. Sci. Eng. C 33 (2013) 3609-3614. https://doi.org/10.1016/j.msec.2013.04.041

[8] U. Anik, Y. Tepeli, M.F. Diouani, Fabrication of electrochemical model influenza a virus biosensor based on the measurements of neuroaminidase enzyme activity, Anal. Chem. 88 (2016) 6151-6153. https://doi.org/10.1021/acs.analchem.6b01720

[9] Oˮ. Erdem, Y. Saylan, N. Cihangir, A. Denizli, Molecularly imprinted nanoparticles based plasmonic sensors for real-time Enterococcus faecalis detection, Biosens. Bioelectron. 126 (2019) 608-614. https://doi.org/10.1016/j.bios.2018.11.030

[10] F.C. Dudak, I. H. Boyaci, Peptide-based surface plasmon resonance biosensor for detection of staphylococcal enterotoxin B, Food Anal. Methods 7 (2014) 506-511. https://doi.org/10.1007/s12161-013-9739-9

[11] S. Cheng, S. Hideshima, S. Kuroiwa, T. Nakanishi, T. Osaka, Label-free detection of tumor markers using field effect transistor (FET)-based biosensors for lung cancer diagnosis, Sens. Actuators B: Chem. 212 (2015) 329-334. https://doi.org/10.1016/j.snb.2015.02.038

[12] Y. Saylan, A. Denizli, Molecular fingerprints of hemoglobin on a nanofilm chip, Sensors 18 (2018) 3016-3029. https://doi.org/10.3390/s18093016

[13] M. Lv, Y. Liu, J. Geng, X. Kou, Z. Xin, D. Yang, Engineering nanomaterials-based biosensors for food safety detection, Biosens. Bioelectron.106 (2018) 122-128. https://doi.org/10.1016/j.bios.2018.01.049

[14] Oˮ. Erdem, Y. Saylan, M. Andac̦, A. Denizli, Molecularly imprinted polymers for removal of metal ions: an alternative treatment method, Biomimetics 3 (2018) 38-53. https://doi.org/10.3390/biomimetics3040038

[15] C. Mao, A. Liu, B. Cao, Virus-based chemical and biological sensing, Angew. Chem. Int. Ed. 48 (2009) 6790-6810. https://doi.org/10.1002/anie.200900231

[16] J. Deng, S. Zhao, Y. Liu, C. Liu, J. Sun, Nanosensors for diagnosis of infectious diseases, ACS Appl. Bio Mater. 4 (5) (2021) 3863-3879. https://doi.org/10.1021/acsabm.0c01247

[17] S.D. Mahapatra, P.C. Mohapatra, A.I. Aria, G. Christie, Y.K. Mishra, S. Hofmann, V. K. Thakur, Piezoelectric Materials for Energy Harvesting and Sensing Applications: Roadmap for Future Smart Materials, Adv. Sci. 8 (2021) 2100864. https://doi.org/10.1002/advs.202100864

[18] B. Ates, S. Koytepe, A. Ulu, C. Gurses, V.K. Thakur, Chemistry, structures, and advanced applications of nanocomposites from biorenewable resources, Chem. Rev. 120 (17) (2020) 9304-9362. https://doi.org/10.1021/acs.chemrev.9b00553

[19] M. Saraf, M.T. Yaraki, Prateek, Y.N. Tan, R.K. Gupta, Insights and perspectives regarding nanostructured fluorescent materials toward tackling COVID-19 and future pandemics, ACS Appl. Nano Mater. 4 (2) (2021) 911-948. https://doi.org/10.1021/acsanm.0c02945

[20] A. Singhal, A. Parihar, N. Kumar, R. Khan, High throughput molecularly imprinted polymers based electrochemical nanosensors for point-of-care diagnostics of COVID-19, Materials Letters 306 (2022) 130898. https://doi.org/10.1016/j.matlet.2021.130898

[21] X.T. Zheng, Y.N. Tan, Recent development of nucleic acid nanosensors to detect sequence-specific binding interactions: From metal ions, small molecules to proteins and pathogens, Sensors Int. 1 (2020), 100034 https://doi.org/10.1016/j.sintl.2020.100034

[22] S. Krishnan, A.K. Narasimhan, D. Gangodkar, S. Dhanasekaran, N.K. Jha, K. Dua, V.K. Thakur, P.K. Gupta, Aptameric nanobiosensors for the diagnosis of COVID-19: An update, Materials Letters 308(B) (2022) 131-237. https://doi.org/10.1016/j.matlet.2021.131237

[23] E. Zavyalova, O. Ambartsumyan, G. Zhdanov, D. Gribanyov, V. Gushchin, A. Tkachuk, E. Rudakova, M. Nikiforova, N. Kuznetsova, L. Popova, B. Verdiev, SERS-Based Aptasensor for Rapid Quantitative Detection of SARS-CoV-2, Nanomaterials 11 (2021) 1394. https://doi.org/10.3390/nano11061394

[24] T. Stanborough, F.M. Given, B. Koch, C.R. Sheen, A.B. Stowers-Hull, M.R. Waterland, D.L. Crittenden, Optical Detection of CoV-SARS-2 Viral Proteins to Sub-Picomolar Concentrations, ACS Omega 6 (9) (2021) 6404-6413. https://doi.org/10.1021/acsomega.1c00008

[25] A. Pramanik, Y. Gao, S. Patibandla, D. Mitra, M.G. McCandless, L.A. Fassero, K. Gates, R. Tandon, P.C. Ray, Aptamer Conjugated Gold Nanostar-Based Distance Dependent Nanoparticle Surface Energy Transfer Spectroscopy for Ultrasensitive Detection and Inactivation of Corona Virus, J. Phys. Chem. Lett. 12 (8) (2021) 2166-2171. https://doi.org/10.1021/acs.jpclett.0c03570

[26] M.A. Tabrizi, L. Nazari, P. Acedo, A photo-electrochemical aptasensor for the determination of severe acute respiratory syndrome coronavirus 2 receptorbinding domain by using graphitic carbon nitride-cadmium sulfide quantum dots nanocomposite, Sens. Actuators B Chem. 345 (2021), 130377. https://doi.org/10.1016/j.snb.2021.130377

[27] J. Tian, Z. Liang, O. Hu, Q. He, D. Sun, Z. Chen, An electrochemical dual-aptamer biosensor based on metal-organic frameworks MIL-53 decorated with Au@ Pt nanoparticles and enzymes for detection of COVID-19 nucleocapsid protein, Electrochim. Acta 387 (2021) 138553. https://doi.org/10.1016/j.electacta.2021.138553

[28] H. Jia, A.o. Zhang, Y. Yang, Y. Cui, J. Xu, H. Jiang, S. Tao, D. Zhang, H. Zeng, Z. Hou, J. Feng, A graphene oxide coated tapered microfiber acting as a supersensor for rapid detection of SARS-CoV-2, Lab Chip 21 (12) (2021) 2398-2406. https://doi.org/10.1039/D0LC01231A

[29] R, Miyazaki, M, Nargis, A.B. Ihsan, N. Nakajima, M. Hamada, Y. Koyama, Effects of glycon and temperature on self-assembly behaviors of α-galactosyl ceramide in water. Langmuir 37 (2021) 7936-7944. https://doi.org/10.1021/acs.langmuir.1c00545

[30] M. Nargis, A.B. Ihsan, Y. Koyama, Thermo-responsive structure and dye-encapsulation of micelles comprising bolaamphiphilic quercetin polyglycoside. Langmuir 36 (2020) 10764-10771. https://doi.org/10.1021/acs.langmuir.0c01564

[31] M. Nargis, A.B. Ihsan, Y. Koyama, Bolaamphiphilic properties and pH-dependent micellization of quercetin polyglycoside. RSC Adv. 9 (2019) 33674-33677. https://doi.org/10.1039/C9RA05711K

[32] A.B. Ihsan, M. Nargis, Y. Koyama, Effects of the hydrophilic-lipophilic balance of alternating peptides on self-assembly and thermo-responsive behaviors. Int. J. Mol. Sci. 20 (2019) 4604-4614. https://doi.org/10.3390/ijms20184604

[33] A.B. Ihsan, Y. Koyama, Impact of polypeptide sequence on thermal properties for diblock, random, and alternating copolymers containing a stoichiometric mixture of glycine and valine. Polymer 161 (2019) 197-204. https://doi.org/10.1016/j.polymer.2018.12.021

[34] A.B. Ihsan, Y. Koyama, T, Taira, T. Imura, Thermo-responsive structure and surface activity of kinetically stabilized micelle composed of fluorinated alternating peptides in orgainic solvent. Chemistry Select 3 (2018) 4173-4178. https://doi.org/10.1002/slct.201800590

[35] Y. Koyama, A.B. Ihsan, T. Taira, T. Imura, Fluorinated polymer surfactants bearing alternating peptide skeleton prepared by three-component polycondensation. RSC Adv 8 (2018) 7509-7513. https://doi.org/10.1039/C8RA00581H

[36] A.B. Ihsan, T.L. Sun, S. Kuroda, M.A. Haque, T. Kurokawa, T. Nakajima, J.P. Gong, A phase diagram of neutral polyampholyte - From solution to tough hydrogel. J. Mat. Chem. B 1 (2013) 4555-4562. https://doi.org/10.1039/c3tb20790k

[37] T.L. Sun, T. Kurokawa, S. Kuroda, A.B. Ihsan, T. Akasaki, K. Sato, M.A. Haque, T. Nakajima, J.P. Gong, Physical Hydrogels Composed of Polyampholytes Demonstrate High Toughness and Viscoelasticity. Nat. Mater. 12 (2013) 932-937. https://doi.org/10.1038/nmat3713

[38] A.B. Ihsan, T.L. Sun, T. Kurokawa, S.N. Karobi, T. Nakajima, T. Nonoyama, C.K. Roy, F. Luo, J.P. Gong, Self-healing behaviors of tough polyampholyte hydrogels. Macromolecules 49 (2016) 4245-4252. https://doi.org/10.1021/acs.macromol.6b00437

[39] F. Luo, T.L Sun, T. Nakajima, T. Kurokawa, Y. Zhao, K. Sato, A.B. Ihsan, X. Li, H. Guo, J.P Gong, oppositely charged polyelectrolytes form tough, self-healing and

Materials Research Forum LLC
https://doi.org/10.21741/9781644902295-10

rebuildable hydrogels. Adv. Mater. 27 (2015) 2722-2727.
https://doi.org/10.1002/adma.201500140

[40] F. Luo, T.L. Sun, T. Nakajima, T. Kurokawa, A.B. Ihsan, X. Li, H. Guo, J.P. Gong, Free reprocessability of tough and self-healing hydrogels based on polyion complex, ACS Macro. Lett. 4 (2015) 961-964. https://doi.org/10.1021/acsmacrolett.5b00501

[41] F. Luo, T.L. Sun, T. Nakajima, D.R. King, T. Kurokawa, Y. Zhao, A.B. Ihsan, X. Li, H. Guo, J.P. Gong, Strong and tough polyion-complex hydrogels from oppositely charged polyelectrolytes: A comparative study with polyampholyte hydrogels. Macromolecules 49 (2016) 2750-2760. https://doi.org/10.1021/acs.macromol.6b00235

[42] F. Luo, T.L. Sun, T. Nakajima, T. Kurokawa, Y. Zhao, A.B. Ihsan, H. Guo, X. Li, J. P. Gong Crack blunting and advancing behaviors of tough and self-healing polyampholyte hydrogel, Macromolecules 47 (2014) 6037-6046. https://doi.org/10.1021/ma5009447

[43] C.K. Roy, H. Guo, T.L. Sun, A.B. Ihsan, T. Kurokawa, M. Takahata, T. Nonoyama, T. Nakajima, J.P. Gong, Self-adjustable adhesion of polyampholyte hydrogels, Adv. Mater. 27 (2015) 7344-7348. https://doi.org/10.1002/adma.201504059

[44] A.B. Ihsan, Y. Tawara, S. Goto, H. Kobayashi, K. Nakajima, A. Fukuoka, Y. Koyama, Effects of 2,5-Furanylene Sulfides in Polymer Main Chain on Polymer Physical Properties, Polymer Journal 51 (2019) 413-422. https://doi.org/10.1038/s41428-018-0140-9

[45] S. Ramanavicius, A. Jagminas, A. Ramanavicius, Advances in Molecularly Imprinted Polymers Based Affinity Sensors (Review), Polymers 13 (2021) 6. https://doi.org/10.3390/polym13060974

[46] R. Gui, H. Jin, H. Guo, Z. Wang, Recent advances and future prospects in molecularly imprinted polymers-based electrochemical biosensors, Biosens. Bioelectron. 100 (2018) 56-70. https://doi.org/10.1016/j.bios.2017.08.058

[47] M. Singh, S. Singh, S.P. Singh, S.S. Patel, Recent advancement of carbon nanomaterials engrained molecular imprinted polymer for environmental matrix, Trends Environ. Anal. Chem. 27 (2020) e00092. https://doi.org/10.1016/j.teac.2020.e00092

[48] F.W. Scheller, X. Zhang, A. Yarman, U. Wollenberger, R.E. Gyurcs'anyi, Molecularly imprinted polymer based electrochemical sensors for Biopolymers, Curr. Opin. Electrochem. 14 (2019) 53-59. https://doi.org/10.1016/j.coelec.2018.12.005

[49] B. Jin, S. Wang, M. Lin, Y. Jin, S. Zhang, X. Cui, Y. Gong, A. Li, F. Xu, T.J. Lu, Upconversion nanoparticles based FRET aptasensor for rapid and ultrasenstive bacteria detection, Biosens. Bioelectron. 90 (2017) 525-533. https://doi.org/10.1016/j.bios.2016.10.029

[50] M. Safavieh, M.U. Ahmed, A. Ng, M. Zourob, High-throughput real-time electrochemical monitoring of LAMP for pathogenic bacteria detection, Biosens. Bioelectron. 58 (2014) 101-106. https://doi.org/10.1016/j.bios.2014.02.002

[51] Y. Liu, H. Zhou, Z. Hu, G. Yu, D. Yang, J. Zhao, Label and label-free based surface-enhanced Raman scattering for pathogen bacteria detection: A review, Biosens. Bioelectron, 94 (2017) 131-140. https://doi.org/10.1016/j.bios.2017.02.032

[52] Y.-S. Lin, P.-J. Tsai, M.-F. Weng, Y.-C. Chen, Anal. Chem. Affinity Capture Using Vancomycin-Bound Magnetic Nanoparticles for the MALDI-MS Analysis of Bacteria, 77 (2005) 1753-1760. https://doi.org/10.1021/ac048990k

[53] S.D. Soelberg, R.C. Stevens, A.P. Limaye, C.E. Furlong, Surface plasmon resonance detection using antibody-linked magnetic nanoparticles for analyte capture, purification, concentration, and signal amplification, Anal. Chem. 81 (2009) 2357-2363. https://doi.org/10.1021/ac900007c

[54] M. Su, S. Li, V.P. Dravid, Microcantilever resonance-based DNA detection with nanoparticle probes, Appl. Phys. Lett. 82 (2003) 3562-3564. https://doi.org/10.1063/1.1576915

[55] Y. Pan, M. Guo, Z. Nie, Y. Huang, C. Pan, K. Zeng, Y. Zhang, S. Yao, Selective collection and detection of leukemia cells on a magnet-quartz crystal microbalance system using aptamer-conjugated magnetic beads. Biosens. Bioelectron. 25 (2010) 1609-1614. https://doi.org/10.1016/j.bios.2009.11.022

[56] Y. Wan, Y. Sun, P. Qi, P. Wang, D. Zhang, Quaternized magnetic nanoparticles-fluorescent polymer system for detection and identification of bacteria, Biosens. Bioelectron. 55 (2014) 289-293. https://doi.org/10.1016/j.bios.2013.11.080

[57] Y. Wan, D. Zhang, B. Hou, Determination of sulphate-reducing bacteria based on vancomycin functionalised magnetic nanoparticles using a modification-free quartz crystal microbalance, Biosens. Bioelectron. 25 (2010) 1847-1850. https://doi.org/10.1016/j.bios.2009.12.028

[58] S.P. Ravindranath, L.J. Mauer, C. Deb-Roy, J. Irudayaraj, Biofunctionalized magnetic nanoparticle integrated mid-infrared pathogen sensor for food matrixes, Anal. Chem. 81 (8) (2009) 2840-2846. https://doi.org/10.1021/ac802158y

[59] https://www.who.int/en/news-room/fact-sheets/detail/hiv-aids.

[60] B. Babamiri, A. Salimi, R. Hallaj, A molecularly imprinted electrochemiluminescence sensor for ultrasensitive HIV-1 gene detection using EuS nanocrystals as luminophore, Biosens. Bioelectron. 117 (2018) 332-339. https://doi.org/10.1016/j.bios.2018.06.003

[61] C.H. Lu, Y. Zhang, S.F. Tang, Z.B. Fang, H.H. Yang, X. Chen, G. N. Chen, Sensing HIV related protein using epitope imprinted hydrophilic polymer coated quartz crystal microbalance, Biosens. Bioelectron. 31 (1) (2012) 439-444. https://doi.org/10.1016/j.bios.2011.11.008

[62] H. Shafiee, E.A. Lidstone, M. Jahangir, F. Inci, E. Hanhauser, T.J. Henrich, D.R. Kuritzkes, B.T. Cunningham, U. Demirci, Nanostructured optical photonic crystal biosensor for HIV viral load measurement, Sci. Rep. 4 (2014) 4116-4123. https://doi.org/10.1038/srep04116

[63] C. Seeger, W.S. Mason, Molecular biology of hepatitis B virus infection, Virology 479 (2015) 672-686. https://doi.org/10.1016/j.virol.2015.02.031

[64] D. Lavanchy, M. Kane, Global epidemiology of hepatitis B virus infection, Hepatitis B Virus in Human Diseases, Humana Press, Cham. (2016) pp. 187-203. https://doi.org/10.1007/978-3-319-22330-8_9

[65] W.M. Hassen, C. Chaix, A. Abdelghani, F. Bessueille, D. Leonard, N. Jaffrezic-Renault, An impedimetric DNA sensor based on functionalized magnetic nanoparticles for HIV and HBV detection, Sens. Actuators B: Chem. 134 (2) (2008) 755-760. https://doi.org/10.1016/j.snb.2008.06.020

[66] L. Uzun, R. Say, S. U¨ nal, A. Denizli, Production of surface plasmon resonance based assay kit for hepatitis diagnosis, Biosens. Bioelectron. 24 (9) (2009) 2878-2884. https://doi.org/10.1016/j.bios.2009.02.021

[67] M.M. Istek, M.M. Erdem, A.E. Gu¨rsan, Impedimetric nanobiosensor for the detection of sequence-selective DNA hybridization, Hacettepe J. Biol. Chem. 46(4) (2019) 495-503. https://doi.org/10.15671/HJBC.2018.257

[68] S. Tasoglu, H.C. Tekin, F. Inci, S. Knowlton, S. Wang, F. Wang-Johanning, G. Johanning, D. Colevas, U. Demirci, Advances in nanotechnology and microfluidics for human papillomavirus diagnostics, Proc. IEEE 103 (2) (2015) 161-178. https://doi.org/10.1109/JPROC.2014.2384836

[69] S. Bedford, Cervical cancer: physiology, risk factors, vaccination and treatment, Br. J. Nurs. 18 (2009) 80-84. https://doi.org/10.12968/bjon.2009.18.2.37874

[70] Y. Saylan, A. Denizli, Virus detection using nanosensors, B. Han, V. K. Tomer, T. A. Nguyen, A. Farmani, P.K. Singh (Eds.), In Micro and Nano Technologies, Nanosensors for Smart Cities, Elsevier 2020, 501-511. https://doi.org/10.1016/B978-0-12-819870-4.00038-4

[71] P. Teengam, W. Siangproh, A. Tuantranont, T. Vilaivan, O. Chailapakul, C.S. Henry, Multiplex paper-based colorimetric DNA sensor using pyrrolidinyl peptide nucleic acid-induced AgNPs aggregation for detecting MERS-CoV, MTB, and HPV oligonucleotides, Anal. Chem. 89 (2017) 5428-5435. https://doi.org/10.1021/acs.analchem.7b00255

[72] X. Peng, Y. Zhang, D. Lu, Y. Guo, S. Guo, Ultrathin Ti3C2 nanosheets based "off-on" fluorescent nanoprobe for rapid and sensitive detection of HPV infection, Sens. Actuators B: Chem. 286 (2019) 222-229. https://doi.org/10.1016/j.snb.2019.01.158

[73] X. Qiu, G. Wong, J. Audet, A. Bello, L. Fernando, J.B. Alimonti, H. Fausther-Bovendo, H. Wei, J. Aviles, E. Hiatt, A. Johnson, J. Morton, K. Swope, O. Bohorov, N. Bohorova, C. Goodman, D. Kim, M.H. Pauly, J. Velasco, J. Pettitt, G.G. Olinger, K. Whaley, B. Xu, J.E. Strong, L. Zeitlin, G.P. Kobinger, Reversion of advanced Ebola virus disease in nonhuman primates with ZMapp, Nature 514 (7520) (2014) 47-53. https://doi.org/10.1038/nature13777

[74] J.T. Baca, V. Severns, D. Lovato, D.W. Branch, R.S. Larson, Rapid detection of Ebola virus with a reagent-free, point-of-care biosensor, Sensors 15 (4) (2015) 8605-8614. https://doi.org/10.3390/s150408605

[75] H. Ilkhani, S. Farhad, A novel electrochemical DNA biosensor for Ebola virus detection, Anal. Biochem. 557 (2018) 151-155. https://doi.org/10.1016/j.ab.2018.06.010

[76] A.A. Yanik, M. Huang, O. Kamohara, A. Artar, T.W. Geisbert, J.H. Connor, H. Altug, An optofluidic nanoplasmonic biosensor for direct detection of live viruses from biological media, Nano. Lett. 10 (12) (2010) 4962-4969. https://doi.org/10.1021/nl103025u

[77] R.J. Meagher, O.A. Negrete, K.K. Van Rompay, Engineering paper-based sensors for Zika virus, Trends Mol. Med. 22 (7) (2016) 529-530. https://doi.org/10.1016/j.molmed.2016.05.009

[78] S. Afsahi, M.B. Lerner, J.M. Goldstein, J. Lee, X. Tang, D.A. Bagarozzi Jr, D. Pan, L. Locascio, A.Walker, F. Barron, B.R.Goldsmith, Novel graphene-based biosensor for early detection of Zika virus infection, Biosens. Bioelectron. 100 (2018) 85-88. https://doi.org/10.1016/j.bios.2017.08.051

[79] A. Kaushik, A. Yndart, S. Kumar, R.D. Jayant, A. Vashist, A.N. Brown, C-Z. Li, M. Nair, A sensitive electrochemical immunosensor for label-free detection of Zika-virus protein, Sci. Rep. 8 (2018) 9700-9705. https://doi.org/10.1038/s41598-018-28035-3

[80] J. Song, M.G. Mauk, B.A. Hackett, S. Cherry, H.H. Bau, C. Liu, Instrument-free point-of-care molecular detection of Zika virus, Anal. Chem. 88 (14) (2016) 7289-7294. https://doi.org/10.1021/acs.analchem.6b01632

[81] F. Krammer, P. Palese, Advances in the development of influenza virus vaccines, Nat. Rev. Drug Discovery. 14 (3) (2015) 167-182. https://doi.org/10.1038/nrd4529

[82] A. Moulick, L. Richtera, V. Milosavljevic, N. Cernei, Y. Haddad, O. Zitka, P. Kopel, Z. Heger, V. Adam, Advanced nanotechnologies in avian influenza: current status and future trendsa review, Anal. Chim. Acta. 983 (2017) 42-53. https://doi.org/10.1016/j.aca.2017.06.045

[83] P.D. Tam, N. Van Hieu, N.D. Chien, A.T. Le, M.A. Tuan, DNA sensor development based on multi-wall carbon nanotubes for label-free influenza virus (type A) detection, J. Immunol. Methods 350 (2009) 118-124. https://doi.org/10.1016/j.jim.2009.08.002

[84] F. Vollmer, S. Arnold, D. Keng, Single virus detection from the reactive shift of a whispering-gallery mode, Proc. Natl Acad. Sci. U.S.A. 105 (52) (2008) 20701-20704. https://doi.org/10.1073/pnas.0808988106

[85] H. Bai, R. Wang, B. Hargis, H. Lu, Y. Li, A SPR aptasensor for detection of avian influenza virus H5N1, Sensors 12 (2012) 12506-12518. https://doi.org/10.3390/s120912506

[86] S. Emir Diltemiz, A. Erso¨z, D. Hu¨r, R. Kec¸ili, R. Say, 4-Aminophenyl boronic acid modified gold platforms for influena diagnosis, Mater. Sci. Eng., C 33 (2013) 824-830. https://doi.org/10.1016/j.msec.2012.11.007

[87] C.E. Jin, T.Y. Lee, B. Koo, H. Sung, S.H. Kim, Y. Shin, Rapid virus diagnostic system using bio-optical sensor and microfluidic sample processing, Sens. Actuators B: Chem. 255 (2018) 2399-2406. https://doi.org/10.1016/j.snb.2017.08.197

[88] M.R. Guerreiro, D.F. Freitas, P.M. Alves, A.S. Coroadinha, Detection and quantification of label-free infectious adenovirus using a switch-on cell-based fluorescent biosensor, ACS Sens. 4 (6) (2019) 1654-1661. https://doi.org/10.1021/acssensors.9b00489

[89] H. Kim, M. Park, J. Hwang, J.H. Kim, D.R. Chung, K.S. Lee, M. Kang, Development of label-free colorimetric assay for MERS-CoV using gold nanoparticles, ACS Sens. 4 (2019) 1306-1312. https://doi.org/10.1021/acssensors.9b00175

[90] M. Trzaskowskia, A. Napio'rkowska, E. Augustynowicz-Kopec, T. Ciach, Detection of tuberculosis in patients with the use of portable SPR device, Sens. Actuators B: Chem. 260 (2018) 786-792. https://doi.org/10.1016/j.snb.2017.12.183

[91] P. Weerathunge, R. Ramanathan, V. Torok, K. Hodgson, Y. Xu, R. Goodacre, B.K. Behera, V. Bansal, Ultrasensitive colorimetric detection of murine norovirus using NanoZyme aptasensor, Anal. Chem. 91 (5) (2019) 3270-3276. https://doi.org/10.1021/acs.analchem.8b03300

[92] H. Kearns, R. Goodacre, L. Jamieson, D. Graham, K. Faulds, SERS Detection of Multiple Anti-microbial Resistant Pathogens using Nanosensors, Anal. Chem. 89 (23) (2017) 12666-12673. https://doi.org/10.1021/acs.analchem.7b02653

[93] R. Nißler, O. Bader, M. Dohmen, S. G. Walter, C. Noll, G. Selvaggio, U. Groß, S. Kruss, Remote near infrared identification of pathogens with multiplexed nanosensors, Nat. Commun. 11 (1) (2020) 5995. https://doi.org/10.1038/s41467-020-19718-5

[94] J. Chen, S.R. Nugen, Detection of protease and engineered phage-infected bacteria using peptide-graphene oxide nanosensors, Anal. Bioanal. Chem. 411 (12) (2019) 2487-2492 https://doi.org/10.1007/s00216-019-01766-6

[95] G. Santopolo, A. Domenech-Sanchez, S.M. Russell, R. de la Rica, Ultrafast and Ultrasensitive Naked-Eye Detection of Urease-Positive bacteria with Plasmonic Nanosensors, ACS Sens. 4 (4) (2019) 961-967 https://doi.org/10.1021/acssensors.9b00063

[96] T. Banerjee, S. Sulthana, T. Shelby, B. Heckert, J. Jewell, K. Woody,V. Karimnia, J. McAfee, S. Santra, Multiparametric Magneto-fluorescent Nanosensors for the Ultrasensitive Detection of Escherichia coli O157:H7, ACS Infect. Dis. 2 (10) (2016) 667-673. https://doi.org/10.1021/acsinfecdis.6b00108

[97] A. Jyoti, R.S. Tomar, R. Shanker, Nanosensors for the Detection of Pathogenic Bacteria. In: Ranjan, S., Dasgupta, N., Lichtfouse, E. (eds) Nanoscience in Food and

Materials Research Foundations 141 (2023) 246-269   https://doi.org/10.21741/9781644902295-10

Agriculture 1. Sustainable Agriculture Reviews, Springer, Cham. 20 (2016) 129-150. https://doi.org/10.1007/978-3-319-39303-2_5

[98] H. Pu, Y. Xu, D.W. Sun, Q. Wei, X. Li, Optical nanosensors for biofilm detection in the food industry: principles, applications and challenges, Crit Rev Food Sci Nutr. 61 (13) (2020) 2107-2124. https://doi.org/10.1080/10408398.2020.1808877

[99] G. Fuertes, I. Soto, R. Carrasco, M. Vargas, J. Sabattin, C. Lagos, Intelligent Packaging Systems: Sensors and Nanosensors to Monitor Food Quality and Safety, Journal of Sensors, vol. 2016 (2016) 4046061. https://doi.org/10.1155/2016/4046061

[100] A. Khiyami, H. Almoammar, Y.M. Awad, M.A. Alghuthaymi, K.A. Abd-Elsalam, Plant pathogen nanodiagnostic techniques: forthcoming changes?, Biotechnology & Biotechnological Equipment, 28 (5) (2014) 775-785. https://doi.org/10.1080/13102818.2014.960739

[101] T. Banerjee, T. Shelby, S. Santra, How can nanosensors detect bacterial contamination before it ever reaches the dinner table, Future Microbiol. 12 (2) (2017) https://doi.org/10.2217/fmb-2016-0202

[102] Y. Saylan, F. Yılmaz, E. O¨ zgu¨r, A. Derazshamshir, H. Yavuz, A. Denizli, Molecularly imprinting of macromolecules for sensors applications, Sensors 17 (2017) 898-928. https://doi.org/10.3390/s17040898

[103] K.H. Cho, D.H. Shin, J. Oh, J.H. An, J.S. Lee, J. Jang, Multidimensional conductive nanofilm-based flexible aptasensor for ultrasensitive and selective HBsAg detection, ACS Appl. Mater. Interfaces 10 (2018) 28412-28419. https://doi.org/10.1021/acsami.8b09918

[104] Y. Pang, J. Jian, T. Tu, Z. Yang, J. Ling, Y. Li, et al., Wearable humidity sensor based on porous graphene network for respiration monitoring, Biosens. Bioelectron. 116 (2018) 123-129. https://doi.org/10.1016/j.bios.2018.05.038

[105] L. La Spada, L. Vegni, Electromagnetic nanoparticles for sensing and medical diagnostic applications, Materials 11 (2018) 603-624. https://doi.org/10.3390/ma11040603

[106] J. Wang, X. Qu, Recent progress in nanosensors for sensitive detection of biomolecules, Nanoscale 5 (2013) 3589-3600 https://doi.org/10.1039/c3nr00084b

[107] N.H. Anh, M.Q. Doan, N.X. Dinh, T.Q. Huy, D.Q. Tri, L.T.N. Loan, B.V. Hao, A.T. Le, Gold nanoparticle-based optical nanosensors for food and health safety monitoring: recent advances and future perspectives, RSC Adv. 12 (18) (2022) 10950-10988. https://doi.org/10.1039/D1RA08311B

Emerging Applications of Nanomaterials   Materials Research Forum LLC
Materials Research Foundations 141 (2023) 270-293   https://doi.org/10.21741/9781644902295-11

Chapter 11

Nanomaterials for Self-Healing Hydrogels

Md. Mahamudul Hasan Rumon, Stephen Don Sarkar, Md. Mahbub Alam and
Chanchal Kumar Roy*

Department of Chemistry, Bangladesh University of Engineering and Technology (BUET),
Dhaka-1000

*ckroy@chem.buet.ac.bd

Abstract

Materials with self-healing property are considered as the smart materials for various advanced technological applications. Previously only living substances such as tissues were thought to have self-healing properties that are required for their survival and adaptation to environmental changes. Recently, novel synthetic materials such as hydrogels have been prepared and developed which demonstrated self-healing performances. Exceptional self-healing has been observed in hydrogels introducing different kinds of nanoparticles. These hydrogels have exceptional functionality and mechanical properties. The techniques of the introduction, exploration, and measurement of the properties related to self-healing are challenging due to the wide variety of substances. In this chapter, efforts have been made to discuss the mechanism, types, and drawbacks of nanomaterial-modified self-healing hydrogels.

Keywords

Self-healing, Hydrogels, Nanomaterials, Nanocomposites, Characterizations

Contents

1.   Introduction

Hydrogels are prepared with polymers that have the exceptional capability to store abundant of water in their integrated polymeric network which triggered attention in different branches of material science [1, 2]. Especially in biomedical engineering, they are being used as one of the base materials for the development of artificial tissues [3-5]. Different kinds of hydrogels have been prepared to represent the properties of natural tissues [6, 7]. The self-healing and mechanical toughness are the primary requirements for the development of artificial tissues from hydrogels [8]. The combination of good mechanical toughness and high self-healing capability is often desired for tissue applications [9]. However, the prepared conventional hydrogels have demonstrated an inverse relationship between the mechanical toughness and self-healing properties [10]. The rheology or movement of polymers in hydrogels plays a dominating role in determining such combinations. Highly covalently cross-linked hydrogels such as double network hydrogels are very tough, however, they demonstrate very weak or negligible self-healing properties [11-13]. On the other hand, physically crosslinked hydrogels with supramolecular interactions such as electrostatic interaction, hydrogen bonding, polar interactions, and host-guest interactions showed weak to moderate self-healing efficiency but compromised mechanical toughness [14-17]. Here the fusion of chemical and physical cross-linking is the viable approach to preparing mechanically tough hydrogels with good self-healing capability. The introduction of nanoparticles for the crosslinking of polymer hydrogels is a novel strategy to develop these hydrogels. Recent nano-chemistry has become a blessing in the synthesis of self-healing hydrogels [18]. Through the modification of nanoparticles, it is possible to introduce physical and covalent crosslinking in the same

Emerging Applications of Nanomaterials   Materials Research Forum LLC
Materials Research Foundations 141 (2023) 270-293   https://doi.org/10.21741/9781644902295-11

hydrogel. Here, in this chapter, the processing, testing techniques, mechanisms and drawbacks of different self-healing nanocomposite hydrogels have been discussed.

2.   Self-healing mechanism in nanocomposite hydrogels

The hydrogel self-healing mechanism possesses similarity with biological systems [19]. Hydrogel self-healing is classified into extrinsic and intrinsic types [20-22]. The extrinsic self-healing occurs with the introduction of an external healing agent, whereas the intrinsic self-healing is achieved spontaneously or in presence of certain stimuli at the inherent molecular network [23-25]. Both processes of self-healing phenomena can be described in a single schematic diagram which is shown in Fig. 1. The healing process progresses through a series of five steps: surface rearrangement, surface approach, wetting, diffusion, and randomization [26, 27]. The first step is the rearrangement of polymers on the cracked surface that take place following an incident of damage that leads to the tendency of recovery of the material to its original form [26]. Then the surface approach step is the one that decides whether the healing procedure is going to occur or not [28]. If the damaged surfaces are properly approached, wetting of the surfaces will begin. It may occur by solvent present in the original material or by a healing agent. This step ensures the chain mobility of the polymeric network [29, 30]. This step may be improved by raising the temperature or by the addition of solvents [30]. The enhanced attractive interaction of polymer and nanoparticles can promote adhesion, which increases the restoring ability of the mechanical properties of the polymeric network [31]. Then the random movement is caused by diffusion. The mobile polymer chains are more likely to get entangled close to the nanoparticle surface, which then leads to the interpenetration into the matrix material [32]. During the randomization step the polymers and nanoparticles moved due to their entropic freedom [33]. However, during the randomization step, failure of the initial crack interfaces may be noticed for heterogeneous distribution of stress.

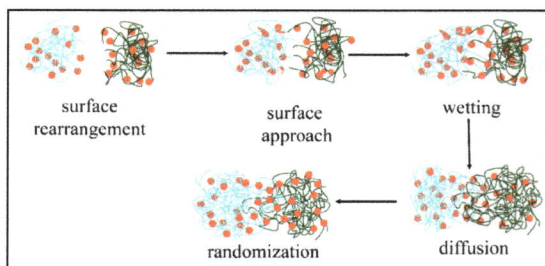

Figure 1. Schematic representation of self-healing mechanism in polymeric hydrogel networks

Emerging Applications of Nanomaterials   Materials Research Forum LLC
Materials Research Foundations 141 (2023) 270-293   https://doi.org/10.21741/9781644902295-11

3.   Recent progress of nanocomposite hydrogels for self-healing

Hydrogels crosslinked by conventional covalent crosslinkers have demonstrated improved mechanical properties in modulus, strength, stress at break, etc. [34, 35] However, these crosslinkers failed to exhibit self-healing characteristics due to the limitation of reversible bond formation ability [36]. Recently, the introduction of nanomaterials in polymer matrix as filler materials and crosslinkers has improved the mechanical properties of nanocomposite hydrogels without compromising its self-healing tendency [10, 37]. The ease of functionality with the large surface area of nanomaterials plays a beneficial role in this case [38]. Additionally, specific and non-specific adsorptions of polymers through different chemical bonds are also responsible for self-healing [39]. This section contains brief discussions of the improvement of the self-healing performance of hydrogels containing various nanomaterials.

3.1   Hydrogel self-healing with metal nanoparticles

Metallic nanoparticles can be directly used as filler and crosslinker in hydrogels [40, 41]. These nanoparticles were able to improve self-healing by the adsorption of polymers in a high dielectric water medium[42]. The required gain in free energy is achieved through the adsorption of polymer on the surface of nanoparticle[43, 44]. On a cracked surface, the exposed nanoparticles and the polymer chains exhibit an affinity to each other as shown in Fig. 2 [45]. In an aqueous medium, the hydrophilic polymeric materials tend to form rod-shaped structures [46]. It does not favour the adsorption of polymer strands on the nanomaterials' surface [47, 48]. Thermal energy, kT per monomer strand is lost by a polymer in this adsorption. It is expected that a huge number of monomers are adsorbed on the nanoparticle surface [49, 50]. This favours the spontaneous adsorption of polymers with the loss of entropy [45]. The cracked surface acts like an adhesive layer that connects nanoparticle and polymer chains, and act as bridge between the cracked chains. When the adhesive joined is extended, the tension is relaxed through the detachment of few adsorbed segments. The viscoelastic energy dissipation works and the joints are sustained [51]. The strong attractive forces could be generated for the improvement of self-healing capabilities through the incorporation of other physical interactions such as van der Waals force, hydrogen bonding, etc. [52, 53]

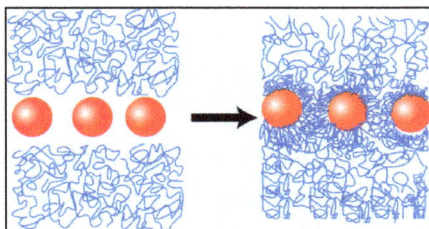

Figure 2. Schematic representation self-healing in metal nanoparticle crosslinked hydrogel

Emerging Applications of Nanomaterials   Materials Research Forum LLC
Materials Research Foundations 141 (2023) 270-293   https://doi.org/10.21741/9781644902295-11

3.2   Self-healing of hydrogels with carbon-based nanoparticles

Recently, carbon-based nanoparticles have become popular for the fabrication of self-healing hydrogel [54, 55]. Carbon-based nanoparticles such as 1D: carbon dots, nanowires, carbon nanotubes (CNT) or fibers, 2D: graphene nanosheets (GNS), 3D: carbon spheres, and other carbon nanomaterials have demonstrated very weak self-recovery in their neat form in hydrogels[56]. The self-healing of hydrogels containing CNT and GNS are depicted in Fig. 3. The modified forms of the carbon materials with different types of functional groups are better than their pristine forms. For example, Jiaqi Liu *et. al.* prepared a self-healing hydrogel using functionalized graphene oxide to crosslink acrylamide hydrogel [57]. It showed a maximum 67% healing efficiency within 2 h. Most of these nanoparticles have been modified to introduce carboxylic acid, thiol, amine, and hydroxyl groups, which are highly capable to provide bonds to facilitate self-healing. The nanoparticles are usually modified for producing van der Waals interactions, polar interactions, hydrogen bonding etc. [58]. These modifications enhance phase compatibility and wetting ability of nanomaterial surface. Sometime, the modified surface of nanoparticles even with alkane chains improves the nanomaterial-polymer interactions in the hydrogel matrix.

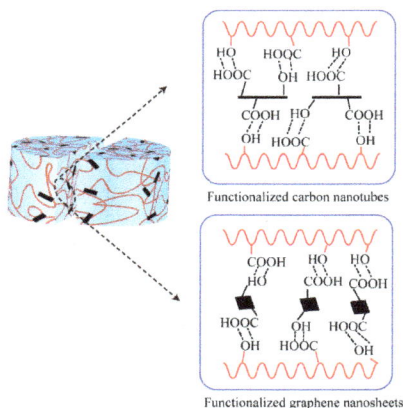

Figure 3. Insight into the self-healing of gels containing carbon nanotubes and graphene nanosheets

3.3   Hydrogel self-healing with polymeric nanoparticles

Nano form of carbon-based polymer have received great attention to develop tough hydrogels consisting of good self-healing capacity [59]. The nano forms of carbon-based polymer possess better chain flexibility and various functionality. The polymeric nano forms of carbon-based materials may consists of a large number of different functional

Emerging Applications of Nanomaterials   Materials Research Forum LLC
Materials Research Foundations 141 (2023) 270-293   https://doi.org/10.21741/9781644902295-11

groups on their structure[60]. For the biomedical applications that requires resistance to physiological conditions, these self-healing systems possess a high potential that can extend their lifetime [61, 62]. Depending on the functional groups present in the cellulose and its derivatives, they can introduce different types of bonding within the hydrogel matrix [63-65]. For instance, nanocrystal cellulose and cellulose nanorods are used to introduce the self-recovery of a nanocomposite hydrogel [66, 67]. The vast number of hydroxyl groups facilitated the formation of hydrogen bonds within the polymer network which provides self-healing properties to the prepared hydrogels [36]. The polymer-nanocrystalline cellulose interactions are demonstrated in Fig. 4. Besides this, carboxymethyl cellulose and carboxylic functionalized nanorod undergo electrostatic interaction with metal ions to provide self-healing properties [68, 69]. The stability of these bonds can be improved by modification. For example, the condensation reaction of cellulose carbonyl group and amine group of polymer results in a acylhydrazone reversible chemical bond that is very stable in ambient condition [62]. Xiao *et al.* designed a nanocomposite hydrogel by crosslinking dialdehyde cellulose nanocrystals with acyl hydrazine-terminated polyethene glycol [70]. The obtained hydrogel demonstrated 97.5% self-healing efficiency with short gelation time. This hydrogel showed good biocompatibility with almost 100% cell viability.

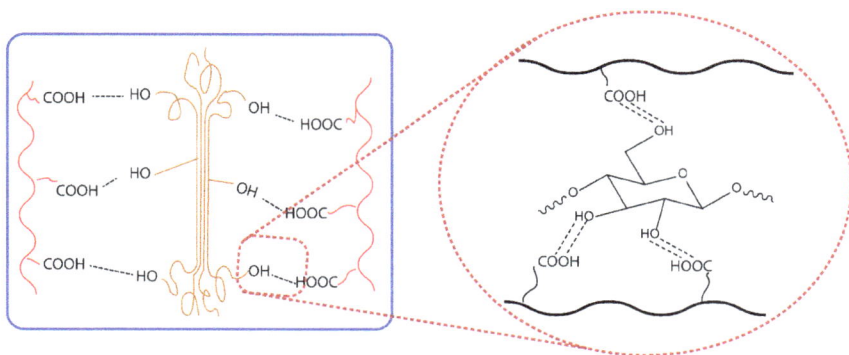

Figure 4. Interactions in self-healing hydrogels with nanocrystalline cellulose

3.4   Hydrogel self-healing with biobased nanoparticles

Synthetic materials with their biobased counterpart are required to intelligently integrated that can replicate biological systems [71]. A number of bio-based nanoparticles have already been used as fillers or crosslinkers in hydrogel materials and widely applied as biomedical materials because of their excellent tuneable functionalities [72, 73]. Bio-based nanomaterials include chitin, its derivatives like chitosan and hyaluronic acid are wildly used. Chitosan is a chitin derivative that has been *N*-deacetylated showed improved biological acceptancy in tissue engineering [74, 75]. Depending on the functional group present on the chitin structure, it may either form chemical crosslinking or physical crosslinking [64]. The NH_2 group forms chemical interactions using Schiff-based linkage of CH=N- with the polymer chains [76-78]. This results in a nanocomposite hydrogel that endowed with a self-healing using Schiff base linkages. The construction of Schiff base linkage with chitin derivatives are illustrated in Fig. 5. Moreover, the physical bonding such as hydrophobic interaction and the hydrogen bond interactions, can also be useful to provide self-healing character onto the chitin-based materials [79, 80]. The OH and NH_2 groups of chitin can undergo reversible hydrogen bond formation with the hydrophilic groups present on the hydrogel polymer chains, and show very fast healing [81, 82].

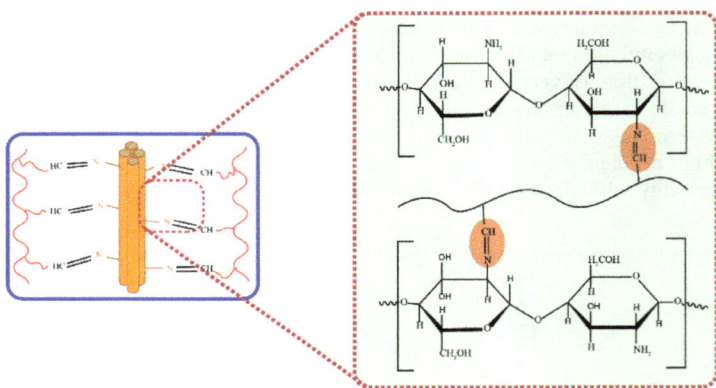

Figure 5. The formation of Schiff base-linkage in self-healable hydrogel with chitin nanofiber

Uses of enzyme like host-guest interactions is another notable way to prepare self-healable hydrogels [83]. In most cases, cyclodextrins (CDs) are employed as the host units to keep the biocompatibility with a suitable guest unit [84, 85]. Two freshly cut hydrogel surfaces are immediately joined using selective adhesion of CD-guest supramolecular interactions [16]. The complexation of the CD with guest unit on a cracked hydrogel surface is demonstrated in Fig. 6. Here, the self-healing efficiency depends on the strength of CD-guest complexation.

Emerging Applications of Nanomaterials   Materials Research Forum LLC

Materials Research Foundations 141 (2023) 270-293   https://doi.org/10.21741/9781644902295-11

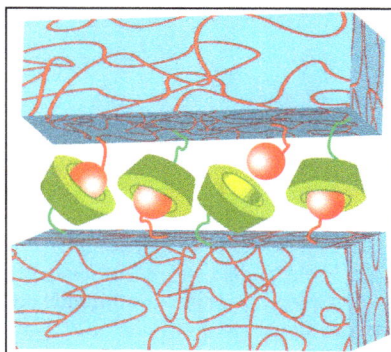

Figure 6. Complexation of CD-guest in self-healing hydrogel system.

4. Self-healing characterization methods

4.1 Microscopic analysis

Optical and different microscopic technologies have been used to investigate the hydrogel self-healing process [86]. This includes scanning electron microscopy (SEM), transmission electron microscopy (TEM) and atomic force microscopy (AFM), which provide information of the healing process on cracked surfaces. These methods often do not provide quantitative information. Some of these methods bring information for qualitative use only. Hence, these techniques are separately usable to investigate the variation of healed surface nanostructure. For example, SEM provides good qualitative information about the microstructure of the healed joined of hydrogels but actual images cannot be obtained for hydrogels due to the vacuum dried sample preparation process [87]. The real images of water containing healed joined can be obtained from TEM observations. The three-dimensional nanostructures in of polymeric system are revealed from TEM images. In this method, very thin samples are required to prepare for analysis. The AFM could compare the rupture and healed surfaces of hydrogel in molecular level. It could provide information of molecular complexations [88]. Sometime, the AFM tip is used to intentionally damage the hydrogel surface and observe the process of healing on nanoscale level. Matyjaszewski's group used AFM to track the healing of hydrogel containing poly (n-butyl acrylate) grafted star polymeric with disulphide functional groups [89]. From the AFM observations a self-healing mechanism was developed. The movement and reaction ability of cut polymer can also be tracked from AFM monitoring. It also provides the information to separate the elastic and non-elastic energy contribution of the adhesive joints of hydrogel systems. In addition, AFM can provide topographical information at the nanoscale, however the chemical composition of the healing joint cannot be obtained from here [89].

4.2   Spectroscopic analysis

Spectroscopic characterizations are utilized to monitor the changes in bonding interactions during hydrogel self-healing [90]. Ultra-Violet, Fourier transform infrared (FTIR) and Raman spectroscopies are commonly used. They reveal the nature of bonding from weak bond to strong bonds of materials. The chemical mechanism of healing can be monitored by assessing the generation of new bonds and interactions in the hydrogel polymer matrix [91]. From the time dependent intensity data, the rate of self-healing is quantified. Coherent anti-Stokes Raman scattering (CARS) is another useful method to determine the location of a molecule and it's in situ functional groups. Sometime CARS is utilized in conjunction with morphological laser reflection imaging to track the restoration of an polymer network [92]. Raman spectroscopy reveals the bonding nature such as chemical bonds or other weak interactions present in materials [93]. The healing process can be determined by observing the spatial distribution of vibrational frequencies generated during the recovery process after inducing damage [91]. Several modern technologies, addition to the classical spectroscopy, can be utilized for the better understanding of the healing process of hydrogels.

4.3   Mechanical test

Self-healing efficiency of hydrogels is commonly evaluated by comparing the mechanical features of the fresh and healed samples [94]. Usually, a cylindrical or dumbbell-shaped hydrogel sample is cut into two similar pieces and joined them using finger pressure. Then the joined samples are left for healing with time. The healing environment must be controlled to prevent the hydrogel from changes due to temperature, pressure, and moisture effects. After certain time duration tensile tests are performed with the healed gels to determine their healing efficiencies [10]. The mechanical performances of the fresh samples are treated as references. The different healing efficiency including working healing efficiency (η_w), stress healing efficiency (η_σ), and the strain healing efficiency (η_ε) are calculated using the following equations [10]:

$$\eta_w = \frac{W_{healed}}{W_{fresh}} \times 100 \ldots\ldots\ldots \quad (1)$$

$$\eta_\sigma = \frac{\sigma_{healed}}{\sigma_{fresh}} \times 100 \ldots\ldots\ldots \quad (2)$$

and,

$$\eta_\varepsilon = \frac{\varepsilon_{healed}}{\varepsilon_{fresh}} \times 100 \dots\dots\dots\dots \quad (3)$$

Here, the work of extension, fracture stress, and fracture strain of the healed samples are termed as W_{healed}, σ_{healed}, and ε_{healed} respectively. The terms W_{fresh}, σ_{fresh}, and ε_{fresh} have been used to denote the corresponding parameters of fresh samples respectively.

Oscillation rheology is another convenient technique to determine the self-healing performance [95]. The amount and type of bonding recovered from healing of hydrogels are investigated using this method. The information on the possibility of generation of self-healed bonds at different time scales are obtained by the application of sinusoidal shear deformation with different frequencies [96, 97]. The shear stress values, storage modulus (G'), and loss modulus (G'') are obtained as a function of time. The G' indicates to the elastic property where G'' to the viscous property of the gel matrix. The improvement of storage moduli after self-healing indicates the recovery of strong chemical bonds. However, the enhancement in the values of G'' corresponds to the weakening of chemical bonds after healing [98].

5. Factors affecting self-healing with nanomaterials

The self-healing of materials is a time-dependent process [99]. A crack in a material takes longer time to recover in its initial stage. From the molecular point of view, a crack in a material is not completely recoverable. The molecules on the crack do not get their initial position and orientation after the crack recovery [100]. These facts are corelated to the self-healing of hydrogels containing nanomaterials. However, the crack recovery with nanomaterials provides advantages for nonspecific molecular adsorption through the surface. The extent of recovery of the damage influences the mechanical performance and stability of the polymeric material. Several factors influence the extent of recoveries of hydrogels such as compositions, functional groups and their density, network of polymer, solvent and temperature, etc., [101, 102]. In this section, the factors that affect the self-recovery of hydrogels are discussed.

5.1 Dimension of components

The presence of small number of nanoparticles can significantly change the viscosity of hydrogels by the introduction of various crosslinking bonds. The healing process of hydrogel are affected by the nanomaterial size and shape [103, 104]. When the size of the nanoparticles is concerned, smaller nanoparticles get greater diffusion feasibility over the damage interfaces. Nanoparticles with a size close to the polymer mesh size have good diffusion rate within the polymer network [105]. This optimal size for maximum self-healing comes as the result of two factors that works in opposite way. The smaller size nanoparticles have higher diffusion rate with lower reinforcing capability [106]. However, the larger sized nanoparticles with lower diffusion rate have greater contribution to damage surface reinforcement. The spherical and rod shaped nanoparticle surfaces provide poor surface adsorption compared to the pellet nanoparticles with larger surface area [107].

Maria C. Arno *et. al.* reported that positively charged pellets particles proved better molecular diffusion then compared with particles with the spherical and cylindrical morphology [107]. Furthermore, the platelet nanoparticles have maximum adhesive performance that helped in quick healing process. The healing performance of a nanomaterial-based polymer also depends on the average molecular weight [108]. When the molecular weight is very high the probability of adsorption on the nanomaterial surface is very high with high the recovery ability resulting shorter time for healing.

5.2   Chemical groups

The presence of functional groups on the surface of the nanomaterials significantly influences the self-healing ability of nanocomposite hydrogels [109]. As instance, nanomaterials with charged surfaces containing functional groups such as COO^-, NH_4^+ etc., can form physical bonds, and play a vital role in the healing process. This functionalization can be either homogeneous or heterogeneous. Chemical functionalization is possible both on solid type metallic nanoparticles as well as in polymeric nanomaterials, such as bio-derived cellulose nanocrystals can be decorated with COO^- and OH^- by the oxidation process [110]. The functionalization enhances the number of H-bonding interaction sites in the polymeric networks of hydrogels, and improves the healing process [85]. This type of functionalization is also possible for chitosan nanoparticles with a high degree of modification with amino, acetamide, and hydroxyl groups[111]. GO has large number of functional groups on it surface, and it undergoes the reversible hydrogen bond formation to facilitate the self-healing phenomena in a hydrogel matrix [112].

5.3   Temperature

The self-healing performance of hydrogel is significantly related with its viscoelastic behaviour [113]. In generally, an increase in temperature enhances polymer diffusion and improves the self-healing [114, 115] because the enhanced diffusion of polymer is correlated to the reduction of friction of hydrogels. Wang *et. al.* have shown that the reduction of equilibrium healing time of nanocomposite hydrogels is associated to the decrease in the Rouse friction coefficient [102]. Here, another fact is also required to be considered, which is deswelling of hydrogels at high temperatures. Nanocomposite hydrogels shrink when deswelled at elevated temperature. It results the formation of strong bonding interaction because the average inter-crosslink distance becomes smaller than before, which could positively affect the self-healing process. This type of evidence has been demonstrated by Kazutoshi Haraguchi *et. al.,* [116]. The effect of temperature on the friction coefficient and healing performance are shown in Fig. 7. However, there could exist an optimum temperature for the improvement of self-healing performance due to the involvement of the adsorption process. An elevated temperature could regulate the desorption of polymers from the surface of nanomaterials, leading decreasing in self-healing performance.

Emerging Applications of Nanomaterials   Materials Research Forum LLC
Materials Research Foundations 141 (2023) 270-293   https://doi.org/10.21741/9781644902295-11

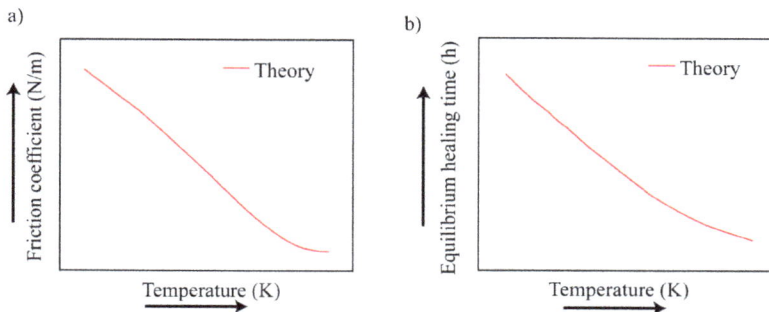

Figure 7. Effects of temperature on the friction coefficient (a), and equilibrium healing time, (b).

5.4   Aging effect

In real-life self-healing applications, the cracked or fractured hydrogels may not be brought into contact immediately but rather after some delaying. During this time, the fractured surfaces with the bulk nanocomposites are remain exposed to the surrounding environment. The healing performance of hydrogels is dependent on this delay due to the changes occurred on the exposed damaged surface during time [117]. The relation between equilibrium healing strength and delay time is presented in Fig. 8, which demonstrates that the maximum healing strength decreases with increasing delaying time [102]. The healing performance is reduced in a long-time delay. During the exposure time, the free end groups of polymer chains those plays the key role in self-healing turn out to be inactive when encountered with molecular species of the environment.

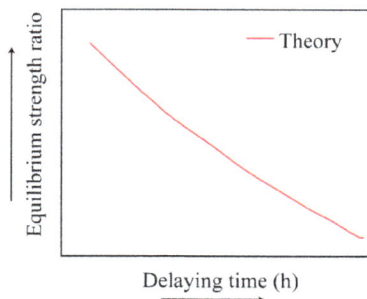

Figure 8. Effects of waiting time on the equilibrium healing strength ratio of nanocomposite hydrogel.

5.5 Water content

Crack healing is sensitive to water content regulation. The presence of water molecules reduces the diffusivity of the polymer matrix, leading to the adhesive failure of the damaged zone [118]. Therefore, highly swollen nanocomposite hydrogels require high energy for intermolecular diffusion, and take a long time for healing[119]. Moreover, the presence of moisture in the environment may lead to structural deformation, where the damage cannot be repaired. On the other hand, if the water contains a Lewis acid or basic solution, it shows a strong affinity to the H-atoms, resulting to the hydrogen bond formation, and enhancing the healing performance [102].

Conclusion and future aspects

Self-healing hydrogels provide several advantages over traditional hydrogels, including endurance and improved mechanical performance. The self-healing performance of hydrogels has reached a new height with the introduction of nanomaterials resulted in the simultaneous improvement of self-healing performance and mechanical toughness. Nanocomposite hydrogels with different kinds of interaction such as host-guest, metal-ion complexation, electrostatic interactions etc. have been tested for self-healing ability. In addition, biobased polymeric nanoparticles have been introduced to offer biocompatibility of self-healing hydrogels. However, there are lots of scope for future study. Recently, two-dimensional nano-sheets have demonstrated their superior flexibility in modification, stability, and preparation of self-healing hydrogels. Functionalized nanosheets can also improve the viscoelastic properties of the hydrogel matrix, leading to enhanced molecular chain diffusion for improved healing. Nevertheless, these materials are so far to be designed for biocompatibility and genuine applications in soft robotics and artificial tissues, which requires further efforts to widen the pertinence of hydrogels.

References

[1] A. Ahsan, W.-X. Tian, M. A. Farooq, D. H. Khan, An overview of hydrogels and their role in transdermal drug delivery, International Journal of Polymeric Materials and Polymeric Biomaterials 70 (2021) 574-584. https://doi.org/10.1080/00914037.2020.1740989

[2] P. B. Mohite, S. S. Adhav, A hydrogels: Methods of preparation and applications, Int. J. Adv. Pharm 6 (2017) 79-85.

[3] A. S. Hoffman, Hydrogels for biomedical applications, Advanced Drug Delivery Reviews 64 (2012) 18-23. https://doi.org/10.1016/j.addr.2012.09.010

[4] R. Mohammadinejad, H. Maleki, E. Larraneta, A. R. Fajardo, A. B. Nik, A. Shavandi, A. Sheikhi, M. Ghorbanpour, M. Farokhi, and P. Govindh, Status and future scope of plant-based green hydrogels in biomedical engineering, Applied Materials Today 16 (2019) 213-246. https://doi.org/10.1016/j.apmt.2019.04.010

[5] S. J. Song, J. Choi, Y. D. Park, J. J. Lee, S. Y. Hong, K. Sun, A three-dimensional bioprinting system for use with a hydrogel-based biomaterial and printing parameter characterization, Artificial Organs 34 (2010) 1044-1048. https://doi.org/10.1111/j.1525-1594.2010.01143.x

[6] M. C. Catoira, L. Fusaro, D. Di Francesco, M. Ramella, F. Boccafoschi, Overview of natural hydrogels for regenerative medicine applications, Journal of Materials Science: Materials in Medicine 30 (2019) 1-10. https://doi.org/10.1007/s10856-019-6318-7

[7] P. Matricardi, C. Di Meo, T. Coviello, W. E. Hennink, F. Alhaique, Interpenetrating polymer networks polysaccharide hydrogels for drug delivery and tissue engineering, Advanced Drug Delivery Reviews 65 (2013) 1172-1187. https://doi.org/10.1016/j.addr.2013.04.002

[8] Q. Chai, Y. Jiao, X. Yu, Hydrogels for biomedical applications: their characteristics and the mechanisms behind them, Gels 3 (2017) 6. https://doi.org/10.3390/gels3010006

[9] G. Su, S. Yin, Y. Guo, F. Zhao, Q. Guo, X. Zhang, T. Zhou, G. Yu, Balancing the mechanical, electronic, and self-healing properties in conductive self-healing hydrogel for wearable sensor applications, Materials Horizons 8 (2021) 1795-1804. https://doi.org/10.1039/D1MH00085C

[10] M. M. H. Rumon, S. D. Sarkar, M. M. Uddin, M. M. Alam, S. N. Karobi, A. Ayfar, M. S. Azam, C. K. Roy, Graphene oxide based crosslinker for simultaneous enhancement of mechanical toughness and self-healing capability of conventional hydrogels, RSC Advances 12 (2022) 7453-7463. https://doi.org/10.1039/D2RA00122E

[11] M. Diba, S. Spaans, K. Ning, B. D. Ippel, F. Yang, B. Loomans, P. Y. W. Dankers, S. C. G. Leeuwenburgh, Self-healing biomaterials: from molecular concepts to clinical applications, Advanced Materials Interfaces 5 (2018) 1800118. https://doi.org/10.1002/admi.201800118

[12] L. Saunders, P. X. Ma, Self-healing supramolecular hydrogels for tissue engineering applications, Macromolecular Bioscience 19 (2019) 1800313. https://doi.org/10.1002/mabi.201800313

[13] Q. Chen, L. Zhu, H. Chen, H. Yan, L. Huang, J. Yang, J. Zheng, A novel design strategy for fully physically linked double network hydrogels with tough, fatigue resistant, and self-healing properties, Advanced Functional Materials 25 (2015) 1598-1607. https://doi.org/10.1002/adfm.201404357

[14] T. Cheng, Y. Z. Zhang, S. Wang, Y. L. Chen, S. Y. Gao, F. Wang, W. Y. Lai, W., Huang, Conductive Hydrogel-Based Electrodes and Electrolytes for Stretchable and Self-Healable Supercapacitors, Advanced Functional Materials 31 (2021) 2101303. https://doi.org/10.1002/adfm.202101303

[15] S. Strandman, X. X. Zhu, Self-healing supramolecular hydrogels based on reversible physical interactions, Gels 2 (2016) 16. https://doi.org/10.3390/gels2020016

[16] Y. Yang, M. W. Urban, Self-healing of polymers via supramolecular chemistry, Advanced Materials Interfaces 5 (2018) 1800384. https://doi.org/10.1002/admi.201800384

[17] Q. Li, C. Liu, J. Wen, Y. Wu, Y. Shan, J. Liao, The design, mechanism and biomedical application of self-healing hydrogels, Chinese Chemical Letters 28 (2017) 1857-1874. https://doi.org/10.1016/j.cclet.2017.05.007

[18] Z. Wei, J. H. Yang, Z. Q. Liu, F. Xu, J. X. Zhou, M. Zrínyi, Y. Osada, Y. M. Chen, Self-Healing Materials: Novel Biocompatible Polysaccharide-Based Self-Healing Hydrogel, Advanced Functional Materials 25 (2015) 1471-1471. https://doi.org/10.1002/adfm.201570065

[19] Y. Zhong, P. Li, J. Hao, X. Wang, Bioinspired self-healing of kinetically inert hydrogels mediated by chemical nutrient supply, ACS Applied Materials & Interfaces 12 (2020) 6471-6478. https://doi.org/10.1021/acsami.9b20445

[20] M. Chen, D. Fan, S. Liu, Z. Rao, Y. Dong, W. Wang, H. Chen, L. Bai, Z. Cheng, Fabrication of self-healing hydrogels with surface functionalized microcapsules from stellate mesoporous silica, Polymer Chemistry 10 (2019) 503-511. https://doi.org/10.1039/C8PY01402G

[21] S. Liu, Z. Rao, R. Wu, Z. Sun, Z. Yuan, L. Bai, W. Wang, H. Yang, H. Chen, Fabrication of microcapsules by the combination of biomass porous carbon and polydopamine for dual self-healing hydrogels, Journal of Agricultural and Food Chemistry 67 (2019) 1061-1071. https://doi.org/10.1021/acs.jafc.8b06241

[22] W. Peng, L. Han, H. Huang, X. Xuan, G. Pan, L. Wan, T. Lu, M. Xu, L. Pan, A direction-aware and ultrafast self-healing dual network hydrogel for a flexible electronic skin strain sensor, Journal of Materials Chemistry A 8 (2020) 26109-26118. https://doi.org/10.1039/D0TA08987G

[23] Z. Rao, S. Liu, R. Wu, G. Wang, Z. Sun, L. Bai, W. Wang, H. Chen, H. Yang, D. Wei, Fabrication of dual network self-healing alginate/guar gum hydrogels based on polydopamine-type microcapsules from mesoporous silica nanoparticles, International Journal of Biological Macromolecules 129 (2019) 916-926. https://doi.org/10.1016/j.ijbiomac.2019.02.089

[24] M.-M. Song, Y.-M. Wang, X.-Y. Liang, X.-Q. Zhang, S. Zhang, B.-J. Li, Functional materials with self-healing properties: A review, Soft Matter 15 (2019) 6615-6625. https://doi.org/10.1039/C9SM00948E

[25] N. Wen, T. Song, Z. Ji, D. Jiang, Z. Wu, Y. Wang, and Z. Guo, Recent advancements in self-healing materials: Mechanicals, performances and features, Reactive and Functional Polymers 168 (2021) 105041. https://doi.org/10.1016/j.reactfunctpolym.2021.105041

[26] D. Sun, G. Sun, X. Zhu, A. Guarin, B. Li, Z. Dai, J. Ling, A comprehensive review on self-healing of asphalt materials: Mechanism, model, characterization and

enhancement, Advances in Colloid and Interface Science 256 (2018) 65-93.
https://doi.org/10.1016/j.cis.2018.05.003

[27] W. Sun, H. Wang, Self-healing of asphalt binder with cohesive failure: Insights from molecular dynamics simulation, Construction and Building Materials 262 (2020) 120538. https://doi.org/10.1016/j.conbuildmat.2020.120538

[28] C. E. Attinger, J. E. Janis, J. Steinberg, J. Schwartz, A. Al-Attar, K. Couch, Clinical approach to wounds: debridement and wound bed preparation including the use of dressings and wound-healing adjuvants, Plastic and Reconstructive Surgery 117 (2006) 72S-109S. https://doi.org/10.1097/01.prs.0000225470.42514.8f

[29] E. D. Rodriguez, X. Luo, P. T. Mather, Linear/network poly (ε-caprolactone) blends exhibiting shape memory assisted self-healing (SMASH), ACS Applied Materials & Interfaces 3 (2011) 152-161. https://doi.org/10.1021/am101012c

[30] C. Kim, N. Yoshie, Polymers healed autonomously and with the assistance of ubiquitous stimuli: how can we combine mechanical strength and a healing ability in polymers?, Polymer Journal 50 (2018) 919-929. https://doi.org/10.1038/s41428-018-0079-x

[31] H. Mokhtari, M. Kharaziha, F. Karimzadeh, and S. Tavakoli, An injectable mechanically robust hydrogel of Kappa-carrageenan-dopamine functionalized graphene oxide for promoting cell growth, Carbohydrate Polymers 214 (2019) 234-249. https://doi.org/10.1016/j.carbpol.2019.03.030

[32] E. J. Bailey, K. I. Winey, Dynamics of polymer segments, polymer chains, and nanoparticles in polymer nanocomposite melts: A review, Progress in Polymer Science 105 (2020) 101242. https://doi.org/10.1016/j.progpolymsci.2020.101242

[33] R. Weeber, M. Hermes, A. M. Schmidt, C. Holm, Polymer architecture of magnetic gels: a review, Journal of Physics: Condensed Matter 30 (2018) 063002. https://doi.org/10.1088/1361-648X/aaa344

[34] M. I. Sujan, S. D. Sarkar, S. Sultana, L. Bushra, R. Tareq, C. K. Roy, M. S. Azam, Bi-functional silica nanoparticles for simultaneous enhancement of mechanical strength and swelling capacity of hydrogels, RSC Advances 10 (2020) 6213-6222. https://doi.org/10.1039/C9RA09528D

[35] Z. Liu, Y. Faraj, X. J. Ju, W. Wang, R. Xie, L. Y. Chu, Nanocomposite smart hydrogels with improved responsiveness and mechanical properties: A mini review, Journal of Polymer Science Part B: Polymer Physics 56 (2018) 1306-1313. https://doi.org/10.1002/polb.24723

[36] S. Spoljaric, A. Salminen, N. D. Luong, J. Seppälä, Stable, self-healing hydrogels from nanofibrillated cellulose, poly (vinyl alcohol) and borax via reversible crosslinking, European Polymer Journal 56 (2014) 105-117. https://doi.org/10.1016/j.eurpolymj.2014.03.009

[37] A. Zengin, J. P. O. Castro, P. Habibovic, S. H. Van Rijt, Injectable, self-healing mesoporous silica nanocomposite hydrogels with improved mechanical properties, Nanoscale 13 (2021) 1144-1154. https://doi.org/10.1039/D0NR07406C

[38] X. Xue, Y. Hu, Y. Deng, J. Su, Recent advances in design of functional biocompatible hydrogels for bone tissue engineering, Advanced Functional Materials 31 (2021) 2009432. https://doi.org/10.1002/adfm.202009432

[39] E. V. Skorb, D. V. Andreeva, Layer-by-Layer approaches for formation of smart self-healing materials, Polymer Chemistry 4 (2013) 4834-4845. https://doi.org/10.1039/c3py00088e

[40] R. Fuhrer, E. K. Athanassiou, N. A. Luechinger, and W. J. Stark, Crosslinking metal nanoparticles into the polymer backbone of hydrogels enables preparation of soft, magnetic field-driven actuators with muscle-like flexibility, Small 5 (2009) 383-388. https://doi.org/10.1002/smll.200801091

[41] N. Asadi, E. Alizadeh, R. Salehi, B. Khalandi, S. Davaran, and A. Akbarzadeh, Nanocomposite hydrogels for cartilage tissue engineering: a review, Artificial Cells, Nanomedicine, and Biotechnology 46 (2018) 465-471. https://doi.org/10.1080/21691401.2017.1345924

[42] F.-m. Cheng, H.-x. Chen, H.-d. Li, Recent advances in tough and self-healing nanocomposite hydrogels for shape morphing and soft actuators, European Polymer Journal 124 (2020) 109448. https://doi.org/10.1016/j.eurpolymj.2019.109448

[43] I. Pastoriza-Santos, C. Kinnear, J. Pérez-Juste, P. Mulvaney, L. M. Liz-Marzán, Plasmonic polymer nanocomposites, Nature Reviews Materials 3 (2018) 375-391. https://doi.org/10.1038/s41578-018-0050-7

[44] V. Bertolino, G. Cavallaro, G. Lazzara, S. Milioto, F. Parisi, Biopolymer-targeted adsorption onto halloysite nanotubes in aqueous media, Langmuir 33 (2017) 3317-3323. https://doi.org/10.1021/acs.langmuir.7b00600

[45] S. Rose, A. Prevoteau, P. Elzière, D. Hourdet, A. Marcellan, and L. Leibler, Nanoparticle solutions as adhesives for gels and biological tissues, Nature 505 (2014) 382-385. https://doi.org/10.1038/nature12806

[46] Z.-B. Zhang, Z.-G. Shen, J.-X. Wang, H.-X. Zhang, H. Zhao, J.-F. Chen, J. Yun, Micronization of silybin by the emulsion solvent diffusion method, International Journal of Pharmaceutics 376 (2009) 116-122. https://doi.org/10.1016/j.ijpharm.2009.04.028

[47] S. B. Marpu, E. N. Benton, Shining light on chitosan: A review on the usage of chitosan for photonics and nanomaterials research, International Journal of Molecular Sciences 19 (2018) 1795. https://doi.org/10.3390/ijms19061795

[48] S. Bhattacharya, S. K. Samanta, Soft-nanocomposites of nanoparticles and nanocarbons with supramolecular and polymer gels and their applications, Chemical Reviews 116 (2016) 11967-12028. https://doi.org/10.1021/acs.chemrev.6b00221

[49] H. Kawaguchi, Functional polymer microspheres, Progress in Polymer Science 25 (2000) 1171-1210. https://doi.org/10.1016/S0079-6700(00)00024-1

[50] T. E. Gartner Iii, A. Jayaraman, Modeling and simulations of polymers: A roadmap, Macromolecules 52 (2019) 755-786. https://doi.org/10.1021/acs.macromol.8b01836

[51] D. T. Cheung, M. E. Nimni, Mechanism of crosslinking of proteins by glutaraldehyde II. Reaction with monomeric and polymeric collagen, Connective Tissue Research 10 (1982) 201-216. https://doi.org/10.3109/03008208209034419

[52] G. A. Williams, R. Ishige, O. R. Cromwell, J. Chung, A. Takahara, Z. Guan, Mechanically robust and self-healable superlattice nanocomposites by self-assembly of single-component "sticky" polymer-grafted nanoparticles, Advanced Materials 27 (2015) 3934-3941. https://doi.org/10.1002/adma.201500927

[53] O. Goor, P. Y. W. Dankers, Advances in the development of supramolecular polymeric biomaterials, Comprehensive Supramolecular Chemistry II. Advances in the Development of Supramolecular Polymeric Biomaterials 255 (2017) 282. https://doi.org/10.1016/B978-0-12-409547-2.12574-0

[54] R. G. Chaudhary, A. K. Potbhare, P. B. Chouke, A. R. Rai, R. P. Mishra, M. Desimone, A. Abdala, Graphene-based nanomaterials and their nanocomposites with metal oxides: biosynthesis, electrochemical, photocatalytic and antimicrobial applications, Magnetic Oxides and Composites II, Materials Research Forum 83 (2020) 79-116. https://doi.org/10.21741/9781644900970-4

[55] M. S. Umekar, G. S. Bhusari, A. K. Potbhare, A. Mondal, B. P. Kapgate, M. Desimone, R. G. Chaudhary, Bioinspired reduced graphene oxide based nanohybrids for photocatalysis and antibacterial applications, Current Pharmaceutical Biotechnology 22 (2021) 1759-1781. https://doi.org/10.2174/1389201022666201231115826

[56] S. G. Giuffrida, W. Forysiak, P. Cwynar, R. Szweda, Shaping macromolecules for sensing applications-from polymer hydrogels to foldamers, Polymers 14 (2022) 580. https://doi.org/10.3390/polym14030580

[57] J. Liu, G. Song, C. He, H. Wang, Self-healing in tough graphene oxide composite hydrogels, Macromolecular Rapid Communications 34 (2013) 1002-1007. https://doi.org/10.1002/marc.201300242

[58] J. Liu, J. Tang, J. J. Gooding, Strategies for chemical modification of graphene and applications of chemically modified graphene, Journal of Materials Chemistry 22 (2012) 12435-12452. https://doi.org/10.1039/c2jm31218b

[59] D. Zhao, Y. Zhu, W. Cheng, W. Chen, Y. Wu, H. Yu, Cellulose-based flexible functional materials for emerging intelligent electronics, Advanced Materials 33 (2021) 2000619. https://doi.org/10.1002/adma.202000619

[60] B. L. Peng, N. Dhar, H. L. Liu, K. C. Tam, Chemistry and applications of nanocrystalline cellulose and its derivatives: A nanotechnology perspective, The

Canadian Journal of Chemical Engineering 89 (2011) 1191-1206. https://doi.org/10.1002/cjce.20554

[61] H. Rammal, A. GhavamiNejad, A. Erdem, R. Mbeleck, M. Nematollahi, S. E. Diltemiz, H. Alem, M. A. Darabi, Y. N. Ertas, E. J. Caterson, Advances in Biomedical Applications of Self-Healing Hydrogels, Materials Chemistry Frontiers 5 (2021) 4368-4400. https://doi.org/10.1039/D0QM01099E

[62] D. Prado-Audelo, M. Luisa, I. H. Caballero-Florán, N. Mendoza-Muñoz, D. Giraldo-Gomez, J. Sharifi-Rad, J. K. Patra, M. González-Torres, B. Florán, H. Cortes, Current progress of self-healing polymers for medical applications in tissue engineering, Iranian Polymer Journal (2021) 1-23. https://doi.org/10.1007/s13726-021-00943-8

[63] L.-H. Fu, C. Qi, M.-G. Ma, P. Wan, Multifunctional cellulose-based hydrogels for biomedical applications, Journal of Materials Chemistry B 7 (2019) 1541-1562. https://doi.org/10.1039/C8TB02331J

[64] S. M. Kabir, P. P. Sikdar, B. Haque, M. A. Bhuiyan, A. Ali, and M. N. Islam, Cellulose-based hydrogel materials: Chemistry, properties and their prospective applications, Progress in Biomaterials 7 (2018) 153-174. https://doi.org/10.1007/s40204-018-0095-0

[65] C. Chang, L. Zhang, Cellulose-based hydrogels: Present status and application prospects, Carbohydrate Polymers 84 (2011) 40-53. https://doi.org/10.1016/j.carbpol.2010.12.023

[66] B. Li, Y. Zhang, C. Wu, B. Guo, Z. Luo, Fabrication of mechanically tough and self-recoverable nanocomposite hydrogels from polyacrylamide grafted cellulose nanocrystal and poly (acrylic acid), Carbohydrate Polymers 198 (2018) 1-8. https://doi.org/10.1016/j.carbpol.2018.06.047

[67] D. M. Nascimento, Y. L. Nunes, M. C. B. Figueirêdo, H. M. C. de Azeredo, F. A. Aouada, J. P. A. Feitosa, M. F. Rosa, A. Dufresne, Nanocellulose nanocomposite hydrogels: Technological and environmental issues, Green Chemistry 20 (2018) 2428-2448. https://doi.org/10.1039/C8GC00205C

[68] Q. Peng, J. Chen, T. Wang, X. Peng, J. Liu, X. Wang, J. Wang, H. Zeng, Recent advances in designing conductive hydrogels for flexible electronics, InfoMat 2 (2020) 843-865. https://doi.org/10.1002/inf2.12113

[69] C. Jiang, W. Fan, N. Zhang, G. Zhao, W. Wang, L. Bai, H. Chen, H. Yang, Surface engineering of cellulose nanocrystals via SI-AGET ATRP of glycidyl methacrylate and ring-opening reaction for fabricating self-healing nanocomposite hydrogels, Cellulose 28 (2021) 9785-9801. https://doi.org/10.1007/s10570-021-04170-5

[70] G. Xiao, Y. Wang, H. Zhang, L. Chen, S. Fu, Facile strategy to construct a self-healing and biocompatible cellulose nanocomposite hydrogel via reversible acylhydrazone, Carbohydrate Polymers 218 (2019) 68-77. https://doi.org/10.1016/j.carbpol.2019.04.080

[71] A. B. W. Brochu, S. L. Craig, W. M. Reichert, Self-healing biomaterials, Journal of Biomedical Materials Research Part A 96 (2011) 492-506. https://doi.org/10.1002/jbm.a.32987

[72] A. U. Chaudhry, A. Abdala, S. P. Lonkar, R. G. Chaudhary, A. Mabrouk, Thermal, electrical, and mechanical properties of highly filled HDPE/graphite nanoplatelets composites, Materials Today: Proceedings 29 (2020) 704-708 https://doi.org/10.1016/j.matpr.2020.04.168

[73] M. Zhu, J. Liu, L. Gan, M. Long, Research progress in bio-based self-healing materials, European Polymer Journal 129 (2020) 109651. https://doi.org/10.1016/j.eurpolymj.2020.109651

[74] M. E. Abd El-Hack, M. T. El-Saadony, M. E. Shafi, N. M. Zabermawi, M. Arif, G. E. Batiha, A. F. Khafaga, Y. M. Abd El-Hakim, A. A. Al-Sagheer, Antimicrobial and antioxidant properties of chitosan and its derivatives and their applications: A review, International Journal of Biological Macromolecules 164 (2020) 2726-2744. https://doi.org/10.1016/j.ijbiomac.2020.08.153

[75] L. Phil, M. Naveed, I. S. Mohammad, L. Bo, D. Bin, Chitooligosaccharide: An evaluation of physicochemical and biological properties with the proposition for determination of thermal degradation products, Biomedicine & Pharmacotherapy 102 (2018) 438-451. https://doi.org/10.1016/j.biopha.2018.03.108

[76] V.-D. Mai, S.-R. Shin, D.-S. Lee, I. Kang, Thermal healing, reshaping and ecofriendly recycling of epoxy resin crosslinked with Schiff base of vanillin and hexane-1, 6-diamine, Polymers 11 (2019) 293. https://doi.org/10.3390/polym11020293

[77] D. D. Steppan, M. F. Doherty, M. F. Malone, A simplified degradation model for nylon 6, 6 polymerization, Journal of Applied Polymer Science 42 (1991) 1009-1021. https://doi.org/10.1002/app.1991.070420415

[78] E. Troschke, M. Oschatz, and I. K. Ilic, Schiff-bases for sustainable battery and supercapacitor electrodes, Wiley Online Library, 2021. https://doi.org/10.1002/EXP.20210128

[79] H. Xu, L. Zhang, J. Cai, Injectable, self-healing, β-chitin-based hydrogels with excellent cytocompatibility, antibacterial activity, and potential as drug/cell carriers, ACS Applied Bio Materials 2 (2018) 196-204. https://doi.org/10.1021/acsabm.8b00548

[80] J. Qu, X. Zhao, Y. Liang, T. Zhang, P. X. Ma, B. Guo, Antibacterial adhesive injectable hydrogels with rapid self-healing, extensibility and compressibility as wound dressing for joints skin wound healing, Biomaterials 183 (2018), 185-199. https://doi.org/10.1016/j.biomaterials.2018.08.044

[81] Z. Wang, L. Yang, W. Fang, Chitosan-based hydrogels, Chitin and chitosan: Properties and applications (2019) 97-144. https://doi.org/10.1002/9781119450467.ch5

Materials Research Forum LLC
https://doi.org/10.21741/9781644902295-11

[82] F. Sami El-banna, M. E. Mahfouz, S. Leporatti, M. El-Kemary, N. An Hanafy, Chitosan as a natural copolymer with unique properties for the development of hydrogels, Applied Sciences 9 (2019) 2193. https://doi.org/10.3390/app9112193

[83] J. Jin, L. Cai, Y.-G. Jia, S. Liu, Y. Chen, L. Ren, Progress in self-healing hydrogels assembled by host-guest interactions: Preparation and biomedical applications, Journal of Materials Chemistry B 7 (2019) 1637-1651. https://doi.org/10.1039/C8TB02547A

[84] M. Mohamadhoseini, Z. Mohamadnia, Supramolecular self-healing materials via host-guest strategy between cyclodextrin and specific types of guest molecules, Coordination Chemistry Reviews 432 (2021) 213711. https://doi.org/10.1016/j.ccr.2020.213711

[85] Y. Zhou, Y. Zhang, Z. Dai, F. Jiang, J. Tian, W. Zhang, A super-stretchable, self-healing and injectable supramolecular hydrogel constructed by a host-guest crosslinker, Biomaterials Science 8 (2020) 3359-3369. https://doi.org/10.1039/D0BM00290A

[86] F. Jiang, Z. Tang, Y. Zhang, Y. Ju, H. Gao, N. Sun, F. Liu, P. Gu, W. Zhang, Enhanced proliferation and differentiation of retinal progenitor cells through a self-healing injectable hydrogel, Biomaterials Science, 7 (2019) 2335-2347. https://doi.org/10.1039/C8BM01579A

[87] V. S. Raghuwanshi, G. Garnier, Characterisation of hydrogels: Linking the nano to the microscale, Advances in Colloid and Interface Science 274 (2019) 102044. https://doi.org/10.1016/j.cis.2019.102044

[88] J. C. G. Jeynes, Nanotubes for Biotechnology, Handbook of Nanophysics: Nanomedicine and Nanorobotics, 2010.

[89] J. A. Yoon, J. Kamada, K. Koynov, J. Mohin, R. Nicolaÿ, Y. Zhang, A. C. Balazs, T. Kowalewski, K. Matyjaszewski, Self-healing polymer films based on thiol-disulfide exchange reactions and self-healing kinetics measured using atomic force microscopy, Macromolecules 45 (2012) 142-149. https://doi.org/10.1021/ma2015134

[90] L. Zedler, M. D. Hager, U. S. Schubert, M. J. Harrington, M. Schmitt, J. Popp, B. Dietzek, Monitoring the chemistry of self-healing by vibrational spectroscopy-current state and perspectives, Materials Today 17 (2014) 57-69. https://doi.org/10.1016/j.mattod.2014.01.020

[91] R. Geitner, F. B. Legesse, N. Kuhl, T. W. Bocklitz, S. Zechel, J. Vitz, M. Hager, U. S. Schubert, B. Dietzek, M. Schmitt, Do you get what you see? Understanding molecular self-healing, Chemistry-A European Journal 24 (2018) 2493-2502. https://doi.org/10.1002/chem.201705836

[92] F. El-Diasty, Coherent anti-Stokes Raman scattering: Spectroscopy and microscopy, Vibrational Spectroscopy 55 (2011) 1-37. https://doi.org/10.1016/j.vibspec.2010.09.008

Emerging Applications of Nanomaterials   Materials Research Forum LLC
Materials Research Foundations 141 (2023) 270-293   https://doi.org/10.21741/9781644902295-11

[93] X. Tong, L. Du, Q. Xu, Tough, adhesive and self-healing conductive 3D network hydrogel of physically linked functionalized-boron nitride/clay/poly (N-isopropylacrylamide), Journal of Materials Chemistry A 6 (2018) 3091-3099. https://doi.org/10.1039/C7TA10898B

[94] M. Caprioli, I. Roppolo, A. Chiappone, L. Larush, C. F. Pirri, S. Magdassi, 3D-printed self-healing hydrogels via digital light processing, Nature Communications 12 (2021) 1-9. https://doi.org/10.1038/s41467-021-22802-z

[95] J. Zhao, D. Diaz-Dussan, M. Wu, Y.-Y. Peng, J. Wang, H. Zeng, W. Duan, L. Kong, X. Hao, R. Narain, Dual-cross-linked network hydrogels with multiresponsive, self-healing, and shear strengthening properties, Biomacromolecules 22 (2020) 800-810. https://doi.org/10.1021/acs.biomac.0c01548

[96] S. Bashir, M. Hina, J. Iqbal, A. H. Rajpar, M. A. Mujtaba, N. A. Alghamdi, S. Wageh, K. Ramesh, S. Ramesh, Fundamental concepts of hydrogels: Synthesis, properties, and their applications, Polymers 12 (2020) 2702. https://doi.org/10.3390/polym12112702

[97] M. Vázquez-González, I. Willner, Stimuli-responsive biomolecule-based hydrogels and their applications, Angewandte Chemie International Edition 59 (2020) 15342-15377. https://doi.org/10.1002/anie.201907670

[98] A. B. Ihsan, T. L. Sun, T. Kurokawa, S. N. Karobi, T. Nakajima, T. Nonoyama, C. K. Roy, F. Luo, J. P. Gong, Self-healing behaviors of tough polyampholyte hydrogels, Macromolecules 49 (2016) 4245-4252. https://doi.org/10.1021/acs.macromol.6b00437

[99] Q. Geng, C. Zhang, K. Zheng, J. Zhang, J. Cheng, W. Yang, Preparation and properties of a self-healing, multiresponsive color-change hydrogel, Industrial & Engineering Chemistry Research 59 (2020) 10689-10696. https://doi.org/10.1021/acs.iecr.0c00219

[100] G. Li, D. Nettles, Thermomechanical characterization of a shape memory polymer based self-repairing syntactic foam, Polymer 51 (2010) 755-762. https://doi.org/10.1016/j.polymer.2009.12.002

[101] T. L. Sun, F. Luo, W. Hong, K. Cui, Y. Huang, H. J. Zhang, D. R. King, T. Kurokawa, T. Nakajima, J. P. Gong, Bulk energy dissipation mechanism for the fracture of tough and self-healing hydrogels, Macromolecules 50 (2017) 2923-2931. https://doi.org/10.1021/acs.macromol.7b00162

[102] Q. Wang, Z. Gao, K. Yu, Interfacial self-healing of nanocomposite hydrogels: Theory and experiment, Journal of the Mechanics and Physics of Solids 109 (2017) 288-306. https://doi.org/10.1016/j.jmps.2017.08.004

[103] Y. Jiang, N. Krishnan, J. Heo, R. H. Fang, L. Zhang, Nanoparticle-hydrogel superstructures for biomedical applications, Journal of Controlled Release 324 (2020) 505-521. https://doi.org/10.1016/j.jconrel.2020.05.041

[104] D. Chimene, R. Kaunas, A. K. Gaharwar, Hydrogel bioink reinforcement for additive manufacturing: A focused review of emerging strategies, Advanced Materials 32 (2020) 1902026. https://doi.org/10.1002/adma.201902026

[105] N. Molinari, G. Jung, S. Angioletti-Uberti, Designing nanoparticles as glues for hydrogels: Insights from a microscopic model, Macromolecules 54 (2021) 1992-2000. https://doi.org/10.1021/acs.macromol.0c02353

[106] J. S. Chen, X. W. Lou, The superior lithium storage capabilities of ultra-fine rutile TiO2 nanoparticles, Journal of Power Sources 195 (2010) 2905-2908. https://doi.org/10.1016/j.jpowsour.2009.11.040

[107] M. C. Arno, M. Inam, A. C. Weems, Z. Li, A. L. A. Binch, C. I. Platt, S. M. Richardson, J. A. Hoyland, A. P. Dove, and R. K. O'Reilly, Exploiting the role of nanoparticle shape in enhancing hydrogel adhesive and mechanical properties, Nature Communications 11 (2020) 1-9. https://doi.org/10.1038/s41467-020-15206-y

[108] A. D. Valino, J. R. C. Dizon, A. H. Espera Jr, Q. Chen, J. Messman, R. C. Advincula, Advances in 3D printing of thermoplastic polymer composites and nanocomposites, Progress in Polymer Science 98 (2019) 101162. https://doi.org/10.1016/j.progpolymsci.2019.101162

[109] D. Fan, G. Wang, A. Ma, W. Wang, H. Chen, L. Bai, H. Yang, D. Wei, L. Yang, Surface engineering of porous carbon for self-healing nanocomposite hydrogels by mussel-inspired chemistry and PET-ATRP, ACS Applied Materials & Interfaces 11 (2019) 38126-38135. https://doi.org/10.1021/acsami.9b12264

[110] M. Zhou, J. He, L. Wang, S. Zhao, Q. Wang, S. Cui, X. Qin, R. Wang, Synthesis of carbonized-cellulose nanowhisker/FeS2@ reduced graphene oxide composite for highly efficient counter electrodes in dye-sensitized solar cells, Solar Energy 166 (2018) 71-79. https://doi.org/10.1016/j.solener.2018.01.089

[111] S. Kim, Competitive biological activities of chitosan and its derivatives: Antimicrobial, antioxidant, anticancer, and anti-inflammatory activities, International Journal of Polymer Science, 2018. https://doi.org/10.1155/2018/1708172

[112] Y. Tu, N. Chen, C. Li, H. Liu, R. Zhu, S. Chen, Q. Xiao, J. Liu, S. Ramakrishna, L. He, Advances in injectable self-healing biomedical hydrogels, Acta Biomaterialia 90 (2019) 1-20. https://doi.org/10.1016/j.actbio.2019.03.057

[113] M. Diba, J. An, S. Schmidt, M. Hembury, D. Ossipov, A. R. Boccaccini, S. C. G. Leeuwenburgh, Exploiting bisphosphonate-bioactive-glass interactions for the development of self-healing and bioactive composite hydrogels, Macromolecular Rapid Communications 37 (2016) 1952-1959. https://doi.org/10.1002/marc.201600353

[114] Y. M. Malinskii, V. V. Prokopenko, N. A. Ivanova, V. A. Kargin, Investigation of self-healing of cracks in polymers, Polymer Mechanics 6 (1970) 240-244. https://doi.org/10.1007/BF00859196

[115] E. Zhang, T. Wang, L. Zhao, W. Sun, X. Liu, Z. Tong, Fast self-healing of graphene oxide-hectorite clay-poly (N, N-dimethylacrylamide) hybrid hydrogels realized by near-infrared irradiation, ACS Applied Materials & Interfaces 6 (2014) 22855-22861. https://doi.org/10.1021/am507100m

[116] K. Haraguchi, K. Uyama, H. Tanimoto, Self-healing in nanocomposite hydrogels, Macromolecular Rapid Communications 32 (2011) 1253-1258. https://doi.org/10.1002/marc.201100248

[117] A. Dev, S. J. Mohanbhai, A. C. Kushwaha, A. Sood, M. N. Sardoiwala, S. R. Choudhury, S. Karmakar, κ-carrageenan-C-phycocyanin based smart injectable hydrogels for accelerated wound recovery and real-time monitoring, Acta Biomaterialia 109 (2020) 121-131. https://doi.org/10.1016/j.actbio.2020.03.023

[118] Y. Li, C. P. Wong, Recent advances of conductive adhesives as a lead-free alternative in electronic packaging: Materials, processing, reliability and applications, Materials Science and Engineering: R: Reports 51 (2006) 1-35. https://doi.org/10.1016/j.mser.2006.01.001

[119] S. Awasthi, J. K. Gaur, S. K. Pandey, M. S. Bobji, and C. Srivastava, High-strength, strongly bonded nanocomposite hydrogels for cartilage repair, ACS Applied Materials & Interfaces 13 (2021) 24505-24523. https://doi.org/10.1021/acsami.1c05394

Emerging Applications of Nanomaterials   Materials Research Forum LLC
Materials Research Foundations 141 (2023) 294-326   https://doi.org/10.21741/9781644902295-12

Chapter 12

Emerging Nanomaterials in Energy Storage

Aniruddha Mondal[1*], Himadri Tanaya Das[2], Sudip Mondal[3], Vaishali N. Sonkusare[4], and Ratiram Gomaji Chaudhary[3*]

[1]Division of Materials Science, Department of Engineering Sciences and Mathematics, Luleå University of Technology, Luleå, SE-971 87 Sweden

[2]Centre of Excellence for Advance Materials and Applications, Utkal University, Bhubaneswar, Odisha, India

[3]Post Graduate Department of Chemistry, Seth Kesarimal Porwal College of Arts, Science and Commerce, Kamptee 441 001 MS, India

[4]Post Graduate Teaching Department of Chemistry, Rashtrasant Tukdoji Maharaj Nagpur University Nagpur-440033, India

aniruddha.mondal@ltu.se; chaudhary_rati@yahoo.com

Abstract

Continuous renewable technologies can only be adequate when coupled with efficient nanomaterial based energy storage systems. These gadgets can reliably provide electricity even on overcast days or at night. To power the majority of consumer devices regardless of environmental conditions, the battery business is thriving. Among electrochemical energy storage devices (EESD), lithium-ion batteries (LiBs) have been a popular option for many eras. Even LiBs with a greater energy density (ED) and strong charge-discharge behaviour still have safety, durability problems and are expensive. Thus, various battery technologies, have attracted the attention of scientists all around the globe. However, both main and secondary batteries are used to power numerous electronic equipment. Focus will be placed on optimising battery performance, cost, and mass manufacturing in order to commercialize the batteries. This chapter will explore several battery kinds with different nanomaterials and their characteristics. Extensive details will be provided on the regulating criteria for battery performance, its fundamental design, and the operating principle of energy storage. In addition to diverse electrodes and electrolytes, this chapter provides information on the benefits and downsides of various batteries as well as ideas for future advancements in smart electronics battery systems.

Keywords

Energy Storage, Nanomaterials, Batteries, Supercapacitors, Energy Density, Power Density

Contents

1.   Introduction

Energy storage systems will be utilized in the upcoming days to integrate variable renewable energy (RE) production with a variety of load situations since they can decouple the time of production and consumption [1, 2]. Electrochemical energy storage (EES) systems are receiving a great deal of interest in the power sector because of ample advantageous characteristics, such as rapid reaction, modular design, and simple assimilation [3–5]. In addition, they aid as an energy buffer to compensate for a mismatch between the amount of energy produced and the amount of power used, in addition to offering a variety of additional functions [6, 7].

The energy storage batteries (ESB) would be seen as a crucial component in diversifying current energy sources. Utilizing an ESB is a practical solution to reduce the recurring nature of RE sources, which means that the energy generated does not always match the energy required. When energy is stored and processed in an efficient and eco-friendly manner, it has a noteworthy impact on the global economy as well as the environment. There are a lot of different types of EES and conversion systems that can be used instead of fossil fuels, which comprise of typical zinc-manganese dioxide (Zn-Mn) [8] and metal-air (Mg/Al/Zn-air) batteries [9], rechargeable nickel-zinc oxide [10], lithium-ion (Li-ion) [11, 12] and magnesium-ion (Mg-ion) [13] batteries, and fuel cells, as favorable substitutes [14–16].

It is not just electric autos and other non-grid-dependent applications that need batteries. The majority of these devices are composed of the lighter elements in the upper rows of the periodic table. The lightest metal for its limited density and specific gravity (0.53 g cm^{-3}) is Li. It is mostly used in LiBs, which can store 3860 mA h g^{-1} [15–17]. A similar light-weight metal, magnesium, may also be found. Mg is utilized in primary and secondary Mg-air and Mg-ion batteries in part due to its utilities of less cost, less toxicity, and great safety. There are more valence electrons in the lesser elements on the periodic table than there are in the heavier ones. There is a Pauling ions radius of 0.60 for Li$^+$, 0.65 for Mg^{2+}, and 8.7 x 10^{-6} for H$^+$. Because they are smaller than bigger ions, they are more easily transported through the electrode and electrolyte [11, 13, 18]. Because of their low weight, a wide variety of objects are suitable for use with various types of batteries [2–4, 19].

Despite advancements, many batteries still have restrictions that hinder broad usage. Poor electron conductivity occurs in lithium and magnesium batteries due to delayed reaction kinetics, high polarisation, and limited ion-diffusion/migration. Limited ED, low capacity, and short durability are the results. Because nanomaterials used as battery electrodes and

Emerging Applications of Nanomaterials   Materials Research Forum LLC

Materials Research Foundations 141 (2023) 294-326   https://doi.org/10.21741/9781644902295-12

catalysts are prohibitively costly, researchers are inventing alternative nanomaterials which are easily accessible, less expensive and efficient. The traditional resources are old. Even under ideal circumstances, most real batteries have lower specific energy and capacity than anticipated by theory. The battery's efficacy is heavily impacted by its functional materials. Fig. 1 shows that the ED of the most common rechargeable battery systems has grown. Volume and weight show this. Less bulky, higher ED, less harmful chemicals, and improved cycle efficiency are trending and consistent pattern [20].

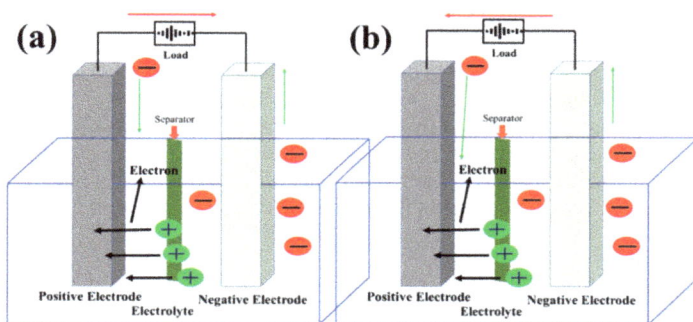

Figure 1. Illustrate the basic mechanism of energy storage system (a) Charging and (b) Discharging

2. Basics of energy storage system (batteries)

Through the conversion of chemical energy, a battery is able to create electrical energy. A battery is made up of two electrodes, an anode (+ve electrode) and a cathode (-ve electrode), which are linked to one another via an electrolyte. Half-cell by half-cell, an electrochemical reaction takes place in each electrode [21]. A diverse selection of electrical equipment may be found in our immediate environment. Only major electric equipment such as televisions and refrigerators need to be plugged into an outlet in order to function. The remaining ones are powered by rechargeable batteries. Smartphones, tablets, and portable music players are all examples of this category of technology. Particularly for big vehicles, such as electric automobiles, TV remote controls, air conditioners, and many other equipment need a significant quantity of batteries. Solar panels are installed to produce energy in the daytime that is then used to power the electrical system of a house during the evening. For usage in residential settings, a particularly big battery with storage capacity has been created [3, 4].These goods contain a wide variety of primary batteries (PB) and secondary/rechargeable batteries (SB), each of which may be used in a variety of various ways. The batteries are used in accordance with the application that was planned for them. The many varieties of batteries, the issues that are associated with them, and the solutions that are offered in this chapter may assist in making well-informed judgments on the purchase of batteries.

To begin, let's go through the many kinds of batteries that are available. There are many distinct varieties of primary cells, including as dry cells (which do not need recharging) and secondary cells (which function as rechargeable batteries), as well as redox cells that generate electrical power by means of chemical processes. In general, it works by producing electrical energy via a chemical reaction that takes place on electrodes in the presence of an electrolyte medium during the charging/ discharging. Where the operations of rechargeable batteries may be reversed [3, 22, 23].

Rechargeable batteries, in contrast to the ones being used here, have the capability of being repeatedly charged and discharged. Despite the fact that it produces energy via chemical processes, it may be distinguished from typical primary batteries due to the fact that the chemical reactions can be reversed (non-rechargeable). The substance does not go through any kind of chemical change as it is being discharged. Therefore, it is a process that causes the material to return to the condition it was in before the reaction took place.

2.1 Charging

Wh en the cell potential is exhausted, one of the options available is to recharge the battery. When an opposing current is delivered to the cell, the anode and cathode eventually switch places with one another. In conclusion, this is the overall model for the anode and cathode,

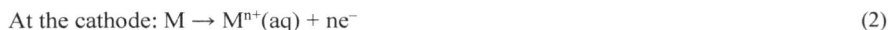

At the anode: $M^{n+}(aq) + ne^- \rightarrow M$ (1)

At the cathode: $M \rightarrow M^{n+}(aq) + ne^-$ (2)

Considering cell reactions where copper (anode) and zinc (cathode):

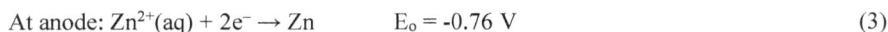

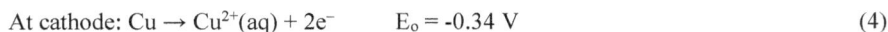

At anode: $Zn^{2+}(aq) + 2e^- \rightarrow Zn$ $E_o = -0.76$ V (3)

At cathode: $Cu \rightarrow Cu^{2+}(aq) + 2e^-$ $E_o = -0.34$ V (4)

Because 1.10 V is the potential of the battery, it is necessary to reach at least that level in order to fully charge the cell. In all likelihood, a far larger potential will be necessary in order to triumph over this polarization.

Polarisation effect: Polarization degrades the performance of batteries by disturbing the electrodes equilibrium potential [24, 25].

2.2 Discharging

During discharge, at full charge, there is a surplus of electrons on the anode (which becomes negative) and a deficiency of electrons on the cathode (which turns positive).

Because of this, a current is produced during discharge operations by flow of electrons from the cathode to the anode via external circuit. Electromotive force (emf), refers to the difference in potential that exists between an anode and a cathode, which has the ability to drive electrons in an external circuit. Once all materials cathode gets reduced and anode is oxidised, the battery is dead. The battery may then be thrown away, recharged, or recycled if at all feasible. This decision is based on the kind of battery, of course (secondary cells) [3, 10, 16].

The general model is

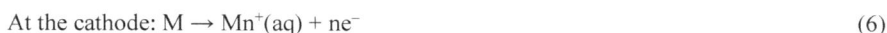

At the anode: $Mn^+(aq) + ne^- \rightarrow M$ (5)

At the cathode: $M \rightarrow Mn^+(aq) + ne^-$ (6)

In actual circumstances, the anode would be copper and the cathode would be zinc, and the half reactions would happen in the following manner:

At anode side: $Zn^{2+}(aq) + 2e^- \rightarrow Zn$ $\quad E_o = -0.76V$ (7)

At cathode side: $Cu \rightarrow Cu^{2+}(aq) + 2e^-$ $\quad E_o = -0.34V$ (8)

So, 1.10 V is the overall potential of the cell.

3.   Different categories of batteries

Generally, batteries are two types: (1) Primary Batteries and (2) Secondary Batteries

3.1   Primary batteries

After usage, PBs must be discarded. Dry cells have neither free nor liquid electrolyte since primary cell electrolytes are generally retained in an absorbent material or separator. The primary battery is a dry cell. Non-rechargeable batteries can't be charged. Gassner of Munich modified Leclanche's cell in 1888. When primary batteries were originally invented, zinc-carbon batteries were the most cost-effective. Its output voltage is 1.5 V and its capacity is 65 W h kg^{-1}. In 1859, Georges Leclanche created zinc-carbon. Even at normal temperature, a dry battery won't work. The battery's contents aren't spillable when it's moving or in any posture. Gels/absorbents may prevent electrolyte spills. Leaking electrolyte threatens the batteries' stability and lifetime. We must distinguish between primary dry cells and leak-proof rechargeable sealed batteries. Some backup batteries aren't charged until used. Solid electrolyte batteries are spill-proof yet operate best when heated [3].

Urry came up with the idea for alkaline cells in 1949 when he was employed with the Eveready Battery Company in Ohio, United States. They are an enhanced form that is

Emerging Applications of Nanomaterials   Materials Research Forum LLC
Materials Research Foundations 141 (2023) 294-326   https://doi.org/10.21741/9781644902295-12

known as Alkaline-manganese cell, it provides 1.6 V and have low rates of self-discharge but do not leak electrolytes on their exhaust. For over sixty years, the Zn-MnO_2 main alkaline battery (particularly AA) with KOH liquid electrolyte has been the most common kind of battery used in portable power sources. Electrolytic manganese dioxide (EMD) is used at the cathode of alkaline batteries, while micron-scale zinc particles are used at the anode. Both of these components are suspended in a gelled electrolyte that is made up of concentrated KOH in water [2]. The capacity of commercial cells is limited by the active mass of Zn in the anode of the cell. When the battery is drained, the MnO_2 interacts with one electron, and when the Zn anode is exhausted, the MnO_2 reacts with two electrons (Equation 9).

$$\text{Anodic reaction: } Zn + 2OH^- \longrightarrow ZnO + H_2O + 2e^- \tag{9}$$

Anode responses have two stages during discharge. OH^- anions oxidise Zn, forming solvated divalent zincate. When zincate concentration nears solubility limit, zinc oxide forms on electrode (Equation 9). World War II saw the creation of zinc-mercury oxide military battery technologies. First button cell was mercury oxide (HgO), used in scientific equipment. Mercuric oxide is the cathode. ZnO takes about 7% of the cathode's weight in the electrolyte. Shelf life was another high-performance feature. 1.35 V battery cathode is mercuric oxide; 1.4 V battery cathode is mercuric oxide with manganese dioxide. Cd/HgO cells have a longer storage life than Zn/HgO cells. Its prolonged storage life is due to Cd's lower solubility in the electrolyte. Despite having a low voltage of 0.9 V, the cells work effectively from -55 to 80 °C. Cd/HgO cells are costly compared to Zn/HgO cells, and Cd is very hazardous, thus they are only used in limited applications. Common Zn/HgO cells [3]

Electrochemical reactions during discharge

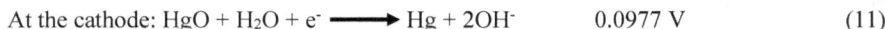

$$\text{At the anode: } Zn + 2OH^- \longrightarrow ZnO + H_2O + 2e^- \qquad 1.25 \text{ V} \tag{10}$$

$$\text{At the cathode: } HgO + H_2O + e^- \longrightarrow Hg + 2OH^- \qquad 0.0977 \text{ V} \tag{11}$$

$$\text{Overall reactions: } Zn + HgO \longrightarrow ZnO + Hg \tag{12}$$

An exhaustible amount of a practically weightless reactive component is carried by the cathode whenever a metal anode is electrochemically connected with an air (oxygen) cathode. At the cathode, there is no loss of material, the air is swiftly refilled by the environment around it, and while oxygen is lost at the cathode, it is soon restored by the air around it. Electrochemically, oxygen is depleted by the cell's porous cathode, which allows it to diffuse through the cell:

$$O_2 + 2H_2O + 4e^- \longrightarrow 4OH^- \qquad 0.401 \text{ V} \tag{13}$$

Materials Research Forum LLC

https://doi.org/10.21741/9781644902295-12

The metallic anode oxidises when OH^- ions diffuse into the electrolyte, which is neutral or alkaline. Theoretical ED is high, and discharge voltage curves are flat. Anodic capacity and parasitic responses may be used to measure capacity. Primary cells are those whose anodes cannot be electrochemically recharged.

With air electrodes, this area has made significant progress. Cathode and current collector are held in place by metal mesh. Through an air access hole, one may blow from nearby air. A diffusion membrane on the cathode surface controls oxygen transport. Cathode oxygen consumption affects cell current. By increasing porosity, current may be increased until the rate-limiting step is the previous reaction. MnO_2 and a carbon matrix (and platinum may be included) are combined into tiny Teflon liner to create a hydrophobic catalyst layer. Platinum may be act as catalyst layer. A dry barrier separates aqueous electrolyte from gas phase. Metal-air cells employ various metallic electrodes. Polarization at the electrode limits potential.

In the 1930s, G. André devised a zinc-and-silver-oxide battery. It had a separator that made the two electrodes clear. With a steady voltage of 1.6 V and 150 W h kg^{-1}, these cells can supply a lot of current quickly. Due to the high cost of creating components, these small batteries, called button cells, have a long lifetime and are appropriate for toys or watches. Graphite may increase the conductivity of monovalent silver oxide cathodes. The resulting liquid is then mixed with an alkaline electrolyte. A membrane of grafted plastic, treated cellophane, and unwoven fibres separates the electrodes. The +ve site is made of nickel-plated steel, while the –ve site is made of copper, tin, steel, or nickel. A gasket prevents short circuits by separating the two surfaces [3, 4, 26].

Electrochemical reactions

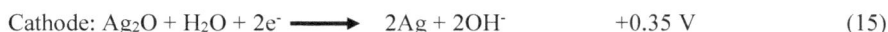

Anode: $Zn + 2OH^- \longrightarrow Zn(OH)_2 + 2e^-$ +1.25 V   (14)

Cathode: $Ag_2O + H_2O + 2e^- \longrightarrow 2Ag + 2OH^-$ +0.35 V   (15)

Overall Reactions: $Ag_2O + H_2O + Zn \longrightarrow 2Ag + Zn(OH)_2$ +1.6 V   (16)

Zinc/air has gained attention as a possible electric vehicle technology in recent years (EVs). The cell is also eco-friendly. This battery was used in military radios from the 1930s until WWII. The 1970s-developed commercial variant, which comes in N and D sizes, is used in paging systems. Watches and hearing aids now use button cells. Cathode holes allow electrochemical reaction to occur. These holes let oxygen into cells. Commercially available battery options include 1.4 V single-cell and 5.6 V and 8 V multicell batteries. 1.4 V batteries provide 442 to 970 W h dm^{-3} energy density. This range represents the batteries' several sizes. These batteries outperform zinc-silver batteries and are more environmentally friendly. Flat cell Zn/air batteries are also produced in big quantities [3].

3.1.2 Lithium (Li) as primary batteries

The lightweight and small device needed high ED lithium batteries. Lightest anode metal is Li. Carbon-based current collectors couple with MnO_2 or thionyl chloride cathodes. Li floats and is light. Li's reduction potential is 3.05 V at 298 K, therefore it can yield 3.860 A h g^{-1}. Zinc is weaker than sodium (1.16 A h g^{-1} against 0.82 A h g^{-1}, respectively). Lithium interacts aggressively with water and air, prohibiting it from operating in aquatic environments. Organic electrolytes are preferred in LiBs due of lithium's moisture sensitivity. Low conductivity makes long-lasting batteries in biomedicine. Propylene carbonate, liquid sulphur dioxide, and thionyl chloride protect Li against corrosion. The safe metal salts and organic liquids for battery electrolytes took years to determine.

In 1975, Sanyo Electric Company began selling cylinder-shaped lithium-ion batteries. First used in cameras and other electronics. Iron sulphide may also be a cathode. A 1.5 V battery may replace an alkaline or Leclanche (carbon-zinc) battery using this procedure [3].

Several lithium delivery methods have been developed in the previous three decades. Lithium-copper fluoride, lithium-silver chromate, lithium-lead bismuthate, lithium–polycarbonate monofluoride, and lithium–iodine primary lithium battery systems are also available. Lithium primary batteries are the same size as alkaline batteries, but their manufacture needs more energy and poses safety issues. High-tech companies produce lithium cells and batteries. Assembly requires relative humidity between 1 and 3%, with 1% as the top limit. Major battery makers have caused dozens of factory fires by mishandling materials. Owing to that, efficient nanomaterials have been used for the efficient energy storage materials.

3.1.3 Lithium/thionyl chloride batteries (Li/SOCL₂)

In the beginning of 1968, Li-primary cell that had a longer service life of 15–20 years and a higher energy density in comparison to earlier Li cells [21]. Both electrodes are submerged in organic lithium-ion electrolyte (Li(AlCl₄)). The lithium foil anode and thionyl chloride cathode (SOCl₂). These cells offer a 450 A h kg^{-1} capacity and 330 W h kg^{-1} energy density. Thionyl chloride is combined with carbon to generate an anode that doesn't require a separator in aqueous batteries. This forms an anode-cathode. The cell's internal resistance is minimal despite its high energy content. When a cell's discharge is halted, a solid electrolyte insulating covering regenerates on its surface. Lithium batteries can sustain high current for a long period. When assembling cells, exercise care to prevent overheating accidents. Li/thionyl discharge reaction explanation:

When it is mixed with a lithium salt such as LiClO₄ or LiAlCl₄, thionyl chloride not only functions as a cathode material, but it also performs the function of an electrolyte in the reaction. The carbon black electrode that makes up the cathode is what makes up the cathode. Because its discharge curve remains flat for an extended period of time, this cell is helpful for computers back-up. The battery cell has a high energy density, a long durability, as well as an excellent fit for medical equipment with medium power needs such as pacemakers, medication pumps, and neuro-stimulators. When it comes to medical equipment, having a discharge curve that flattens off and then has quick voltage drop at the end might be a major issue.

Table 1. Comparison of the different types of primary batteries and their applications

System	Nominal Cell Voltage (V)	Capacity (W h kg^{-1})	Advantages	Disadvantages	Applications
C/Zn	1.50	65	Lowest cost; variety of shapes and sizes	Low energy density; poor low-temperature performance	Torches; radios; electronic toys and games
Mg/MnO$_2$	1.60	105	Higher capacity than C/Zn; good shelf life	High gassing on discharge; delayed voltage	Military and aircraft receiver-transmitters
Zn/Alk/ MO$_2$	1.50	95	Higher capacity than C/Zn; good low-temperature performance	Moderate cost	Personal stereos; calculators; radio; TV
Zn/HgO	1.35	105	High Energy density; flat discharge; stable voltage	Expensive; energy density only moderate	Hearing aids; pacemakers; photography; military sensors/detectors
Cd/HgO	0.90	45	Good high and low-temperature performance; good shelf life	Expensive; low energy density	
Zn/Ag$_2$O	1.60	130	High Energy density, good high-rate performance	Expensive (but cost effective)	Watches; photography; missiles; Larger space applications
Zn/Air	1.50	290	High Energy density; long shelf life	Dependent on environment; limited power output	Watches; hearing aids; railway signals; electric fences
Li/SOCl$_2$	3.60	300	High Energy density; long shelf life	Only low to moderate rate applications	Memory devices; standby electrical power devices
Li/SO$_2$	3.00	280	High energy density; best low-temperature performance; long shelf life	High-cost pressurized system	Military and special industrial needs

Emerging Applications of Nanomaterials   Materials Research Forum LLC

Materials Research Foundations 141 (2023) 294-326   https://doi.org/10.21741/9781644902295-12

Table 2. Advantages and disadvantages of the primary battery system

Advantages	Disadvantages
Minimum initial cost	Higher life-cycle cost
Disposable	-
Replacement readily available	Performance is not optimal for heavy loads and high discharge rates
Because it is lighter, more compact, and, as a result, more portable, it is historically suitable for use in portable applications.	This is not the right choice for application such as load balancing, emergency backup, hybrid batteries, and expensive military applications.
Decent charge retention and longer service per charge	Traditionally, applications are limited to specific areas

3.2 Secondary batteries

In SBs, the electric current that flows through cells during discharge travels in the opposite direction of the current that flows through them during recharge. This allows secondary batteries to be electrically refilled to the condition they were in before they were discharged. Reversible chemical processes may be found inside the cells that make up storage batteries, secondary batteries, and charge accumulators. This indicates that it may be possible to return the device to its initial chemical state by introducing power from the outside, in other words, by running a current through the device. These SBs are implemented in many applications including transportation, emergency lighting sources, and telephone exchanges. Plante invented lead-acid batteries in 1859, which are the primary kind of secondary battery used today. In 1960, Andre was the first to market a secondary battery with a silver-zinc structure [2, 3, 21, 26–29].

In a perfect world, battery storage would be simple, easy to build, low-cost, and resistant. This battery is reliable, inexpensive, and long-lasting. Poor weight-to-volume ratio and mechanical strength. Despite its high cost and capacity, nickel-cadmium batteries have a long shelf life and are quite durable. Not all responses are reversible. Silver-zinc batteries provide great energy density while being pricey. When talking about batteries, efficiency is often measured in terms of ampere hours, volt hours, or watt hours, as seen below

A h efficiency = economic A h output during discharge/all input to fully charged   (17)

V efficiency = average V during discharge/ average V during charge   (18)

W h efficiency = A h efficiency X V efficiency

     (19)

The ampere-hour efficiency measures how much of the stored power is used up during the discharge process and how much is lost due to gas evolution, heating, and other activities involving the electrodes. It is essential to keep in mind that voltage efficiency is only one of many ways to assess polarization; hence, this efficiency will be equal to one for a reversible progression when there is an overvoltage (a or c = 0) and IR is at its lowest [3, 28, 29].

Table 3. Advantages and disadvantages of the secondary battery system

Advantages	Disadvantages
	Higher preliminary cost
Using convenient and low-cost charging reduces life-cycle costs	
	Regular preservation is mandatory
Superior high discharge rate performance at heavy loads	
	Periodic recharging compulsory
Ideally suited for load-leveling, emergency backup, hybrid battery and high-cost military applications	
The overall inherent versatility of secondary battery systems allows its use and continuing research for a large spectrum of applications.	While replacement batteries are available, they are not produced in the same quantity as primary batteries. Orders may need to be placed in advance.
The lithium battery technology has recently made secondary batteries smaller, lighter as well as in assembled big sizes. Now a days, It is widely implemented in different areas.	Comparatively to primary battery systems, traditional secondary batteries (especially aqueous secondary batteries) perform less well at retaining charge.

3.2.1 Categories and classification of secondary batteries

The architecture of secondary batteries, such as whether they are sealed or vented, and whether they contain acid or alkaline electrolyte, as well as whether they are fixed or portable determines how they are classed. There is a wide variety of ampere-hour capabilities, from less than one to many thousand.

3.2.1.1 Lead-acid battery

An anode, also known as sponge metallic lead; a cathode, sometimes known as lead dioxide (PbO_2); and an electrolyte, often known as a diluted combination of aqueous sulfuric acid make up a lead-acid battery [30, 31]. Among the several uses that it has, some of the more common ones are in automobiles, trucks, forklifts, construction machinery, recreational boats, and standby or backup systems. The most typical use for these batteries is in engines. These cells are used in the production of more than half of all batteries sold. In addition to having a cheap price point and a lengthy lifetime, these materials are also able to withstand being mistreated. In addition to this, they perform admirably in settings of both hot and low temperatures, as well as in tasks that involve a significant amount of water flow. The battery is denoted by the symbols $Pb/H_2SO_4/PbO_2$, Pb or $Pb/PbSO_4$, $H_2SO_4/PbSO_4/PbO_2/Pb$ respectively. The chemistry of lead acid batteries (LAB), in terms of the processes that occur in half-cells, is as follows:

The overall reaction: $Pb + PbO_2 + 2H_2SO_4 \longrightarrow 2PbSO_4 + 2H_2O$ (20)

The double sulfate theory (as $PbSO_4$ is formed in both reactions)

The OCV is given by: $E_{rev} = E_0 - \frac{RT}{NF}\ln\frac{(a_{H_2O})^2}{(a_{H_2SO_4})^2}$ (21)

Whereas E_0 stand for numerical sum of the two standard potentials of 0.355 V and +1.68 V is equal to 2.042 V at 298 K. The maximum charging and discharging voltage are 2.2 and 1.8 V, respectively.

(a) Design of a lead-acid battery

LAB also adhere to the overall battery design, however there are issues with some components of the design. When subjected to a voltage that is more than 2.39 V, water disassembles into its component parts, hydrogen (H_2), and oxygen (O_2). Therefore, it is of the utmost importance to replenish the water that is contained inside the cell. In addition to this, an explosion will occur if there is an excessively high concentration of hydrogen and oxygen in the cell because hydrogen and oxygen are released from the cell. When an acid or hydroxide solution is used, this method may also release vapors that have the potential to damage the surrounding surroundings in close proximity to the battery. Finally, throughout the 1970s, wrapped cells were readily accessible on the commercial market as a solution to the majority of these issues. The phrase "valve-regulated cells" is more appropriate than "lead-acid cells" due to the fact that lead-acid cells cannot be entirely shut. If those two gases were present, there would be a significant rise in pressure that would exceed what is considered to be acceptable limits. Catalytic gas recombination is a solution that has the potential to significantly address this issue. The conversion of hydrogen and oxygen back into water allows for the possibility of reaching an efficiency level of up to 85%. Despite the fact that H_2 and O_2 gases are not entirely removed, the amount of water that is lost is so little that there is no need for a refill. Batteries that consisting of these cells do not need to be maintained are referred to as maintenance-free batteries. In addition, the production of fumes is stopped by the design of this kind of cell. Cells that have a short cycle life have a quick rate of self-discharge and have lower energy densities (normally 30 to 40 W h kg^{-1}). In spite of the fact that the LAB has a voltage of 2 V and a power density of up to 600 W kg^{-1}, it is an excellent choice for use in automobile batteries, even if it is not ideal.

A design that is rectangular should be used since it is acceptable. The cathode consists of a grid made of lead antimony alloy that is covered with PbO_2, while a negative plate is made up of a rectangular plate of spongy lead (PbO_2). During this step, the electrodes are submerged in sulfuric acid, a substance that has 27-39% H_2SO_4 based on their weight. An insulator made of porous material sits between the +ve and -ve electrodes. Acidic materials are used for the construction of the containers, coverings, and vents because they are resistant to corrosion. The inner working volume is divided into three unique working areas: a gas space, an element space, and a bottom space. Each of the locations is designed to do a certain kind of work. In the element space, you will find supporting plates, as well as electrodes and separators. Separators are entwined with both positively charged and negatively charged electrodes. Electrolyte is poured into the space that may be found between the electrodes and the bottom of the tube.

(b) Electrode plates in a LAB

Castings made of lead alloyed with antimony or calcium are used to create grids, which are then coated with litharge or red lead and diluted sulfuric acid (H_2SO_4). After this step, a basic sulfate is created by drying the solution, which is followed by an electrolysis step in diluted H_2SO_4. Grids are utilized as cathodes in this electrolysis process, and when the oxide paste is decreased, the grids gradually get coated with lead that has the consistency of sponge. These are the components that make up the anodes of the battery. In the process of electrolysis, the anodes get coated with lead oxide as a consequence of the paste's oxidation. Similarly, a two percent concentration of Pb_2SO_4 might be added to similar grids. These materials are used to create the cathodes in batteries. The porosity of the plates significantly improves as the electroforming process continues. When a result of the porosity in the case of the battery, Pb_2SO_4 is more effectively preserved as the battery is discharged. Expanders may also be made out of graphite, Ba_2SO_4, and lampblack when working with high current discharge batteries.

Ebonite tubes that are thin, vertical, and have holes in them so that the electrolyte may enter them are used to hold the lead dioxide in place, which is the positive location. The lifespan of a plate is governed by (1) the amount of PbO_2 that is lost from its surface, and (2) the amount of parasitic corrosion that occurs in the grid material. We use sturdy plates to help reduce the amount of shedding that occurs. Corrosion is caused by local galvanic activity, which results in lead being anodic and hydrogen being produced at cathodic locations when antimony is present. When there is copper present, the presence of silver may also promote the corrosion process in a battery. There are a number of significant benefits that come from using calcium (0.08%) as an alloying agent rather than antimony.

In addition, plates may be formed by charging and discharging sheets of pure lead in an acid solution that is diluted with sulfuric acid. The addition of 8–12% antimony to lead has many benefits, including a low melting point, excellent hardness, and a decreased creep point. If there is antimony present, the plates will discharge more easily. Brittleness and cracking may be caused by casting faults if the amount of antimony in the material is less than 5%. In addition, Grids may be reinforced by the use of heat treatment. Grain may be refined by combining elements such as Ag, Si, Te, Se, and Sn at a concentration of 0.075%.

Lead and calcium precipitation-hardened alloys have a lower rate of corrosion and a longer service life than other types of alloys.

(c) Separators in a LAB

The internal resistance (IR) is lowered on reducing the distance between the electrodes. The plates can not be very closely as: (1) Pb has the propensity to create a dendritic structure on the substrate during charge that led to short-circuit; (2) As a result of buckling caused by overcharged or undercharged plates, Pb_2SO_4 has the tendency to advance the plates under pressure; and (3) Due to the mechanical and thermal shocks caused by the supports, the plates are forced to be exiled. There have been a lot of different materials tested out for this purpose.

(d) Electrolyte in a LAB

The activity of sulfuric acid in a battery is what determines the emf that is produced by the battery. At a temperature of 298 K, the open circuit potential varies from 1.88 V when the acid concentration is 5% to 2.3 V when the acid concentration is 40%. During the process of discharge, the acid is watered down. The specific gravity of a charged battery is comparable to that of diluted acid. The rate of dilution increases in tandem with the volume of discharge. For this reason, it is very necessary to begin with a surplus of sulfuric acid in order to keep the resistance at a low level.

(e) Capacity of LABs

LABs have a maximum capacity that is determined by the quantity of active material that they keep within. The sole material that is employed is lead sulfate, which is a substance that does not carry electricity. The sole material that is employed is lead sulfate, which is a substance that does not carry electricity. At the rate of 3.65 g of water produced per hour, the transformation of sulfuric acid into water takes place concurrently. The outcome of this is that IR will go up. Cathodic redox reactions are responsible for the transformation of lead oxide into lead sulfate. When a result, a rise in IR may be seen as the battery is discharged. When compared to the theoretical need, the amount of active material that is actually utilized in practice is three to four times more.

The following data have been gained through years of experience:

1. The Pb_2SO_4 blocks the pores on the plates, preventing acid from reaching the active material and causing the pores to become clogged.

2. The positive plate develops buckles and bulges as a result of the deposition of lead sulfate (Pb_2SO_4) (cathode).

3. Plates with a larger surface area have a lower mechanical strength than plates with a smaller surface area because of the larger surface area.

4. High rates of current cannot be sustained on thick plates if the acid is diluted, hence this limits their use.

(f) Effectiveness of LABs

Due to the fact that only one-third of the material is consumed during the discharge process, a lead acid battery has a relatively low overall share of its potential efficacy. The active substance undergoes a full transformation while it is being charged. The ampere-hour efficiency is typically between 90 and 99%, although it may go as high as 90%. The lower the numbers, the more heat is produced when charging, the more gas is produced during discharging, and the lower the temperature is at which the battery is working.

It is possible that the following factors may lead to a decline in the Labs' performances:

1. If the battery is overcharged, there will be a significant amount of gassing.

2. Plates that have been bent and deformed as a result of their physical degradation.

3. Filling the bottom region with Pb_2SO_4 and other shed products, resulting in a lower active volume

4. The weaker lead grid is a direct consequence of the overoxidation that occurred.

5. Sulfating caused by over discharge operations, which is generating the development of hard crystalline $PbSO4$, which is essentially nonconducting and clogs pores).

6. A little quantity of discharge from the self each day (estimated at 1%)

7. As a result of inverse or reverse charging, the incorrect plates are contaminated with lead or lead oxide (accidental).

Dry charged batteries

The procedure for charging dry lead-acid batteries is very similar to the procedure for charging wet lead-acid batteries, with the exception of the extra stages of cleaning the batteries with water and drying them in an inert environment. Since the battery contains spongy Pb negative plates and PbO positive plates, dry separators are a necessary component of the battery. In order for these components to become active, sulfuric acid of the necessary concentration must be added to them. Dry-charged batteries have an advantage over wet batteries during storage since wet batteries degrade over time. Regular charging slows the process of deterioration. Additionally, lead-acid batteries that cannot leak are manufactured on a large scale [3, 4].

3.2.1.3 Alkaline storage batteries

In 1919, alkaline batteries were first made available to the public. There are two different ways that Edison cells may be constructed: using NiO and Fe, or NiO and Cd [32]. An alloy of Ni and steel is used to construct the cathodes, and this alloy is supported by a grid made of nickel. Iron powder adhered to steel supports makes up the nickel-iron battery's anodes, which are supported by the nickel-iron. A solution of 20-25% KOH may be found contained inside an inert steel container.

As a consequence of this, the emf is independent of the amount of OH^- ions present, and it may be seen to lie anywhere between 1.33 and 1.35 V.

3.2.1.4 Nickel cadmium (Ni-Cd) cells

Cadmium makes up the anodes, whereas nickel oxyhydroxide $Ni(OH)_2$ makes up the cathodes. Cadmium and aqueous potassium hydroxide make up the electrolytes (KOH). Electro-optical technology finds use in a wide variety of products, such as digital cameras, pagers, laptops, tape recorders, spotlights, and medical equipment [33].

The cathode is a weave that has been plated with nickel, and the anode is a net that has been coated with cadmium. Because of the cadmium coating on this cell, the potential for it to have a detrimental influence on the surrounding environment may be exaggerated. The electrolyte of this cell, KOH, serves merely as an ion conductor and has no impact on the process inside the cell. This results in a reduced need for electrolytes, which in turn leads to a reduction in overall body mass. In contrast to KOH, NaOH does not carry electricity very well yet it does a good job of minimizing leaks.

Because of the processes that take place in the cells, we are able to write.

Their high-discharge performance and their ability to operate at low temperatures are only two of their many benefits. In addition to this, they have a lengthy shelf life and may be used for a considerable amount of time. The disadvantages of lead acid batteries are their higher price tag and lesser power density compared to other types of batteries. The memory effect, in which the cell preserves traits from earlier cycles, is the constraint that is perhaps the most well-known. It is a brief decrease in the capacity of a cell that occurs when the cell is recharged before it has been completely emptied of its previous charge. Cadmium hydroxide has the potential to result in the passivation of electrodes or the deterioration of batteries. In the first scenario, the issue can be fixed by performing a few cycles of draining and charging the battery, although this might reduce the battery's overall lifespan. A memory effect that might be considered "real" was detected. When it comes to "training" a Ni-Cd battery, performing a deep discharge is an essential step.

Construction of the electrodes

$Ni(OH)_2$ precipitates out as a byproduct of the reaction between $NiSO_4$ and NaOH. After that, a Ni plate is used to apply pressure on the material. To create the punched steel cylinders, about 300 layers of Ni and $Ni(OH)_2$ in alternating layers are created and assembled. $Ni(OH)_2$ is first charged, which results in the formation of Ni_2O_3 or NiO_2 + $Ni(OH)_3$. NiO_2 then crumbles to produce oxygen, and the active cathode is produced. In the anode section, hydrogen gas is used to convert ferric oxide into a powdery combination of iron and iron oxide (Fe_2O_3) Powder is often packed using steel grids that have holes bored into them. With a specific gravity of 1.21, a nickel iron cell can utilize KOH at a concentration of 22 percent, whereas a nickel cadmium cell can use KOH at a concentration of 20%. No matter how much KOH is introduced into the cell, the outcome will be the same. The volume of the electrolyte must be adequate for the active region of the plate to be completely immersed in order for the reaction to take place. Since $Fe(OH)_2$ and $Cd(OH)_2$ are the results of oxidation and are nonconductors, more of the Fe powder needs to be taken into account during the discharge process in order to verify the high conductivity of the material. In addition, NiO and Ni(OH)$_2$ have a low electrical conductivity. Therefore, it is very necessary to include metallic nickel into the cathode. Because of the reduction in volute size, there is no blockage that occurs during the discharge of the acid cell.

The capacity of these cells may be affected by the following factors:

i. Whether or not the emf drops with the rate of discharge (Iamp), the ampere-hour capacity remains the same. This is true regardless of idis. Because of its increased accessibility, KOH provides easy accessibility for electroactive areas in plates.

ii. As the temperature in the circuit drops, the capacitance drops as a result of the passive nature of the Fe anodes; this is because capacitance is proportional to temperature. Cd anodes do not display this characteristic in any way.

iii. The rate of self-discharge is relatively low (20% in a year), and as a result, there is no need for trickle charging.

Merits of the system: i. The damage won't be irreversible during the first5 cycles of charge-cycling, even if the vehicle has been idle for an extended period of time.

ii. Overcharging does not have an adverse effect on performance.

iii. The product's original expense is somewhat compensated by its low cost of maintenance.

iv. Alternating current may be used to facilitate the charging process.

Efficiency of the system: As a result, the efficiency is just 25% when measured in terms of the quantity of metal, but it is 80% when measured in terms of amps per hour. The decrease in value may be attributed to the gassing that takes place during the charge process. Efficiency measured in terms of watt-hours varies between 55 and 65%. The discrepancy in the charging/discharging curves contributes to the voltage efficiency being just 70% effective.

3.2.1.5 Sealed storage batteries

In order to accomplish sealing or non-spillability, it is feasible to either completely eliminate or significantly minimize the generation of gases during discharge, or to electrochemically transform these gases into other products. Another option is to maintain the electrolyte in the form of a gel, which would then be absorbed by a conducting media. These types of batteries provide a number of benefits, including the following:

1) It can be positioned in any orientation and still be utilized.

2) As a consequence, there is no leakage, and as a result, the chemicals in the battery won't have any effect on the operational instruments.

3) There is no way for vapors or sprays to go out.

4) In most cases, the topping is not necessary.

Lead acid batteries may be engineered to be unspillable. In order to achieve this goal, please follow these steps:

1) A separator is responsible for the absorption of electrolytes;

2) Vent traps are included to prevent leaks from occurring in any location. However, this does not prevent the spraying of acid;

3) A sponge-like lead is utilized to soak up oxygen while the battery is being charged;

4) By tripping the charger when excessive voltages are detected, gassing may be prevented.

In 1951, the nickel-cadmium penlight cell that had been sealed was made public for the first time. Among the requirements for sealing are the following:

1) the extraction of gases from the electrodes of the cell.

2) The rate of gas evolution slows down while the vehicle is being charged and while it is idling.

3) Regulation of the battery's voltage at the completion of the charging procedure

It is possible to achieve the aforementioned three objectives by:

I. There are specialized closures utilized for the valves

II. O_2 is transported throughout the cell and combines with the energetic components of the -ve electrodes. There is an abnormally high concentration of cadmium oxide present. Through diffusion beginning at the positive electrode, a reaction between cadmium sponge and oxygen takes place.

III. There is no production of hydrogen from the interaction between cadmium oxide and KOH;

IV. Making use of a charging method that has a regulated voltage

There are three different shapes of sealed cells: rectangle cells (50-400 mA h), flat button cells (50-400 mA h), cylindrical cells (AA) (450 mA h to 8 or even 50 A h), and button cells (5-25 A h).

It wasn't until 1960 when sealed silver-zinc cells made their debut. Their one-of-a-kind characteristics include a regulated charging voltage, which stops before 1.97 V. In certain cells, hydrogen evolution is prevented by severely amalgamating zinc, and a 3rd electrode is used to either oxidize hydrogen into hydrogen plus or decrease oxygen into OH$^-$ ions.

3.2.1.6 Nickel-metal hydride (NiMH) cells

In NiMH cells, it is normal practice to make the anodes out of rare-earth or nickel alloys that include a variety of metals. The cathode is comprised of nickel oxyhydroxide [34]. An electrolyte composed of KOH is used. The applications include things like electric cars, cell phones, camcorders, emergency lights, power tools, computers, and portables, as well as other things.

The concentration of the electrolyte does not change when the battery is going through the charging and discharging cycle. The nominal cell voltage is 1.2 V. When compared to Ni-H_2 batteries, this battery has the advantage of not necessitating the use of high pressures for the storage of hydrogen. This sealed cell is formed by the combination of Ni-Cd and Ni-H_2 cells. In spite of the fact that hydrogen has outstanding anodic qualities, it calls for the pressurization of the cell, which in the past rendered this battery unfit for commercial usage. The anodes of contemporary NiMH batteries are made up of metal alloys such as V, Ti, Zr, Ni, Co, and Fe as a consequence of recent developments in the field of materials science. Except for the anode, the construction of a NiMH cell is quite similar to that of a Ni-Cd cell. Because the voltage is so close to being constant at 1.2 V, the cells may often be substituted for one another in a variety of applications.

Some of the anodes in these cells include complex alloys of a variety of metals in addition to V, Ti, Zr, Ni, Cr, Co, and Fe. These metals are found in the anodes. Because research into the chemistry that lies underneath these alloys and the reasons for their higher performance is not yet finished in its entirety, the composition of these alloys is decided via an empirical process. A fascinating characteristic about these alloys is that, when they take in hydrogen, some metals undergo a process in which they simultaneously absorb heat and release heat. Since the movement of hydrogen between the electrodes quickly and without the transfer of energy is beneficial to a battery, the presence of any of these is undesirable. In order to do this, an alloy has to include both an exothermic and an endothermic metal within its composition. The electrolyte of a standard NiMH commercial battery is 6 M KOH. Although it is more expensive and has a lifespan that is only half as long as that of a Ni-Cd cell, it has a capacity that is 30% greater and a higher power density (theoretically 50% higher, practically 25% higher). The memory effect will always be present in NiMH batteries so long as they are cared for properly. Prior to this discovery, it was thought that NiMH batteries did not possess the memory effect. To prevent the memory effect, it should be completely discharged around once every 30 cycles. It is difficult to choose which alternative is the superior one. The primary properties of a battery are what should be considered when evaluating its quality for a certain application.

3.2.1.7 Lithium-ion cells

The anodes of LiBs are made up of carbon compounds, whereas the cathodes are graphite electrodes that are infused with lithium oxide [3, 16, 19, 26]. There are a variety of uses, some of which include laptops, mobile phones, and electric vehicles. Batteries that are based on lithium metal might be dangerous when used as a secondary (rechargeable) energy source. As a direct consequence of this, various cell chemistries have resulted in the development of lithium compounds as an alternative to Li metal. The cathode is made up of layered crystals (graphite), and lithium is intercalated into those crystal layers. In addition, lithiated metal oxides such as $LiCoO_2$, $NiNi_{0.3}Co_{0.7}O_2$, $LiNiO_2$, LiV_2O_5, LiV_6O_{13}, $LiMn_4O_9$, and $LiNiO_{0.2}CoO_2$ have been used in experimental cells. The vast majority of electrolytes are made from $LiPF_6$, however aluminum corrosion is an issue that these electrolytes have, thus researchers are looking at other potential replacements. In this context, $LiBF_4$ is a contender that may be considered. Batteries used in the generation of current typically have an electrolyte that is a liquid organic solvent. Membranes are necessary in order to accomplish the separation of electrons and ions. The majority of today's batteries make use of polyethylene membranes that include microscopic pores. An intercalation procedure, which has been the subject of research for a significant amount of time, has recently been applicable in the real world. Interstitial gaps in graphite crystals may accommodate a variety of ions of varying sizes. Graphite is an excellent option for use in battery construction because to its low cost, structural integrity, and compatibility with lithium (which prevents dendrite development during charge and discharge cycles) [3, 19, 26].

3.2.1.8 Manganese-titanium (lithium) cells

These cells employ the more common lithium-titanium oxide anodes in their construction. The cathode is formed from manganese dioxide that has been intercalated with lithium. Watches and other applications that need a very low discharge are a possibility. In this technique, lithium is used as the metal instead of manganese titanium, which may be a more familiar term. Because it has a voltage range that is comparable to that of the lithium-ion cell (1.5-1.22 V), it may be used in circumstances that previously required the use of main coin batteries. In contrast to cells based on lithium, it can withstand continuous overcharging at voltages between 1.6 and 2.6 V without being harmed. Even though it has a life certified for just 500 cycles, it has the potential to live more than 15 years even when subjected to shallow discharging since its annual rate of self-discharge is under 10%. These cells only generate a very little amount of amp-hours and available current between them. For instance, Panasonic has a selection of capacities ranging from 0.9 to 14 mA h. It is feasible for drain currents to be in the range of 0.1 to 0.5 mA [3, 35].

3.2.1.9 Rechargeable alkaline manganese cells

Rechargeable Zinc serves as the positive electrode in alkaline manganese (RAM) rechargeable batteries. As the cathode, it is composed of MnO_2. The electrolyte is a solution of KOH that has been dissolved in water. Applications may be downloaded into consumer electronics. This is an alkaline battery, which has been engineered from the ground up to be rechargeable. The process of charging a battery requires the application of direct current electrical power in order to bring about a transformation in the charge states of the chemical components that make up the battery. During the discharge process, the MnOOH transforms into MnO_2, while the ZnO transforms into Zn metal. It is vital to take precautions to evade overcharging the cell in order to avoid electrolysis and charging at voltages higher than 1.65 V (temperature dependant), all of which are necessary in order to avoid the formation of additional MnO_2 [36].

3.2.1.10 Redox (liquid electrode) cells

A semipermeable membrane is wrapped over a variety of liquids to create a liquid electrode, also known as a redox electrode. Ions can pass across the membrane, but liquids cannot mix together as they try to cross it. When an electrical contact is created, the liquids perform the role of passive conductors. When ions pass across a membrane, conductors produce an electric current because of the movement of the ions. There are two different approaches that may be used in order to recharge these cells and batteries. In the traditional method, the stream flows in the other direction. There are also other options, such as refilling the liquids in the cell so that they may be utilized again. It is feasible to recharge an extensive quantity of liquid stored outside of the cell using a very tiny devices. Despite having a poor volumetric efficiency, batteries have a very long lifespan. Electrochemical systems may be constructed with $FeCl_3$ serving as the cathode and $TiCl_3$ or $CrCl_2$ serving as the anode [37].

3.2.1.11 Vanadium redox cells

An uncommon kind of oxygen reduction cell, known as vanadium redox cells, use vanadium oxides in a variety of stages of oxidation as the anode and cathode, respectively. Because they have the ability to be charged as either oxidizing or reducing components, these solutions will not be harmed in the event that the membrane springs a leak [38, 39].

3.2.1.12 Solid-state batteries

Solid-state batteries convert chemical energy to electrical energy by transporting electrons across their layers. Ions generated at the anode obtain their charges from the cathode, called the electron exchanger. Effective electronic insulator, or separator, transfers positive ion through dielectric. Solid-state batteries should be manufactured from a single, unique material with three different sections separated by homojunctions. Ion source, separator, and electron exchanger. Fig. 2 shows the present battery structure. An electrochemical cell has an anode, a separator, and cathodes. Anodes emit positive ions, which allow oxidised electrons to travel to a separator and subsequently an external circuit. Positive ions can only pass via ion conductors. Electrons and +ve ions from the external circuit enter the electron exchanger, continuing the reduction process. An external load including negative metal ions and a positive intercalation chemical can drain a battery. During spontaneous oxidation and reduction, oxidation-reduction reactions may create electrical energy [40, 41].

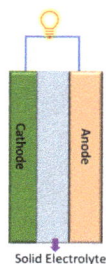

Figure 2. Represents of architecture of solid-state battery

Here is an illustration of the progression of solid-state batteries (primary as well as secondary) that use solid electrolytes, including, but not limited to, silver ion batteries, primary lithium batteries, sodium batteries, lithium iron sulfide batteries, polymeric batteries, lithium halogen batteries, and lead-cupric fluoride thin-layer batteries.

3.3.1.12 Other batteries

Reserve batteries, often called postponed action batteries, are easy to activate. Reserve batteries are sometimes called postponed action batteries, one-shot batteries, and dry-charged batteries. Reserve batteries may be activated in various ways before use; while not in use, they are dormant. One critical battery component must wait until the others are done. This triggers battery activation. Reserve batteries give certain benefits: Long durability; high consistent performance; flexibility to choose a system that allows highly reactive chemistry; and diverse designs. Reserve batteries last forever. You may rely on the backup battery's durability, but not its performance. Depending on their storage material, batteries are liquid, gaseous, thermally activated, or fusion salt. Manual or automatic activation is possible.

4. Manufacturing process

The process of manufacturing involves a number of stages, including the following: (1) Adding a binder and conductive additive to the materials that will be used for the cathode or anode; (2) painting ink on the current collector (metallic foil); (3) allowing the ink to dry; and (4) applying pressure to the finished product. The subsequent step includes (1) assembling the cathode and anode and simultaneously rolling them up with the separator, (2) inserting the electrode, (3) injecting electrolyte into the battery box, and (4) sealing, as illustrated in Fig. 2.

Active materials dissolved in polyvinylidene fluoride and n-methyl-2-pyrrolidinone are mixed with conductive additives to make an electrode. Production uses this paste. After applying paste on both sides of aluminium foil, it is dried and rolled. Then it is cut to size. After the gel electrolyte sheet is immediately polymerized in polymer lithium battery manufacture, the cathode and anode sheets are rolled, compressed, wrapped with aluminium lamination film, and heated to ensure their quality. In two to four weeks, the battery will be checked for shorts. During anode-carbon contact repair, a thin protective layer forms (SEI). Items must be inspected before shipment. Putting cathode and anode together while rolling them with separator [3, 14, 19, 23, 28, 41].

5. Battery economics

When compared to primary power, the cost of battery power is much higher. In addition to it, the following things need to be taken into consideration:

a) The cost of the ac-dc converters that are required in order to charge the battery should be included.

b) In order for the operational instruments to function properly while using AC, dc-ac converters need to be installed in them.

c) Although the loss of energy cannot be completely prevented, it may be reduced to a minimum.

d) When using a multicell assembly, it is essential to gather an extremely large quantity of the relevant material.

e) Taking into account the potential risks of pollution are necessary in order to shield the environment from the damage done by old batteries.

6. Nanomaterials in energy storage system

Materials with nanostructured electrodes and separators that can be used in supercapacitors and batteries

1. The creation of new kinds of batteries as well as the optimization of LiBs for use in automotive drives and stationary mass storage via nanostructuring and the use of nanocomposites for electrode materials and separator/electrolytic systems; (lithium-air, lithium-sulphur etc.)

2. Electrode materials based on carbon nanotubes (CNT) that have large charge capacities and can be used in supercapacitors and batteries, Nanoporous adsorption storage for mobile gas storage, incorporating metallo-organic cage-like structures (hydrogen, natural gas)

3. Hydrogen storage based on nanocrystalline metal hydrids. Products developed on the basis of nanotechnology are anticipated to achieve significant levels of market volume within the energy storage sector in the years to come.

Nanotechnology-based energy storage is primarily used for decentralised energy storage and electromobility. Decentralized energy storage (e.g., rooftop solar panels with storage in homes or businesses) is projected to become more important in the near future. Future technologies include lithium-ion batteries and supercapacitors.

The batteries used in today's externally charging electric cars are too heavy, large, and expensive for a major industry shift. Work is underway to construct a more powerful, lighter, mass-manufacturable energy storage device. Lithium-ion batteries for cars are expected to triple by 2020. All-electric automobiles will be used for specialized purposes or short distances until a solution to the long charging period is found, such as nanotechnology (several hours as a rule). When charging an electric car takes fewer than 10 minutes, there will be a demand for the corresponding infrastructure. Below are some types of energy storage.

Electrode materials and separators with nanostructures for supercapacitors and batteries:

Nanostructuring and the usage of nanocomposites for electrode materials and separator/electrolytic systems to optimise lithium-ion batteries for automotive motors and stationary mass storage; the creation of novel battery types (lithium-air, lithium-sulphur).

Materials Research Forum LLC
https://doi.org/10.21741/9781644902295-12

Carbon based nanomaterials such as nanotube (CNT), graphene nanocomposites based electrode materials for supercapacitors and batteries with high charge capacities.

The primary applications for nanotechnology-based energy storage are believed to be in the storage of decentralised energy and in electromobility; as a result, the storage of decentralised energy (e.g., through a combination of rooftop solar panels and storage in private homes or businesses) will become increasingly significant in the near future. Particularly, lithium-ion batteries and supercapacitors are recognised as future-critical technology.

Current research and development focuses specifically on rechargeable lithium batteries. This sort of battery will thus be described in further depth below. Lithium may be used to manufacture high-performance batteries. Through enlarging the electrode surfaces and optimising the separators, the energy density and stability of batteries are boosted, allowing for new applications in electric and hybrid cars as well as stationary power storage.

In batteries, lithium provides a variety of advantages:

- Its significant specific charge

- Its very negative standard potential (-3.05 V)

- Its stability in several organic (and some inorganic) electrolytes.

Lithium-ion batteries are now the most prevalent (rechargeable) mobile energy storage device: five billion lithium-ion batteries were purchased by customers in last 10 years for usage in computers, cameras, mobile phones, and electric vehicles. To be acceptable for such applications, the batteries must have a high voltage, a large capacity, a long lifespan, and a high level of safety and dependability. Lithium-ion batteries have an energy density that is rather high. Using novel cathode materials (such as $LiNi_{0.8}Co_{0.15}Al_{0.05}O_2$; NCA), the energy density of lithium-ion batteries might almost double over the next decade. Depending on which subsystem of lithium-ion batteries is studied, current energy densities vary from around 80 W h kg^{-1} (lithium iron phosphate, LFP) to approximately 250 W h kg^{-1} (nickel cobalt acetate, NCA). In contrast, the energy density of lead-acid batteries is around 40 W h kg^{-1}.

7.   Lithium-ion battery

Recent R&D has emphasized lithium-ion batteries. We will discuss this battery type below. Lithium makes high-performance batteries. Increasing electrode surface size and changing separators enhances battery stability. This allows battery usage in electric and hybrid cars and stationary power storage. Organic electrolyte stability. Most rechargeable cellphone batteries are lithium-ion. In 2013, 5 billion lithium-ion batteries were purchased for computers, phones, and electric cars. High voltage, a large capacity, and a long-life cycle are needed for these purposes. High-energy-density lithium-ion batteries. Lithium-ion battery energy density might treble in 10 years if $LiNi_{0.8}Co_{0.15}Al_{0.05}O_2$; NCA is used. Lithium-ion battery energy densities vary from 80 to 250 W h kg^{-1}. Lead-acid batteries pack 40 W h kg^{-1}. High-voltage lithium-ion batteries are stable. 2017 might see 400 W h kg^{-1} energy density. Dimensional electrode stability, high cell voltage (3.2 to 3.7 V), and cycle count (5,000) are maintained. These qualities can be achieved without reducing electrode size. Electrolyte, cathode, anode, and separator comprise lithium-ion batteries. Electrolytes boost charge transfer between electrodes because they contain ionized lithium. Charging moves Li-ions from cathode to anode. Graphite anodes average 370 A h kg^{-1}. Anodes include carbon black, doped carbon, and fullerenes. Nanowires, nanorods, nanotubes, nanoporous particles, and nanocomposites containing nanoscale silicon or tin are utilized or manufactured. Almost every substance can contain crystalline lithium. 700 A h kg^{-1} electrodes. Discharge moves lithium ions from anode to cathode. Modern battery cathodes use $LiCoO_2$. Economic and environmental factors increase the adoption of novel materials. $LiMnPO_4$, $LiCoPO_4$, or $LiFePO_4$ nanomaterials are presented. Nanoparticles save materials. Cathode nanostructure improves energy storage and charge cycle stability. Batteries require separators. Overheating and short circuits are prevented. They transmit ions between electrodes. The separator is constructed of non-woven polymeric fibres that do not transmit electricity and an inorganic material that allows ions pass through. Polymer nanofibers make up separator components. These porous nonwovens have small pores. Sub-100-nanometer particles create ion-conducting material. The separator's high tensile strength and pliability allow it to adapt to electrode shape changes without breaking.

8. Lithium-air battery/sodium-air battery

Although lithium-air battery development is in its early phases, these batteries should offer a high energy density. Lithium-air batteries might have an energy density similar to a gasoline engine, 10 times that of current batteries. Discharging a battery oxidises lithium to lithium oxide, releasing energy. During charging, the lithium anode regenerates and releases oxygen into the environment. Lithium-air batteries require air oxygen. This approach's oxygen cathode doesn't contribute to the battery's weight or volume. In these batteries, nanostructured materials play a vital role in both the carbon cathode and the oxygen-permeable separator membrane. A lithium-air battery charge in an electric car would be prohibitively expensive, hence this kind of battery is unlikely to be used for this purpose. Sodium-air batteries might have five times the energy density of lithium-ion batteries. Sodium is a cheaper battery ingredient than lithium. This gives sodium-air batteries a better chance of success, say scientists. These batteries have only reached 100 charge cycles till now.

9. Lithium-sulphur battery/sodium-sulphur battery

Lithium-sulphur batteries are promising. These batteries have two to five times the energy density of existing lithium-ion batteries. Also, battery components may be cheaper. Soon-to-be-introduced lithium-sulphur batteries will have 500 W h kg^{-1}. Lithium works as both an electrode and a source of lithium ions in lithium-sulphur batteries. Cathode is sulphur. Cathode is sulphur. During discharge, the anode lithium dissolves and combines with the cathode sulphur to form lithium sulphide. During charging, lithium is created on the anode from lithium sulphide. Nanoscale carbon, such as graphene, coats the sulphur electrode to avoid premature degradation. Carbon nanotubes (CNTs) are used to increase battery performance. Silicon and tin are suggested as anode materials to increase cycle life. Electrolytes and anode compositions are further optimization options. The lithium-sulphur battery has many benefits over lithium-ion batteries, the most notable being that it does not require dangerous heavy metals like cobalt or nickel. Sulfur isn't detrimental to the environment, but lithium sulphides are. Lithium sulphides react with acids to form hydrogen sulphides. The compartments must be well-sealed. Several research and development activities are employing sodium instead of lithium since it's more available. Sodium-anode batteries have lower energy densities than others. Cost and cycle stability are vital for stationary applications; therefore they may be used.

10. Printed battery

Printing electrical components onto surfaces using specific inks is another energy storage approach (printed electronics). Because metallic nanoparticles are disseminated in the ink, this method is termed nanotechnology. Due to their superior conductivity, the inks are widely used in the conductor industry. Metallo-organic decomposition inks hinder spray nozzles somewhat (abbreviated as MOD inks). Metals utilised include silver, copper, aluminium, and nickel. Metal oxide, usually zinc oxide, is used to produce ink. Clear and oxidation-resistant, they are great. Organic conductors and metals may be used as electrodes, eliminating the need for expensive indium tin oxide (ITO). Printed batteries aren't very powerful. Industry experts expect this to grow in the near future and be used to make ultra-thin, flexible, rechargeable batteries.

11. Supercapacitor

Supercapacitors bridge the gap between capacitors and batteries. Electrolyte, like battery electrolyte, conducts electricity between electrodes [20, 42,43]. Supercapacitors have the highest capacity per component among capacitors. Supercapacitors can be charged and drained much faster than standard batteries like lithium-ion batteries. Supercapacitors have poor storage capacity [44–46]. This gives them high power density but poor energy density. Due to the supercapacitor's low cell voltage, more elements must be linked in series to attain the required voltage. Work in this subject focuses on boosting performance and energy density [47, 48]. Reduce production costs and employ environmentally friendly materials. Supercapacitors are used in electric automobiles to speed up battery discharge or store energy in regenerative braking systems. This includes improving battery discharge. These applications need rapid power input and output, therefore batteries are not an option. A supercapacitor may last up to 80,000 cycles, depending on temperature and voltage load [49, 50]. Nanotechnology must improve high-capacity capacitors' electrical storage capability to compete with batteries. Carbon nanotubes (both single-walled and multi-walled carbon nanotubes), metal oxides, and conductive polymers are all being considered as main components. Graphene's theoretical and practical usefulness is also growing. Graphene has a vast surface area, storage capacity, and strong conductivity. Graphene-based improvements have resulted to a possible specific capacity of 550 Fg^{-1}, substantially higher than 200-300 Fg^{-1}. Activated carbon capacitors may achieve 100 Fg^{-1}.

Flexible supercapacitors will soon be available. Nanoporous nickel fluoride films (NiF_2) on thermoplastics may replace lithium in supercapacitors. Due to their large surfaces, they can store enormous quantities of electrical charge (66 mF cm^{-2}, 384 Wh kg^{-1}, 112 kW kg^{-1}) without pore structure deterioration after 10,000 charge/discharge cycles.

11. Nanocapacitor

Electrical capacitors known as nanocapacitors have individual structures that are less than 100 nanometers in size per. They have not moved on beyond the research phase yet. The production of a configuration of nanotubes that results in a complete capacitor with a large capacity is one of the advances that are taking place now. This configuration involves electrical connections between the nanotubes. Among the other things that are being researched, a novel method of structuring electrode surfaces is being investigated. Therefore, the power density of a nanocapacitor may be improved by more than ten times that of electrolytic capacitors with the assistance of a nanoporous highly structured aluminum oxide layer that is covered with conductive titanium nitride (TiN).

Further research is now being conducted on topics such as the encapsulation of practical concerns and the optimization of the capacity of the system. First and foremost, prohibitive manufacturing costs continue to be the primary barrier to the widespread usage of nanocapacitors.

12. Metal hydrid storage

Hydrogen's high energy density and clean burning make it a promising vehicle fuel [12]. Several companies have been exploring hydrogen-powered cars for 20 years. Fuel cells need "energy-rich" material, unlike batteries. Fuel cell materials include hydrogen. In a fuel cell, hydrogen-oxygen interactions drive electrons into an external circuit. This circuit drives a motor. Fuel cells solely produce water. Hydrogen-powered cars won't cut GHG emissions in 10 to 15 years. Hydrogen and oxygen fuel cells are costly. Hydrogen requires energy to make, extract from fossil fuels, and separate in water. Cooling hydrogen to -253 °C uses 30 to 40% of its energy. Hydrogen leaks may cause explosions. Novel H_2 storage technologies are being developed. Researchers use nanostructured materials to store a lot of hydrogen in a little area. Ammonia borane, CNTs, metal hydrids, or titanium, iron, or nickel alloys reduce the danger of explosion by reversibly attaching hydrogen. Metalorganic nanocubes provide a large storage surface. Nanocubes have nanoholes. Heat releases hydrogen electrically or chemically. This storage mechanism's metallic link lowers loss. Hydrogen storage efficiency is 29%. Long-term efficiency might reach 46%. Metal hybrid hydrogen storage reduces costs and increases safety (average operating pressure of less than 10 bar). Charge-discharge cycles don't deplete storage media. Without waste, the tanks may be used forever.

Summary and future possibility

Because batteries are a daily requirement, it is important to enhance battery technology. Due to restricted technology in scaled-up manufacturing and effective results, there is a large gap between research labs and the battery industry. Scaled-up production demands more power, which may explain the disparity. Businesses and academic organisations have joined together to establish a unified battery platform. This effort aims to meet customer wants without burdening manufacturing costs or the environment. To keep up with new battery innovations and meet market demands, it is important to grasp the working principles of different battery types, as well as their challenges and potential. The chapter featured an overview of main and secondary batteries, their components, and battery advances. An efficient battery design combines high-performing electrode materials with stable electrolytes to create high-tech, financially viable energy storage systems. Combining high-performance and stable electrolytes achieves this. This highlights the necessity for a robust battery sector to power electric cars and other devices.

References

[1]   J. Wiehe, J. Thiele, A. Walter, A. Hashemifarzad, J. Hingst, C. Haaren, Nothing to regret: Reconciling renewable energies with human wellbeing and nature in the German Energy Transition, Int. J. Energy Res. 45 (2021) 745–758. https://doi.org/10.1002/er.5870

[2]   C. Chen, A. Yang, Power-to-methanol: The role of process flexibility in the integration of variable renewable energy into chemical production, Energy Convers. Manag. 228 (2021) 113673. https://doi.org/10.1016/j.enconman.2020.113673

[3]   Viswanathan, Balasubramanian, Batteries: Energy Sources; Viswanathan, B., Ed.; Elsevier: Amsterdam, The Netherlands (2017) 263-313. https://doi.org/10.1016/B978-0-444-56353-8.00012-5

[4]   A K. Potbhare, R.G. Chaudhary, P.B. Chouke, A. Rai, A. Abdala, R. Mishra, M. Desimone, Graphene-based materials and their nanocomposites with metal oxides: Biosynthesis, electrochemical, photocatalytic and antimicrobial applications. Mater. Res. Forum. 83 (2020) 79-116. http://dx.doi.org/10.21741/9781644900970-4

[5]   Q. Zhang, J. Zhou, Z. Chen, C. Xu, W. Tang, G. Yang, C. Lai, Q. Xu, J. Yang, C. Peng, Direct Ink Writing of Moldable Electrochemical Energy Storage Devices: Ongoing Progress, Challenges, and Prospects, Adv. Eng. Mater. 23 (2021) 2100068. https://doi.org/10.1002/adem.202100068.

[6]   H.T. Das, E.B. T, S. Dutta, N. Das, P. Das, A. Mondal, M. Imran, Recent trend of CeO_2-based nanocomposites electrode in supercapacitor: A review on energy storage applications, J. Energy Storage. 50 (2022) 104643. https://doi.org/10.1016/j.est.2022.104643

[7] Das HT, Balaji TE, Dutta S, Das N, Maiyalagan T. Recent advances in MXene as electrocatalysts for sustainable energy generation: A review on surface engineering and compositing of MXene, Int. J. Energy Res. 46 (2022) 8625–8656. https://doi.org/10.1002/er.7847

[8] N. Zhang, F. Cheng, J. Liu, L. Wang, X. Long, X. Liu, F. Li, J. Chen, Rechargeable aqueous zinc-manganese dioxide batteries with high energy and power densities, Nat. Commun. 8 (2017) 405. https://doi.org/10.1038/s41467-017-00467-x

[9] Y. Li, J. Lu, Metal–Air Batteries: Will They Be the Future Electrochemical Energy Storage Device of Choice?, ACS Energy Lett. 2 (2017) 1370–1377. https://doi.org/10.1021/acsenergylett.7b00119

[10] A. Mondal, PB Chouke, V Sonkusre, T Lambat, AA Abdala, S Mondal, Ni-doped ZnO nanocrystalline material for electrocatalytic oxygen reduction reaction, Materials Today: Proceedings, 2020, 29 (3), 715-719. https://doi.org/10.1016/j.matpr.2020.04.170

[11] L. Zhao, Z. Hu, W. Lai, Y. Tao, J. Peng, Z. Miao, Y. Wang, S. Chou, H. Liu, S. Dou, Hard Carbon Anodes: Fundamental Understanding and Commercial Perspectives for Na-Ion Batteries beyond Li-Ion and K-Ion Counterparts, Adv. Energy Mater. 11 (2021) 2002704. https://doi.org/10.1002/aenm.202002704

[12] A. Mondal, H.T. Das, Energy storage batteries: basic feature and applications, in: Ceram. Sci. Eng., Elsevier (2022) 323–351. https://doi.org/10.1016/B978-0-323-89956-7.00008-5

[13] S. Rasul, S. Suzuki, S. Yamaguchi, M. Miyayama, High capacity positive electrodes for secondary Mg-ion batteries, Electrochim. Acta. 82 (2012) 243–249. https://doi.org/10.1016/j.electacta.2012.03.095

[14] A. Saha, P. Bharmoria, A. Mondal, S.C. Ghosh, S. Mahanty, A.B. Panda, Generalized synthesis and evaluation of formation mechanism of metal oxide/sulphide@C hollow spheres, J. Mater. Chem. A. 3 (2015) 20297–20304. https://doi.org/10.1039/C5TA05613F

[15] C.-T. Chu, A. Mondal, N. V. Kosova, J.-Y. Lin, Improved high-temperature cyclability of AlF_3 modified spinel $LiNi_{0.5}Mn_{1.5}O_4$ cathode for lithium-ion batteries, Appl. Surf. Sci. 530 (2020) 147169. https://doi.org/10.1016/j.apsusc.2020.147169

[16] A. Mondal, S. Maiti, K. Singha, S. Mahanty, A.B. Panda, TiO_2-rGO nanocomposite hollow spheres: large scale synthesis and application as an efficient anode material for lithium-ion batteries, J. Mater. Chem. A. 5 (2017) 23853–23862. https://doi.org/10.1039/C7TA08164B

[17] H.T. Das, K. Mahendraprabhu, T. Maiyalagan, P. Elumalai, Performance of Solid-state Hybrid Energy-storage Device using Reduced Graphene-oxide Anchored Sol-gel Derived Ni/NiO Nanocomposite, Sci. Rep. 7 (2017) 15342.

https://doi.org/10.1038/s41598-017-15444-z

[18] E. Duraisamy, P. Gurunathan, H.T. Das, K. Ramesha, P. Elumalai, [Co(salen)] derived Co/Co_3O_4 nanoparticle@carbon matrix as high-performance electrode for energy storage applications, J. Power Sources. 344 (2017) 103–110. https://doi.org/10.1016/j.jpowsour.2017.01.100

[19] N. Nitta, F. Wu, J.T. Lee, G. Yushin, Li-ion battery materials: present and future, Mater. Today. 18 (2015) 252–264. https://doi.org/10.1016/j.mattod.2014.10.040

[20] M.S. Uddin, H. Tanaya Das, T. Maiyalagan, P. Elumalai, Influence of designed electrode surfaces on double layer capacitance in aqueous electrolyte: Insights from standard models, Appl. Surf. Sci. 449 (2018) 445–453. https://doi.org/10.1016/j.apsusc.2017.12.088

[21] J. Chen, Y. Liu, M. Gao, Z. He, Z. Yu, Battery charging and discharging feature extraction method based on the best u-shapelets, in: 2018 IEEE 16th Int. Conf. Ind. Informatics, IEEE, (2018) 207–211. https://doi.org/10.1109/INDIN.2018.8471940

[22] I. Mexis, G. Todeschini, Battery Energy Storage Systems in the United Kingdom: A Review of Current State-of-the-Art and Future Applications, Energies. 13 (2020) 3616. https://doi.org/10.3390/en13143616

[23] B. Peng, J. Chen, Functional materials with high-efficiency energy storage and conversion for batteries and fuel cells, Coord. Chem. Rev. 253 (2009) 2805–2813. https://doi.org/10.1016/j.ccr.2009.04.008

[24] C. Qiu, G. He, W. Shi, M. Zou, C. Liu, The polarization characteristics of lithium-ion batteries under cyclic charge and discharge, J. Solid State Electrochem. 23 (2019) 1887–1902. https://doi.org/10.1007/s10008-019-04282-w

[25] M. Stern, A.L. Geaby, Electrochemical Polarization, J. Electrochem. Soc. 104 (1957) 56. https://doi.org/10.1149/1.2428496

[26] Pistoia, Gianfranco, Basic Battery Concepts, Batter. Portable Devices (2005) 1-15. https://doi.org/10.1016/B978-044451672-5/50001-6

[27] C. Jin, J. Nai, O. Sheng, H. Yuan, W. Zhang, X. Tao, X.W. (David) Lou, Biomass-based materials for green lithium secondary batteries, Energy Environ. Sci. 14 (2021) 1326–1379. https://doi.org/10.1039/D0EE02848G

[28] Y. Nishi, Lithium ion secondary batteries; past 10 years and the future, J. Power Sources. 100 (2001) 101–106. https://doi.org/10.1016/S0378-7753(01)00887-4

[29] X. Wang, G. Tan, Y. Bai, F. Wu, C. Wu, Multi-electron Reaction Materials for High-Energy-Density Secondary Batteries: Current Status and Prospective, Electrochem. Energy Rev. 4 (2021) 35–66. thttps://doi.org/10.1007/s41918-020-00073-4

[30] J.F. Manwell, J.G. McGowan, Lead acid battery storage model for hybrid energy systems, Sol. Energy. 50 (1993) 399–405. https://doi.org/10.1016/0038-092X(93)90060-2

[31] P. Ruetschi, Review on the lead—acid battery science and technology, J. Power Sources. 2 (1977) 3–120. https://doi.org/10.1016/0378-7753(77)85003-9

[32] N.E. Galushkin, N.N. Yazvinskaya, D.N. Galushkin, Generalized Model for Self-Discharge Processes in Alkaline Batteries, J. Electrochem. Soc. 159 (2012) A1315–A1317. https://doi.org/10.1149/2.081208jes

[33] F. Putois, Market for nickel-cadmium batteries, J. Power Sources. 57 (1995) 67–70. https://doi.org/10.1016/0378-7753(95)02243-0

[34] H.S. Lim, G.R. Zelter, D.U. Allison, R.E. Haun, Characteristics of nickel-metal hydride cells containing metal hydride alloys prepared by an atomization technique, J. Power Sources. 66 (1997) 101–105. https://doi.org/10.1016/S0378-7753(96)02488-3

[35] A.-H. Marincaş, P. Ilea, Enhancing Lithium Manganese Oxide Electrochemical Behavior by Doping and Surface Modifications, Coatings. 11 (2021) 456. https://doi.org/10.3390/coatings11040456

[36] M.B. Lim, T.N. Lambert, B.R. Chalamala, Rechargeable alkaline zinc–manganese oxide batteries for grid storage: Mechanisms, challenges and developments, Mater. Sci. Eng. R Reports. 143 (2021) 100593. https://doi.org/10.1016/j.mser.2020.100593

[37] W. Wang, Q. Luo, B. Li, X. Wei, L. Li, Z. Yang, Recent Progress in Redox Flow Battery Research and Development, Adv. Funct. Mater. 23 (2013) 970–986. https://doi.org/10.1002/adfm.201200694

[38] M. Skyllas-Kazacos, F. Grossmith, Efficient Vanadium Redox Flow Cell, J. Electrochem. Soc. 134 (1987) 2950–2953. https://doi.org/10.1149/1.2100321

[39] M. Rychcik, M. Skyllas-Kazacos, Characteristics of a new all-vanadium redox flow battery, J. Power Sources. 22 (1988) 59–67. https://doi.org/10.1016/0378-7753(88)80005-3

[40] W.D. Richards, L.J. Miara, Y. Wang, J.C. Kim, G. Ceder, Interface Stability in Solid-State Batteries, Chem. Mater. 28 (2016) 266–273. https://doi.org/10.1021/acs.chemmater.5b04082

[41] J.G. Kim, B. Son, S. Mukherjee, N. Schuppert, A. Bates, O. Kwon, M.J. Choi, H.Y. Chung, S. Park, A review of lithium and non-lithium based solid state batteries, J. Power Sources. 282 (2015) 299–322. https://doi.org/10.1016/j.jpowsour.2015.02.054

[42] H.T. Das, P. Barai, S. Dutta, N. Das, P. Das, M. Roy, M. Alauddin, H.R. Barai, Polymer Composites with Quantum Dots as Potential Electrode Materials for Supercapacitors Application: A Review, Polymers (Basel). 14 (2022) 1053. https://doi.org/10.3390/polym14051053

[43] A. Saha, A. Mondal, S. Maiti, S.C. Ghosh, S. Mahanty, A.B. Panda, A facile method for the synthesis of a C@MoO$_2$ hollow yolk–shell structure and its electrochemical properties as a faradaic electrode, Mater. Chem. Front. 1 (2017)

1585–1593. https://doi.org/10.1039/C7QM00006E

[44] Y.-K. Hsu, A. Mondal, Y.-Z. Su, Z. Sofer, K. Shanmugam Anuratha, J.-Y. Lin, Highly hydrophilic electrodeposited NiS/Ni$_3$S$_2$ interlaced nanosheets with surface-enriched Ni^{3+} sites as binder-free flexible cathodes for high-rate hybrid supercapacitors, Appl. Surf. Sci. 579 (2022) 151923. https://doi.org/10.1016/j.apsusc.2021.151923

[45] R.G. Chaudhary, V.N. Sonkusare, G.S. Bhusari, A. Mondal, D.P. Shaik, H.D. Juneja, Microwave-mediated synthesis of spinel CuAl$_2$O$_4$ nanocomposites for enhanced electrochemical and catalytic performance, Res. Chem. Intermed. 44 (2018) 2039–2060. https://doi.org/10.1007/s11164-017-3213-z

[46] A. Mondal, C.-Y. Lee, H. Chang, P. Hasin, C.-R. Yang, J.-Y. Lin, Electrodeposited Co$_{0.85}$Se thin films as free-standing cathode materials for high-performance hybrid supercapacitors, J. Taiwan Inst. Chem. Eng. 121 (2021) 205–216. https://doi.org/10.1016/j.jtice.2021.04.017

[47] S.S. Shah, H.T. Das, H.R. Barai, M.A. Aziz, Boosting the Electrochemical Performance of Polyaniline by One-Step Electrochemical Deposition on Nickel Foam for High-Performance Asymmetric Supercapacitor, Polymers (Basel). 14 (2022) 270. https://doi.org/10.3390/polym14020270

[48] S. Vinoth, H.T. Das, M. Govindasamy, S.-F. Wang, N.S. Alkadhi, M. Ouladsmane, Facile solid-state synthesis of layered molybdenum boride-based electrode for efficient electrochemical aqueous asymmetric supercapacitor, J. Alloys Compd. 877 (2021) 160192. https://doi.org/10.1016/j.jallcom.2021.160192

[49] H.T. Das, S. Saravanya, P. Elumalai, Disposed Dry Cells as Sustainable Source for Generation of Few Layers of Graphene and Manganese Oxide for Solid-State Symmetric and Asymmetric Supercapacitor Applications, ChemistrySelect. 3 (2018) 13275–13283. https://doi.org/10.1002/slct.201803034

[50] E. Duraisamy, H.T. Das, A. Selva Sharma, P. Elumalai, Supercapacitor and photocatalytic performances of hydrothermally-derived Co$_3$O$_4$/CoO@carbon nanocomposite, New J. Chem. 42 (2018) 6114–6124. https://doi.org/10.1039/C7NJ04638C

Emerging Applications of Nanomaterials
Materials Research Foundations 141 (2023) 327-352

Materials Research Forum LLC
https://doi.org/10.21741/9781644902295-13

Chapter 13

Prospective Nanomaterials for Food Packaging and Safety

Mohammad Harun-Ur-Rashid[1], Israt Jahan[2], Abu Bin Imran[3], and Md. Abu Bin Hasan Susan[4]*

[1]Department of Chemistry, International University of Business Agriculture and Technology, Dhaka 1230, Bangladesh

[2]Department of Cell Physiology, Graduate School of Medicine, Nagoya University, Nagoya 464-0813, Japan

[3]Department of Chemistry, Bangladesh University of Engineering and Technology, Dhaka 1000, Bangladesh

[4]Department of Chemistry, University of Dhaka, Dhaka 1000, Bangladesh

*susan@du.ac.bd

Abstract

Nanotechnology is being explored widely to improve food packaging. The development of innovative packaging materials using nanotechnology has had remarkable growth in the last few years. For the last two decades, substantial scientific efforts have been placed into replacing bulk and conventional materials with eco-friendly and biodegradable nanotechnology products or, more specifically, nanostructured materials in the food packaging industry. The advantages of nanotechnology and applications of nanostructured materials in food packaging are overviewed in this chapter. The common, profitable, and marketable acceptance of nanomaterial based food packaging systems and future perspectives are discussed by providing a broad and improved understanding of implementing nanotechnology products in food packaging.

Keywords

Nanotechnology, Nanomaterials, Active Packaging, Intelligent Packaging, ZnO Nanoparticles, Carbon-Based Nanomaterials, Green Polymer Nanocomposites

Contents

1.   Introduction

With the invention of canning in the nineteenth century, modern food packaging has made significant advances due to universal trends, technological and scientific improvements, and buyer preferences. Food casing or packaging has always been a part of constant development and the manufacturing company is constantly under pressure to supply more. Recent food science and technology have broadened, expanded, and refined conventional food casing methods and added new ones. The materials used in food packaging acts as a barrier to external elements and allow reaching food to the customer in a healthy and safe way. In addition to the move toward globalization, safety, protection and longer shelf life are necessary, along with monitoring safety and quality based on global standards. Gradually, the conventional materials have been replaced by smart nanomaterials such as nanocomposite gels [1-3], polymer nanocomposite [4-6], structural colored nanomaterials [7-9], molecular machines [10], nanomaterial based biosensors [11], and so on.

The advent of nanotechnology has had a most important impact on food packaging applications due to unique physicochemical and biological properties of nanomaterials. It has become one of the most up-and-coming technologies to refashion conventional food science applications. The implementation of nanotechnology in the processing, packaging, and safety of food has proved its competence in food industries, illustrated in Fig. 1 [12]. Various preparation techniques could produce nanomaterials with desirable chemical and physical properties, which might be utilized in food trade.

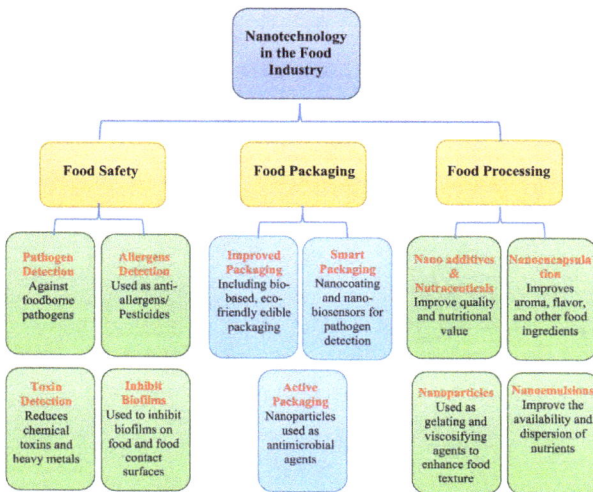

Figure 1. Multifunctional applications of nanotechnology in food processing, packaging, and safety, ensured by food industries. The figure has been reproduced with the permission from [12].

Implementation of nanotechnology by utilizing various nanomaterials such as nanoparticles (NPs), nanocomposites, nanoemulsions, and nanoengineered materials in food trading sectors has proven to be the best solution for addressing many issues and limitations, especially in food packaging and preservations [13, 14]. In this regard, with the help of nanoscale reaction engineering, nanobiotechnology, heat and mass transfer, and molecular synthesis, different nanostructured materials have been synthesized for manufacturing nanotracers and nanosensors for food packaging, preservation, delivery, and development. Biodegradable materials are preferred for the improvement and manufacturing of food packaging and preservation to control environmental pollution [15]. For the last few years, the coating of biodegradable and natural polymers on the surface of food has attracted tremendous attention and has proven prospective impacts in preserving food items [16, 17]. It is noteworthy that the stakeholders of food industries, working in the area of food packaging, are concerned about the introduction of antimicrobial packaging and detection of contamination during the packaging of food material. Modern and advanced food packaging systems may offer superior, easy, less expensive, and cost-effective methods to preserve foods than conventionally practiced techniques such as drying, freezing, canning, and dehydration. Nanotechnology has a wide variety of applications in food trading, as shown in Fig. 2.

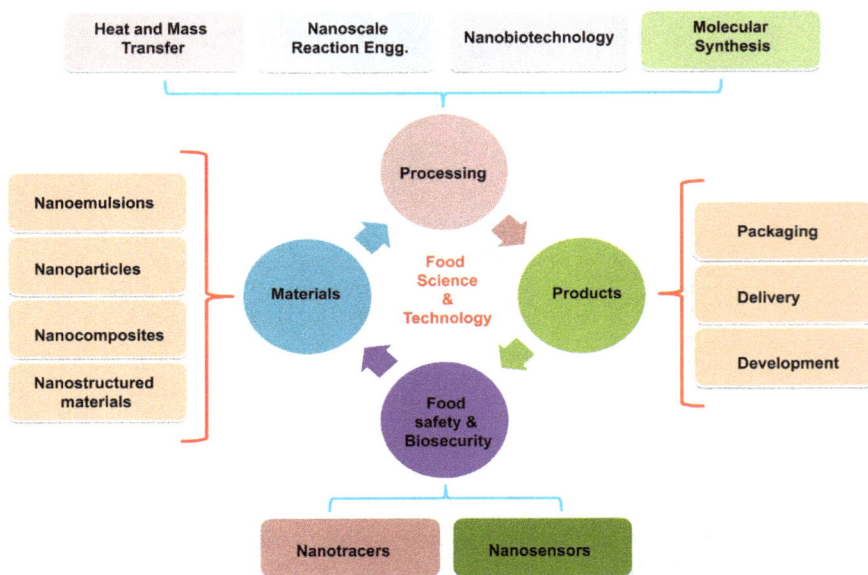

Figure 2. Potential impact of nanotechnologies in food trading through the utilization of nanomaterials such as nanoparticles, nanocomposites, nanoemulsions, and nanoengineered materials. The figure has been reproduced with the permission from [12].

The key importance of packaging is to save the different types of food items from biological, chemical, and physical damage. The food products such as vegetables, fruits, milk and milk based products, dry fruits, raw and dried meat, bakery products, raw and dried fish, sauces, and so on require more attention on the nutrition quality and shelf life [18]. In this connection, diverse techniques of packaging and materials are determined by the types of products packed. The food packaging operations is carried out to encounter several objectives such as physical and chemical protection, handling and transportation, stocking and marketing, information transmission, antitampering, and anticounterfeiting (Fig. 3).

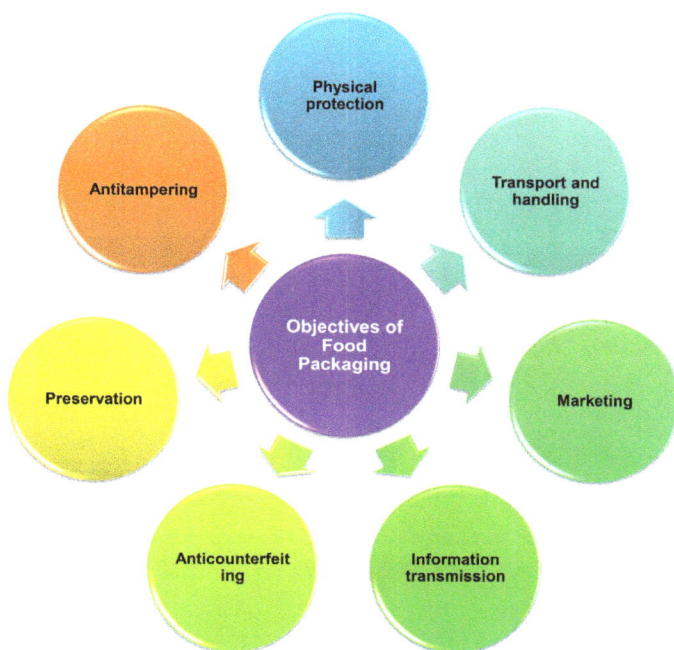

Figure 3. The objectives of food packaging.

Nanomaterials are more beneficial than conventional materials in providing improved preservation and quality maintenance of food products. NPs can alter the physical and mechanical properties of packaging polymers by improving their strength, durability, flexibility, barrier, and reusing properties. Many new food items have been introduced into the market with the requirements of consumers to monitor quality during and before the consumption. Moreover, it is required to minimize and control the food adulterant throughout the preservation period. All these requirements have led to the development of advanced packaging systems, for example, active packaging (AP) and intelligent packaging (IP), illustrated in Fig. 4. The detection of defects and the monitoring of food quality from the manufacturing to consumption stage are the basic functions of IP system by applying different indicators and sensors like gas indicators, time-temperature indicators, and humidity, calorimetric, optical, and electrochemical sensors and biosensors. Active packaging technology helps to maximize the shelf-life of food products by diffusing and absorbing O_2, CO_2, and ethanol [19].

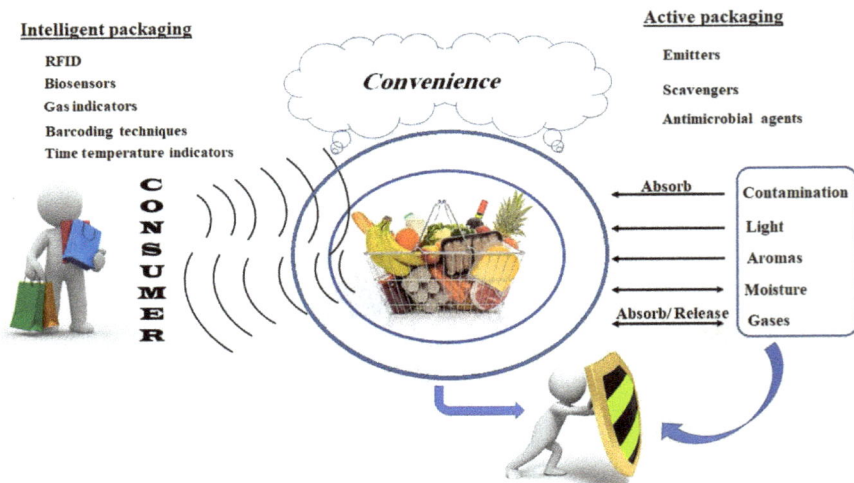

Figure 4. Types of modern food packaging technologies such as intelligent packaging and active packaging. The figure is reproduced with the permission from [19].

2. Prospective nanomaterials in food packaging

At present, nanostructured materials have found multidimensional applications in the food industry. Fig. 5 shows a variety of nanomaterials utilized in food trade. Food-borne illness is a global public health concern. The severity of food-borne diseases affects the economy, global health, trade, and commerce badly by lowering productivity and increasing medical expenses. Consequently, the clamor for new techniques to control food-borne pathogens has increased remarkably in recent years. So food packaging plays an essential role in maintaining food quality, ensuring food safety, and minimizing food-borne diseases. Food packaging having novel functionalities called active packaging can provide food safety by maintaining quality by extending food shelf life, increasing cost-benefit ratio, and eventually improving overall convenience. Functional materials present in active food packaging systems come in contact with food content and alter the environment or change the food composition inside the pack. For instance, antimicrobial packaging is one of the active packaging systems where food surface interacts with the antimicrobial functional materials and inhibits or reduces the growth of microorganisms that spoil the food items, illustrated in Fig. 6.

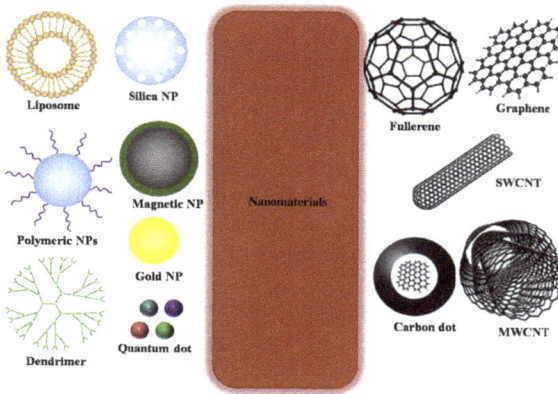

Figure 5. Different types of nanostructured materials utilized in food industries for numerous purposes. The figure has been reproduced with the permission from [20].

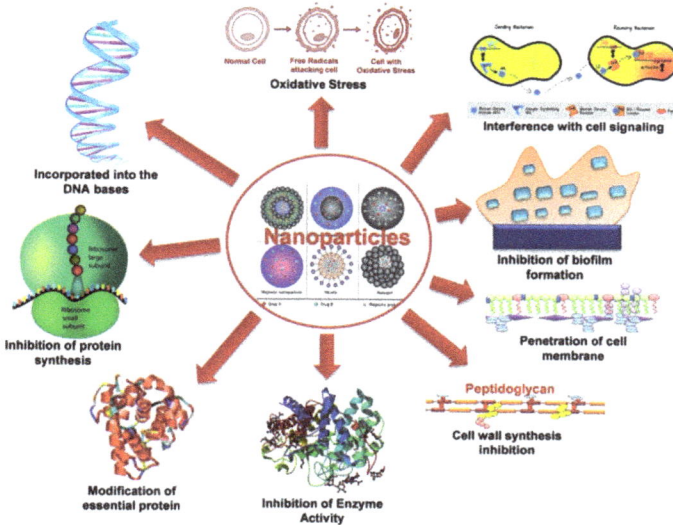

Figure 6. Different concepts and clarifications of antimicrobial activities demonstrated by various nanomaterials. The figure has been reproduced with the permission from [21].

2.1 ZnO nanoparticles

Antimicrobial actions of ZnO NPs either in the form of suspensions (nanofluids) or in the form of intrinsic NPs. ZnO NPs are effective against gram-positive bacteria such as *Staphylococcus aureus* and *Bacillus subtilis* [22-24]. Some gram-negative bacteria, for example *Escherichia coli*, *Campylobacter jejuni*, and *Pseudomonas aeruginosa* are sensitive to ZnO NPs [25-29]. Compared to *S. Aureus, E. coli* exhibits more sensitivity to ZnO NPs [30, 31]. Fig. 7 and Fig. 8 illustrate the structural characteristics of the membranes of gram-positive and gram-negative bacteria and the antimicrobial actions of ZnO NPs by multiple mechanisms towards microorganisms respectively [32].

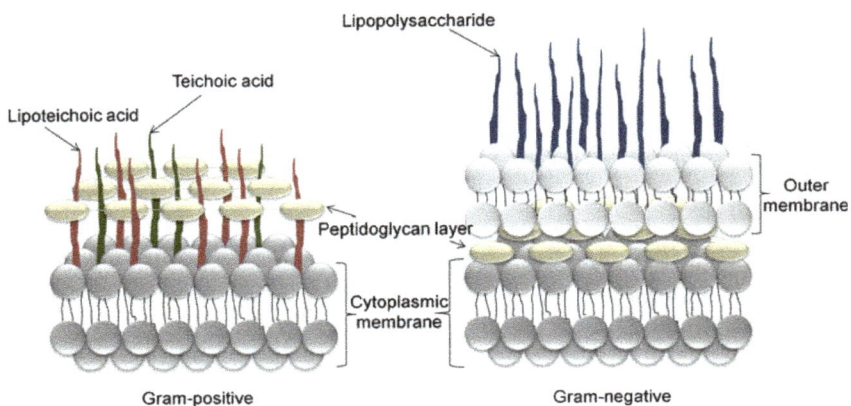

Figure 7. Structural characteristics of the membranes of gram-positive and gram-negative bacteria. The figure has been reproduced with the permission from [32].

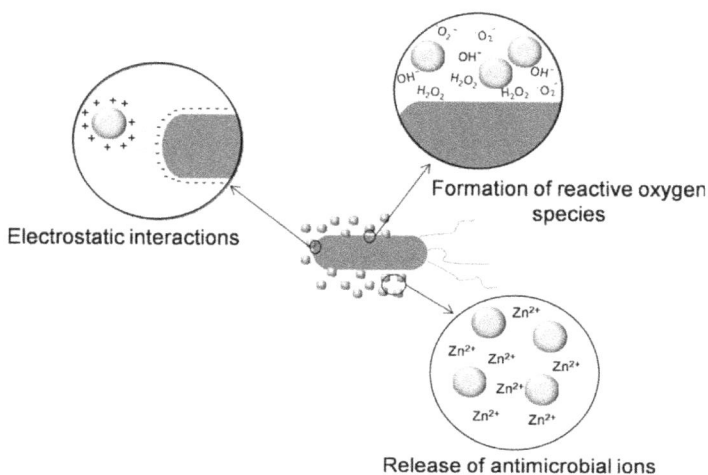

Figure 8. ZnO NPs capable of exhibiting antimicrobial actions through multiple mechanisms. The figure has been reproduced with the permission from [32].

2.2   Carbon-based nanomaterials in food packaging

The learning of carbon-based nanomaterials (CBNs) for food packaging applications has attracted attention due to their exceptional chemical and physical properties, including thermal, mechanical, electrical, optical, and structural diversity. CBNs are a novel class of materials that are widely used in food industry. CBNs, viz., carbon dots (CDs), graphene, activated carbon-based nanocomposites, carbon nanotubes, etc., are environmentally benign and better materials for food packaging. With antibacterial efficiency, they support food preservation and other applications [33]. Applications of CBNs in food trading have been depicted in Fig. 9. CBNs are better resources over the other materials in the ground of nanoscience and material science. They have attracted considerable interest from scientists since their discovery. They can be roughly divided according to their spatial dimensions, into fullerenes (zero-dimensional), carbon nanotubes (one-dimensional), graphene (two-dimensional), graphene coil (multidimensional), etc. They are found to be superior food-packaging materials compared to conventional ones.

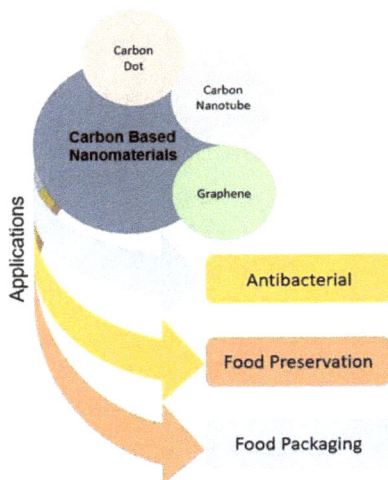

Figure 9. The applications of carbon-based nanomaterials in food preservation and packaging. The figure has been reproduced with the permission from [33].

CDs also known as fluorescent carbon are a new class of fluorescent small-carbon nanomaterials with particle sizes of less than 10 nm. CDs have various sorts of potential applications in food packaging. They have been used as antioxidant, antimicrobial, photoluminescent, and UV-light blocker additives in food packaging materials to reduce chemical deterioration and inhibit the growth of pathogenic and spoilage microorganisms in foods with the capacity to sense the freshness of food. They can be manufactured from environmentally friendly sources at a low price, such as microorganisms, food by-products, and waste streams, or they may be generated in foods during normal processing operations, such as cooking. CDs /polyvinyl alcohol (PVA) films can be used as active food packaging materials [34]. The incorporation of CDs in the PVA film enhances the water resistance properties, antimicrobial and antioxidant properties, mechanical properties, thermal stability, and UV blocking properties, illustrated in Fig. 10. The application of CDs /PVA film as active packaging is capable of notably increasing the shelf life of various foods such as jujube, banana, and fried meatballs.

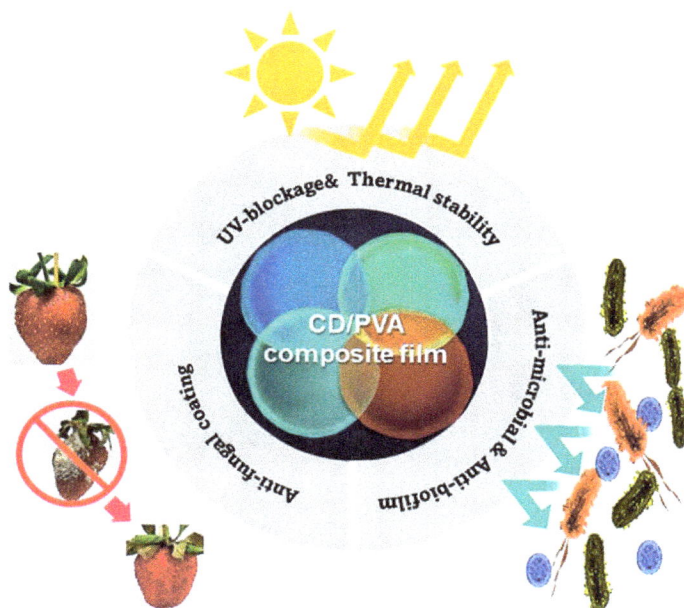

Figure 10. The application of multicolor emitting CDs and PVA composite films as edible food packaging materials and coatings with antimicrobial and UV light blocking properties. The figure has been reproduced with the permission from [35].

Flexible, transparent, and photoluminescent CDs reinforced PVA based nanocomposite films show UV-blocking, mild antioxidant, and good antimicrobial activity against both gram-positive and gram-negative bacteria [35]. This CDs/PVA films significantly extend the shelf-life of strawberries as a model for perishable food products (Fig. 11). Fresh strawberries, dip coated with CDs/PVA, exhibit strong resistance towards fungal growth, spoilage, and weight and moisture loss. The CDs/PVA nanocomposite films are non-cytotoxic, UV light blocker, and antimicrobial. As a result, these nanocomposite films can be used as additives for producing edible coatings and food packaging.

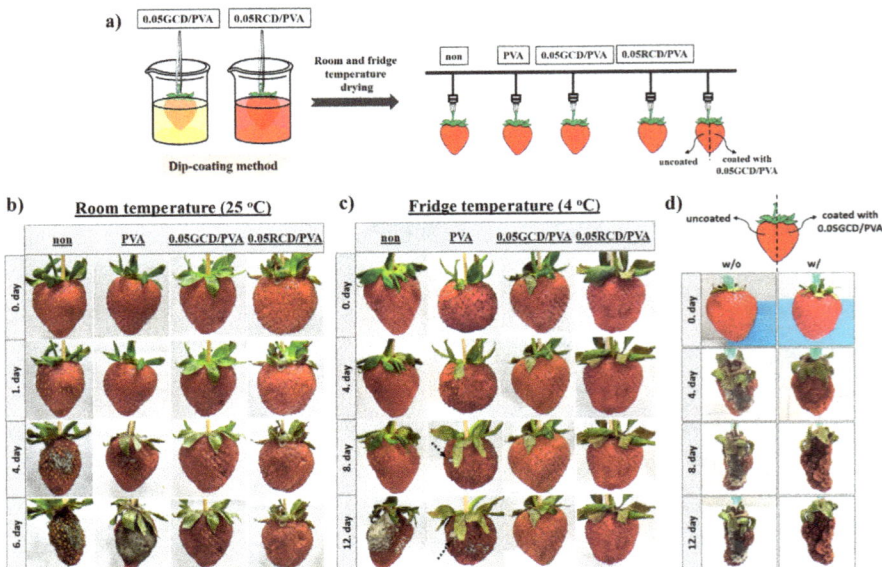

Figure 11. (a) Schematic illustration of strawberry coating with PVA and CD/PVA nanocomposites (green and red-emitting CDs) by the dip-coating method. Digital images of the appearance of non-coated and coated strawberries with PVA, 0.05GCD/PVA, and 0.05RCD/PVA at (b) room temperature, (c) fridge conditions, and (d) room-temperature images of 0.05GCD/PVA-coated and uncoated strawberries at varying storage times. The figure has been reproduced with the permission from [35].

Carbon nanotubes are available as either single wall nanotube (SWNT) or multiwalled nanotubes (MWNT). They are commonly used as low-resistance conductors and catalytic reaction vessels. Globular milk proteins can self-assemble into similarly structured nanotubes in some appropriate environmental conditions. Carbon nanotubes are used to improve the mechanical and antimicrobial properties of the polymer used in packaging. They are also used to form oxygen sensors to monitor their concentration in modified atmospheric packaging. Carbon nanotubes are incorporated into a synthetic polymer matrix used for food packaging to provide antimicrobial properties and intelligent sensors that can detect food spoilage.

Carbon nanotubes reinforced polymer nanocomposite films of PVA modified with ZnO NPs are important candidates as food packaging materials [36]. The synthesis and application of CNT-PVA-ZnO NP polymer nanocomposite films have been illustrated in

Fig. 12. These films have greater thermal stability, hydrophobicity, water vapor transmission rate, and antibacterial activity than PVA films. The packaging system made from CNT-PVA-ZnO NPs polymer nanocomposite films restrict the water loss in vegetables for up to 4 days at room temperature. This packaging material can extend the shelf life of chicken meat, since CNT-PVA-ZnO NP polymer nanocomposite films restrict the growth of natural microorganisms in raw chicken. The experimental results show that CNT-PVA-ZnO NP polymer nanocomposite films having good transparency are suitable for food packaging applications.

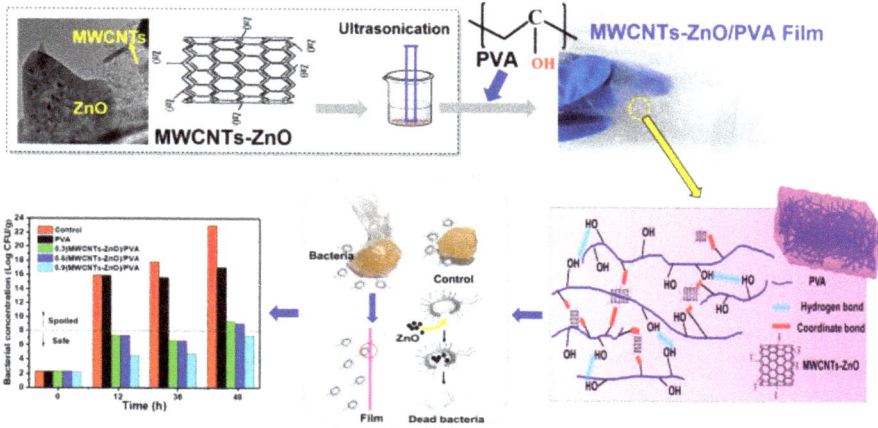

Figure 12. Carbon nanotubes reinforced polymer nanocomposite films of PVA modified with ZnO NPs as food packaging materials. The figure has been reproduced with the permission form [36].

Graphene-based nanocomposites have been used to get better UV resistance as a blockade against gases and still have good thermal, mechanical, and electrical properties compared to their polymeric matrices. Furthermore, graphene-based nanomaterials shared through biodegradable polymers suggest high potentials to be used as antimicrobial and antioxidant in dynamic packaging technology resulting in more excellence, safety, extended shelf-life, and added value [37]. Graphene derivatives in biopolymer-based composites for food packaging applications have been shown in Fig. 13. They have many extraordinary and outstanding properties, which make them ideally effective for use in biosensors to monitor and track the quality of food. These properties include high conductivity, mechanical flexibility, amenability for versatile surface functionalization, ultrahigh surface area, biocompatibility, and so on. [38].

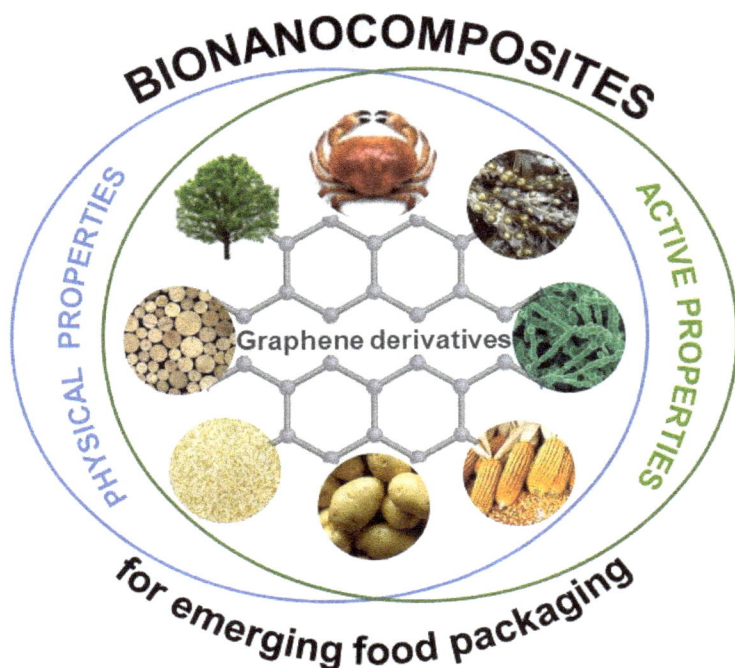

Figure 13. Graphene derivatives in biopolymer-based composites for food packaging applications. The figure has been reproduced with the permission from [37].

Polyhydroxybutyrate (PHB), a natural polymer of microbial origin, is an excellent alternative to petroleum-based food packaging materials. The inferior thermal, mechanical, and barrier properties of PHB have restricted the versatile application for commercial food packaging. Such limitations can be overcome by incorporating graphene Gr-NPs and can be used in food packaging [39]. The Gr-NPs-PHB nanocomposites have been investigated and found with superior thermal, mechanical, cytotoxicity, barrier, and biodegradable properties, which are considered as key factors for selecting any material for food packaging applications (Fig. 14). They can increase the shelf life four times more for oxygen and moisture sensitive food items such as milk produces and potato chips. PVA/graphene-based nanocomposites are just one example of polymeric nanocomposites with considerable potential in the food packaging industry. They have exceptional gas and moisture barrier properties and high-quality thermal resistance. In smart packaging, graphene can be used as a sensor for biochemical or microbial changes in the product packed to detect specific food-borne pathogens or gases. Thus, this type of packaging can serve as oxygen and spoilage indicators for food safety and quality monitoring [40].

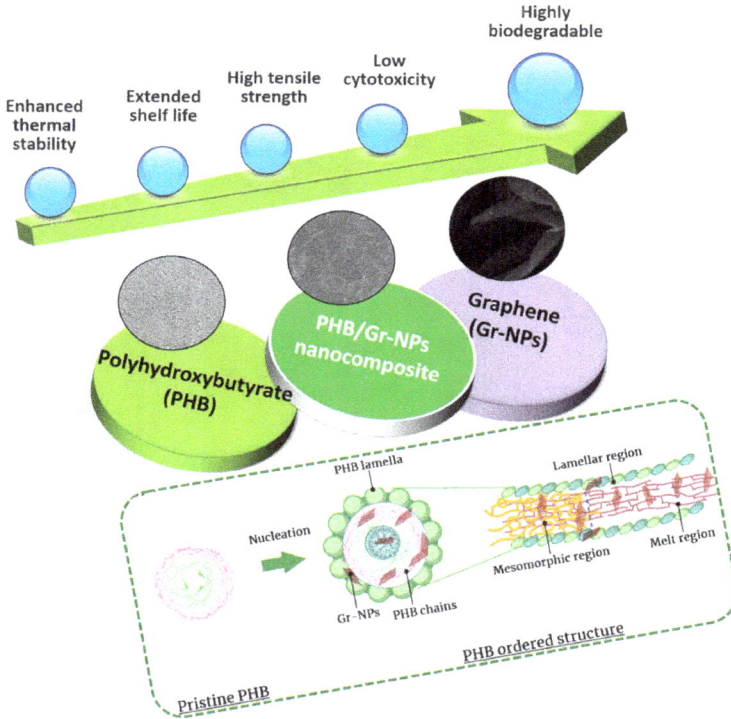

Figure 14. Graphene reinforced highly biodegradable PHB based polymer nanocomposites with antimicrobial activity for food packaging applications. The figure has been reproduced with the permission from [39].

2.3 Bio-based nanomaterials for food packaging

Bio-based nanomaterials are eco-friendly and sustainable packaging materials. These are chosen to make an obstacle for insects and microorganisms that cause degradation of food and spread diseases among the consumers. As a consequence, spoilage of foods and diseases of consumers are completely controlled and minimized [41, 42]. Natural and bio-polymers have found their prospective applications in safe and sustainable food packaging and preservation because of their outstanding, attractive, and unique features, such as biodegradability, chemical stability, and biocompatibility, which make them an essential source of viable materials for commercial production [43]. Generally, protein-based naomaterials, carbohydrate (polysaccharide)-based nanomaterials, lipid-based nanomaterials, antibacterial agents, and bio-based nanocomposites are used in antimicrobial food packing system, illustrated in Fig. 15.

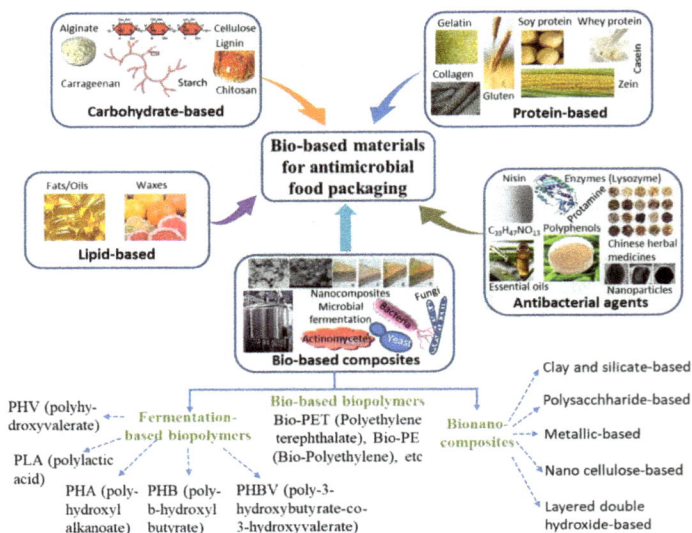

Figure 15. Different types of bio-based nanostructured materials utilized for antimicrobial food packaging applications. The figure has been reproduced with the permission from [42].

Starch, chitosan, and cellulose are common non-toxic and biodegradable polysaccharides to serves as an important source of nanosized reinforcements in antimicrobial food packaging [44]. Chitosan is one of the most popular and effective biopolymers in food coating and packaging since it has excellent film forming, antimicrobial, and biodegradable characteristics [45-47]. Chitosan and its derivatives have been employed as an alternative of natural antioxidant and antibacterial agents in powder form, as coating, casted film, and NPs.

Starch, produced from rice, corn, maize, barley, wheat, vegetables, potato, and soya, is extensively utilized as naturally renewable carbohydrate polymeric nanomaterials for making polymer films to be used in food packaging. Starch-based polymer films are cheap, biodegradable, non-toxic, and naturally abundant. They show excellent oxygen barrier properties but incapable of showing moisture barrier characteristics; however, the inclusion of chitosan in starch can overcome that limitation and improve the mechanical and barrier properties [48]. Starch and its derivatives are very much potential for manufacturing bio-nanocomposite, which could be utilized as suitable materials for cost-effective food spoilage detection and food packaging applications [49]. Nano-starch (NS) is a unique type of starch material with outstanding physiochemical properties. Because of its nano-scale size, NS shows a tendency to agglomeration, a natural process. In addition, due to the presence of a single hydroxyl (OH-) group, NS is unsuitable for hydrophobic environments. That is why modified-NS (MNS) with enhanced hydrophobicity, dispersion property, and stability is more effective to be utilized in many commercial sectors, especially in food packaging applications, illustrated in Fig. 16 [50].

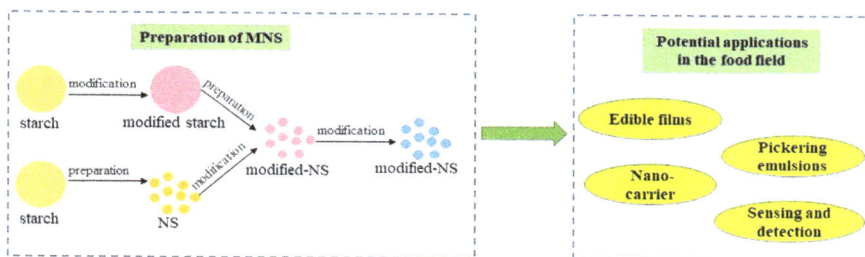

Figure 16. Preparation of modified nano-starch (MNS) particles and their potential applications in food trading. The figure has been reproduced with the permission from [50].

2.4   Green polymer nanocomposites for food packaging

Recent concern on petrochemical-based polymeric materials has created huge interest in green polymer nanocomposites for food packaging industries. Biodegradable green polymer nanocomposites should be effectively sensitive to microbial degradation and decomposed into carbon dioxide, water, and inorganic compounds in the presence of naturally occurring microorganisms like bacteria, fungi, and algae when disposed into soil. Green polymer nanocomposite materials must meet basic requirements such as barrier properties for light, aroma, water vapor and gases, strength, printing properties, optical properties, molding and welding properties, migration resistance, thermal and chemical resistance, disposal requirements, antistatic properties, retaining sensory properties, and above all environmental and health safety. The permeability of polymeric nanocomposites depends on the shape, size, and distribution of nanofillers throughout the polymer matrix. The alignment and dispersion pattern of filler NPs may enhance the tortuosity of the gaseous path and diffusion rate in polymer nanocomposites. The presence of regular dispersion of nanofillers with greater surface area and aspect ratio in the polymer matrix may improve the barrier properties of polymer nanocomposite packaging materials against water vapor, gases, and liquids.

The application of advanced green polymer nanocomposite based packaging materials offers several advantages, such as the protection of food from the attack of various microorganisms and UV radiation, illustrated in Fig. 17. These nano packaging materials can release antimicrobial agents, retain water vapor, absorb ethylene, and remove oxygen. The chitosan/MgO nanocomposite is an essential candidate in food packaging applications, with 86% improved tensile strength and 38% higher elastic modulus compared to pure chitosan [51].

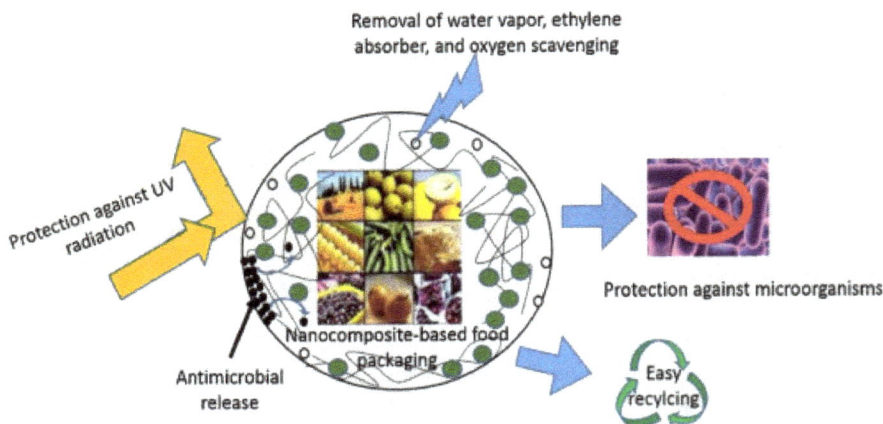

Figure 17. The functionality of green polymer nanocomposites-based food packaging, which is eco-friendly and protects food against microorganisms and UV light. The figure has been reproduced with the permission from [52].

The effective implementation of nanotechnology has provided the green polymer nanocomposites excellent performance as packaging materials concerning health and environmental safety, and economical benefits including a reduction in emissions and wastages, easy transport and storage, decrease in the chance of any attack of food components and human health, decrease in energy factors of production, protection from gases and light, and biodegradability. Polymers employed for preparing green polymer nanocomposites are different types of carbohydrates, including starch, cellulose, chitosan, agar, alginate, and carrageenan, naturally occurring proteins (soy protein, collagen, casein, corn zein, wheat gluten, gelatin whey protein,), and derived components from biodegradable polymers such as poly(L-lactide), poly(glycolic acid), poly(butylene succinate), PVA, poly(ε-caprolactone), and microbial polyesters such as poly(3-hydroxybutyrate-co-3-hydroxyvalerate), poly(hydroxyalkanoates), and poly(ß-hydroxybutyrate). Green polymer nanocomposite thin films made from starch, kefiran, and ZnO NPs can be used as food packaging material [53]. Such polymer nanocomposite has excellent packaging features such as greater mechanical strength, moisture and oxygen barrier, flexibility, rigidity, thermal and chemical stability, and resistance against the attack of food constituents and microorganisms (Fig. 18).

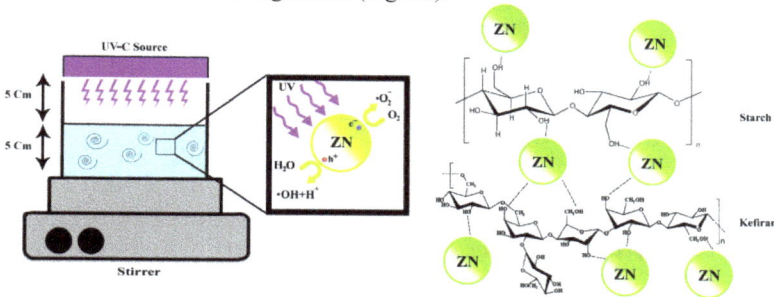

Figure 18. Schematic representation of UV irradiation technique and the activities of ZnO NPs under UV radiation (left side) and the distribution and arrangement of ZnO NPs between the starch and kefiran chains. The figure has been reproduced with the permission from [53].

Green polymer nanocomposites, composed of starch and nanoclay fillers, are suitable for biodegradable packaging materials, as shown in Fig. 19 [54]. Remarkable improvements in barrier properties and mechanical strength might be achieved through the inclusion of nanomaterials as fillers. Extensive research has been conducted to produce environmentally benign green polymer nanocomposites as environmental and health safety are a great concern in the present age.

Soybean polysaccharide based nanocomposites composed of TiO_2 NPs exhibit excellent performance as food packaging materials that extend the food shelf-life up to 6 months. Biodegradable aliphatic polyester polylactide, produced from renewable resources, finds prospective applications in food packaging. Polylactide has been blended with nanofillers like protein, inorganic fillers, starch, and natural fiber flax to prepare a green polymer nanocomposite that represents an effective alternative to conventional polymer composite.

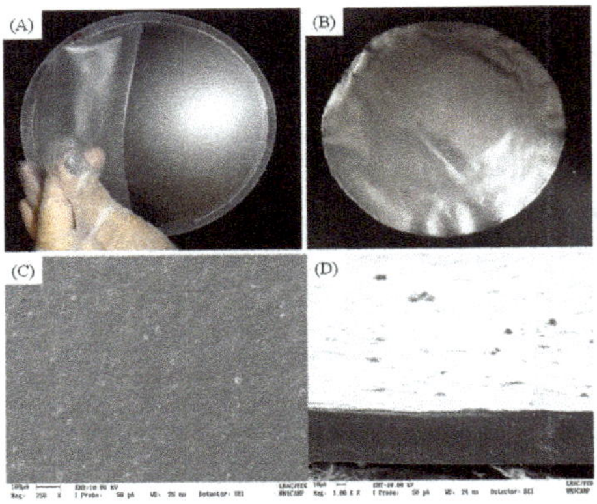

Figure 19. Photographs (a and b) and scanning electron microscopic images (c and d) of the film produced with 4% (mass/mass) arrowroot starch and 17% glycerol (mass/mass of starch). The figure has been reproduced with permission from [54].

3.   Concerns, future scopes, and conclusion

Nanotechnology has a great impact on society through applications in food packaging. It can offer items at a lower price and create more effective food delivery systems. In view of offering safe food, good nutrition, and better health to the present and future world population, novel and advanced technologies are undeniable in the food preservation and packaging industries. For the last few decades, extensive research has been conducted to explore the effective and successful utilization of nanomaterials in different formats by implementing unique and excellent physiochemical characteristics. Nanomaterials have found their modern and advanced applications in the development of novel techniques for food packaging, analysis, preservation, transportation, and safety. The application of nanotechnology and its products in food trading has offered less energy consumption and waste production during any associated process. However, any new technology conveys moral accountability regarding sensible application and the acceptance that there are probable precipitous dangers that might accompany great positive potential. In correspondence to the development and application of nanotechnologies, it is pretty standard that there will be new administrative rules and regulations to oblige nanotechnology products.

Although many research works have been conducted to develop the quality of nanomaterials, investigations on toxicity and freshness are still not adequate or more likely at the initial stage. Substantial studies should be accomplished before the potential application of nanomaterials, especially in the food sector, which is associated with public health and environmental concern. Advanced methods must be implemented so that the properties and the fate of nanomaterials can be identified, characterized, and determined to know the biotransformation in food.

Nanomaterials are progressively being utilized in a broad spectrum of applications. According to present food packaging research, the application of nanostructured materials through the practical application of nanotechnology can provide a variety of alternatives for an active and intelligent packaging system for food preservation and transportation. The approach of utilizing nanomaterials in food packaging is well accepted has become increasingly practical in the food industry in the future due to the increasing demand for the varieties of fascinating foods and the resulting arrangement of safer packaging of food items. Nanotechnology, used to fabricate food packaging, offers remarkable improvements in the characteristics of packaging materials, but more research initiative should be taken to interpret the role of nanotechnology products, their advantages, and disadvantages in the food packaging process.

Acknowledgements

A.B. Imran gratefully acknowledges the support from the Committee for Advanced Studies and Research (CASR) in BUET for funding. MABHs also acknowledges supports from Bose Centre for Advanced Research in Naural Sciences and Semiconductor Technology Research Centre of the University of Dhaka, Bangladesh.

References

[1] M. Harun-Ur-Rashid, T. Seki, Y. Takeoka, Structural colored gels for tunable soft photonic crystals, Chem. Rec. 9 (2009) 87-105. https://doi.org/10.1002/tcr.20169

[2] M. Harun-Ur-Rashid, A.B. Imran, Superabsorbent hydrogels from carboxymethyl cellulose", in Ibrahim H. Mondal (ed.) Carboxymethyl Cellulose. Volume I: Synthesis and Characterization, Nova Science Publishers, New York, 2019, pp. 159-182.

[3] M.R. Karim, Harun-Ur-Rashid, A.B. Imran, Highly stretchable hydrogel using vinyl modified narrow dispersed silica particles as cross-linker, ChemistrySelect, 5 (2020) 10556-10561. https://doi.org/10.1002/slct.202003044

[4] M. Harun-Ur-Rashid, T. Foyez, A.B. Imran, Fabrication of stretchable composite thin film for superconductor applications. In Sensors for Stretchable Electronics in Nanotechnology, CRC Press, 2021, pp. 63-78. https://doi.org/10.1201/9781003123781-5

[5] M. Harun-Ur-Rashid, A.B. Imran, M.A.B.H. Susan, Green Polymer Nanocomposites in Automotive and Packaging Industries, Curr. Pharm. Biotechnol. 6 May (2022). https://doi.org/10.2174/1389201023666220506111027..

[6] A-N. Chowdhury, J. Shapter, and A.B. Imran, Innovations in Nanomaterials, Nova Science Publishers, Inc., NY, USA, 2015

[7] M. Harun-Ur-Rashid, A.B. Imran, T. Seki, Y. Takeoka, M. Ishii, H. Nakamura, Template synthesis for stimuli-responsive angle independent structural colored smart

materials, Trans. Mater. Res. Soc. 34 (2009) 333-337.
https://doi.org/10.14723/tmrsj.34.333

[8] M. Harun-Ur-Rashid, A.B. Imran, T. Seki, M. Ishii, H. Nakamura, Y. Takeoka, Angle-independent structural color in colloidal amorphous arrays, ChemPhysChem, 11 (2010) 579-583. https://doi.org/10.1002/cphc.200900869

[9] Y. Takeoka, S. Yoshioka, M. Teshima, A. Takano, M. Harun-Ur-Rashid, T. Seki, Structurally coloured secondary particles composed of black and white colloidal particles, Sci. Rep. 3 (2013) 1-7. https://doi.org/10.1038/srep02371

[10] A.B. Imran, M. Harun-Ur-Rashid, Y. Takeoka, Polyrotaxane Actuators. In Soft Actuators, Springer, Singapore, 2019, pp. 81-147. https://doi.org/10.1007/978-981-13-6850-9_6

[11] M. Harun-Ur-Rashid, T. Foyez, I. Jahan, K. Pal, A.B. Imran, Rapid diagnosis of COVID-19 via nano-biosensor-implemented biomedical utilization: a systematic review, RSC Advances, 12 (2022) 9445-9465. https://doi.org/10.1039/D2RA01293F

[12] Z.H. Mohammad, F. Ahmad, S.A. Ibrahim, S. Zaidi, Application of nanotechnology in different aspects of the food industry, Discover Food, 2 (2022) 1-21. https://doi.org/10.1007/s44187-022-00013-9

[13] M. Aufan, et al., Towards a defnition of inorganic nanoparticles from an environmental, health and safety perspective. Nat Nanotechnol. 4 (2009) 634-41. https://doi.org/10.1038/nnano.2009.242

[14] L. Rashidi, K. Khosravi-Darani, The applications of nanotechnology in food industry. Crit. Rev. Food Sci. 51 (2011) 723-30. https://doi.org/10.1080/10408391003785417

[15] L.K Ncube et al. Environmental impact of food packaging materials: a review of contemporary development from conventional plastics to polylactic acid-based materials. Mater. 13 (2020) 4994. https://doi.org/10.3390/ma13214994

[16] S.H. Nile et al. Nanotechnologies in food science: applications, recent trends, and future perspectives. Nano-Micro Lett. 12 (2020) 1-34. https://doi.org/10.1007/s40820-020-0383-9

[17] Y. Luo, Q. Wang, Y. Zhang, Biopolymer-based nanotechnology approaches to deliver bioactive compounds for food applications: a perspective on the past, present, and future. J. Agric. Food Chem. 68 (2020) 12993-3000. https://doi.org/10.1021/acs.jafc.0c00277

[18] R. Dobrucka, R. Cierpiszewski Active and intelligent packaging Food-Research and Development-A review, Pol. J. Food Nutri. Sci. 64 (2014) 7-15. https://doi.org/10.2478/v10222-012-0091-3

Materials Research Forum LLC

https://doi.org/10.21741/9781644902295-13

[19] M.S. Firouz, K. Mohi-Alden, M. Omid, A critical review on intelligent and active packaging in the food industry: Research and development. Food Resear. Intl. 141 (2021) 110113. https://doi.org/10.1016/j.foodres.2021.110113

[20] R.P. Singh, Utility of nanomaterials in food safety. In Food Safety and Human Health, Academic Press, 2019, pp. 285-318. https://doi.org/10.1016/B978-0-12-816333-7.00011-4

[21] A. Singh, A.K. Gupta, S. Singh, Molecular mechanisms of drug resistance in Mycobacterium tuberculosis: Role of nanoparticles against multi-drug-resistant tuberculosis (MDR-TB). In NanoBioMedicine, Springer, Singapore, 2020, pp. 285-314. https://doi.org/10.1007/978-981-32-9898-9_12

[22] L.K Adams, D.Y. Lyon, P.J. Alvarez, Comparative eco-toxicity of nanoscale TiO2, SiO2, and ZnO water suspensions, Water Resear. 40 (2006) 3527-3532. https://doi.org/10.1016/j.watres.2006.08.004

[23] T. Gordon, B. Perlstein, O. Houbara, I. Felner, E. Banin, S. Margel, Synthesis and characterization of zinc/iron oxide composite nanoparticles and their antibacterial properties. Colloid. Surf. A: Physicochem. Engg. Aspects, 374 (2011) 1-8. https://doi.org/10.1016/j.colsurfa.2010.10.015

[24] K.M. Reddy, K. Feris, J. Bell, D.G. Wingett, C. Hanley, A. Punnoose, Selective toxicity of zinc oxide nanoparticles to prokaryotic and eukaryotic systems. App. Phy. Lett. 90 (2007) 213902. https://doi.org/10.1063/1.2742324

[25] R. Brayner, R. Ferrari-Iliou, N. Brivois, S. Djediat, M.F. Benedetti, F. Fiévet, Toxicological impact studies based on Escherichia coli bacteria in ultrafine ZnO nanoparticles colloidal medium. Nano Lett. 6 (2006) 866-870. https://doi.org/10.1021/nl052326h

[26] T. Ohira, O. Yamamoto, Y. Iida, Z. Nakagawa, Antibacterial activity of ZnO powder with crystallographic orientation. J. Mater. Sci. Mater. Med. 19 (2008) 1407-1412. https://doi.org/10.1007/s10856-007-3246-8

[27] M. Premanathan, K. Karthikeyan, K. Jeyasubramanian, G. Manivannan, Selective toxicity of ZnO nanoparticles toward Grampositive bacteria and cancer cells by apoptosis through lipid peroxidation. Nanomed.: Nanotechnol. Biol. Med. 7 (2011) 184-192. https://doi.org/10.1016/j.nano.2010.10.001

[28] J. Sawai, Quantitative evaluation of antibacterial activities of metallic oxide powders (ZnO, MgO and CaO) by conductimetric assay. J. Microbiol. Meth. 54 (2003) 177-182. https://doi.org/10.1016/S0167-7012(03)00037-X

[29] Y. Xie, Y. He, P.L. Irwin, T. Jin, X. Shi, Antibacterial activity and mechanism of action of zinc oxide nanoparticles against Campylobacter jejuni, App. Env. Microbiol. 77 (2011) 2325-2331. https://doi.org/10.1128/AEM.02149-10

[30] G. Applerot, A Lipovsky, R. Dror, N. Perkas, Y. Nitzan, R. Lubart, A. Gedanken, Enhanced antibacterial activity of nanocrystalline ZnO due to increased ROS-mediated cell injury. Adv. Funct. Mater. 19 (2009) 842-852. https://doi.org/10.1002/adfm.200801081

[31] O. Yamamoto, Influence of particle size on the antibacterial activity of zinc oxide, Intl. J. Inorg. Mater. 3 (2001), 643-646. https://doi.org/10.1016/S1466-6049(01)00197-0

[32] P.J.P. Espitia, N.D.F.F. Soares, J.S.D.R. Coimbra, N.J. de Andrade, R.S. Cruz, E.A.A. Medeiros, Zinc oxide nanoparticles: synthesis, antimicrobial activity and food packaging applications. Food Biopro. Technol. 5 (2012) 1447-1464. https://doi.org/10.1007/s11947-012-0797-6

[33] P.K Raul, A. Thakuria, B. Das, R.R. Devi, G. Tiwari, C. Yellappa, D.V. Kamboj, Carbon nanostructures as antibacterials and active food-packaging materials: A review, ACS omega, 7 (2022), 11555-11559. https://doi.org/10.1021/acsomega.2c00848

[34] L. Zhao, M. Zhang, A.S. Mujumdar, B. Adhikari, H. Wang, Preparation of a novel carbon dot/polyvinyl alcohol composite film and its application in food preservation. ACS App. Mater. Interf. 14 (2022), 37528-37539. https://doi.org/10.1021/acsami.2c10869

[35] M.O. Alaş, G. Doğan, M.S. Yalcin, S. Ozdemir, R. Genç, Multicolor emitting carbon dot-reinforced pva composites as edible food packaging films and coatings with antimicrobial and uv-blocking properties, ACS omega, 7 (2022) 29967-29983. https://doi.org/10.1021/acsomega.2c02984

[36] Y.H. Wen, C.H. Tsou, M.R. de Guzman, D. Huang, Y.Q. Yu, C. Gao, Z.H. Wang, Antibacterial nanocomposite films of poly (vinyl alcohol) modified with zinc oxide-doped multiwalled carbon nanotubes as food packaging, Polym. Bullet. 79 (2022) 3847-3866. https://doi.org/10.1007/s00289-021-03666-1

[37] A. Barra, J.D. Santos, M.R. Silva, C. Nunes, E. Ruiz-Hitzky, I. Gonçalves, P.A. Marques, Graphene derivatives in biopolymer-based composites for food packaging applications. Nanomaterials 10 (2020) 2077. https://doi.org/10.3390/nano10102077

[38] A.K. Sundramoorthy, T.H.V. Kumar, S. Gunasekaran, Graphene-based nanosensors and smart food packaging systems for food safety and quality monitoring. In Graphene bioelectronics, Elsevier, 2018, pp. 267-306). https://doi.org/10.1016/B978-0-12-813349-1.00012-3

[39] N.A. Manikandan, K. Pakshirajan, G. Pugazhenthi, Preparation and characterization of environmentally safe and highly biodegradable microbial polyhydroxybutyrate (PHB) based graphene nanocomposites for potential food packaging applications, Intl. J. Biol. Macromol. 154 (2020) 866-877. https://doi.org/10.1016/j.ijbiomac.2020.03.084

[40] H. Yilmaz Dogan, Y. Altin, A. Ç. Bedeloğlu, Fabrication and properties of graphene oxide and reduced graphene oxide reinforced Poly (Vinyl alcohol) nanocomposite films for packaging applications. Polym. Polym. Compo. 30 (2022) 09673911221113328. https://doi.org/10.1177/09673911221113328

[41] N.A. Al-Tayyar, A.M. Youssef, R. Al-hindi, Antimicrobial food packaging based on sustainable bio-based materials for reducing food-borne pathogens: A review. Food Chem. 310 (2020) 1-17. https://doi.org/10.1016/j.foodchem.2019.125915

[42] C. Tan, F. Han, S. Zhang, P. Li, N. Shang, Novel bio-based materials and applications in antimicrobial food packaging: Recent advances and future trends. Intl. J. Mole. Sci. 22 (2021) 9663. https://doi.org/10.3390/ijms22189663

[43] H.P.S. Abdul Khalil, Y. Davoudpour, C.K. Saurabh, M.S. Hossain, A.S. Adnan, R. Dungani, M.T. Paridah, M.Z.I. Sarker, M.R.N. Fazita, M.I. Syakir et al., A review on nanocellulosic fibres as new material for sustainable packaging: Process and applications. Renew. Sustain. Energy Rev. 64 (2016) 823-836. https://doi.org/10.1016/j.rser.2016.06.072

[44] T.N. Prabhu, K. Prashantha, A review on present status and future challenges of starch based polymer films and their composites in food packaging applications. Polym. Compos. 39 (2016) 2499-2522. https://doi.org/10.1002/pc.24236

[45] A.M. Youssef, F.M. Assem, M.E. Abdel-Aziz, M. Elaaser, O.A. Ibrahim, M. Mahmoud, M.H. Abd El-Salam, Development of bionanocomposite materials and its use in coating of Ras cheese. Food Chem. 270 (2019) 467-475. https://doi.org/10.1016/j.foodchem.2018.07.114

[46] N.A. Al-Tayyar, A.M. Youssef, R. Al-hindi, Antimicrobial packaging efficiency of ZnO-SiO2 nanocomposites infused into PVA/CS film for enhancing the shelf life of food products. Food Packag. Shelf. 25 (2020) 100523. https://doi.org/10.1016/j.fpsl.2020.100523

[47] H.F. Youssef, M.E. El-Naggar, F.K. Fouda, A.M. Youssef, Antimicrobial packaging film based on biodegradable CMC/PVA-zeolite doped with noble metal cations. Food Packag. Shelf. 22 (2019) 100378. https://doi.org/10.1016/j.fpsl.2019.100378

[48] L. Ren, X. Yan, J. Zhou, J. Tong, X. Su, Influence of chitosan concentration on mechanical and barrier properties of corn starch/chitosan films. Int. J. Biol. Macromol. 105 (2017) 1636-1643. https://doi.org/10.1016/j.ijbiomac.2017.02.008

[49] S.H. Othman, Bio-nanocomposite materials for food packaging applications: Types of bio-based polymer and nanosized filler, Agric. Agric. Sci. Procedia, 2 (2014) 296-303. https://doi.org/10.1016/j.aaspro.2014.11.042

[50] Y. Wang, G. Zhang,. The preparation of modified nano-starch and its application in food industry. Food Resear. Intl. 140 (2021) 110009. https://doi.org/10.1016/j.foodres.2020.110009

[51] T.V. Duncan, Applications of nanotechnology in food packaging and food safety: Barrier materials, antimicrobials and sensors. J. Colloid. Interface. Sci. 363 (2011) 1-24 https://doi.org/10.1016/j.jcis.2011.07.017

[52] S. Kumar, M. Nehra, N. Dilbaghi, K. Tankeshwar, K.H. Kim, Recent advances and remaining challenges for polymeric nanocomposites in healthcare applications. Prog. Polym. Sci. 80 (2018) 1-38. https://doi.org/10.1016/j.progpolymsci.2018.03.001

[53] I. Shahabi-Ghahfarrokhi, A. Babaei-Ghazvini, Using photo-modification to compatibilize nano-ZnO in development of starch-kefiran-ZnO green nanocomposite as food packaging material. Intl. J. Biolo. Macromol. 124 (2019) 922-930. https://doi.org/10.1016/j.ijbiomac.2018.11.241

[54] G.F. Nogueira, F.M. Fakhouri, R. Oliveira, Extraction and characterization of arrowroot (Maranta Arundinaceae L.) starch and its application in edible films. Carbohydr. Polym. 186 (2018) 64-72. https://doi.org/10.1016/j.carbpol.2018.01.024

[55] A. Kausar, A review of high performance polymer nanocomposites for packaging applications in electronics and food industries. J. Plast. Film Sheeting, 36 (2020) 94-112. https://doi.org/10.1177/8756087919849459

Emerging Applications of Nanomaterials   Materials Research Forum LLC
Materials Research Foundations 141 (2023) 353-371   https://doi.org/10.21741/9781644902295-14

Chapter 14

Nanomaterials in Machine Learning Based Prosthesis

Deepali Salwan[1] and Shri Kant[2]

[1]Senior Technology Manager, ITeXchange. India

[2]Centre for Cyber Security and Cryptology, Dept. of Computer science and Engineering, Sharda University, Greater Noida, India

*shrikant.ojha@gmail.com

Abstract

Machine learning has found its application in various fields. With the advent of technology, it has even found its way in the prosthesis field as doctors are using nano materials to conduct surgery in amputees and predicting the results of the surgery with the help of machine learning. The same techniques have also been successfully applied within the medical field, in addressing diverse clinical applications such as interpreting electrocardiograms, detecting dementia, cardiovascular diseases, or predicting the prognosis and survival rates for breast cancer or melanomas. By adopting Machine Learning Techniques (MLT) into medical analysis, diagnostic accuracy is increased, costs are reduced, and human resources are saved. To date, the effectiveness of MLTs for enhancing joint implant design has not been examined. Our goal of this paper is to check the role nanomaterials play during the design of a machine learning based prosthesis.

Keywords

Prosthesis, Machine Learning, Reinforcement Learning, Joint Implant, Hip Surgery, Amputee

Contents

1.   Introduction

The number of medical procedures performed in a population is an indicator of how important a treatment is. A total of one million hip fractures happens every year around the world. In the decade between 2000 and 2009, the number of hip replacements grew by about 25% [1]. Even the number of knee replacements has seen a huge surge with a growth of 5% year on year, most commonly taking place in India and China. This trend is anticipated to continue as the population ages, medical care in developing countries improves, and the average age at which first hip replacements occur decreases [2]. Moreover, these patients often require revision surgery when their life expectancy differs from the lifetime of the implant, as children and young people are among the ones who typically require it [5].

According to general consensus, biomedical innovation has been mainly driven by the appropriate application of materials designed for other purposes than biomedical ones [3]. Many alloys used to make prosthetic implants were developed by the aircraft industry, such as Ti-6Al-4V alloy. The research on new material concepts has already had an impact on healthcare and is increasingly used to manufacture prosthetics and delivery devices for drugs [7]. The future of Biomaterials Science is being shaped by these novel material concepts.

With a total hip replacement, the presence of a rigid stem in the femur changes the bone's mechanical properties considerably compared to the conditions in the healthy bone. With the implant inserted, the flexural rigidity increases, leading to a decrease in mechanical stresses and strains (stress shielding) within the bone, particularly, in the area immediately surrounding the implant. In addition, bone loss is often caused by this erosion of mechanical stresses and strains [3,4], thus decreasing implant stability and longevity [5] and making successful revision surgery more difficult [6].

With nanotechnology's potential to transform this century, along with other emerging technologies, medical device development will be able to open up new concepts in the near future [10]. We have seen rapid advancements in biomedical imaging and diagnosis, drug delivery, gene therapy, anticancer therapy, and other areas by using nanomaterials with diameters of *100 nm. There is an endless range of applications for nanoparticles in these fields. Researchers are still at a very early stage in their investigation of the potential effects of nanoparticles on human health. Further patient investigation is needed. Some of the nanoparticles need to undergo rigorous testing before they can be used on humans [8].

Researchers and developers are continuously researching and developing hip prostheses so, they can extend their life span, replicate the skin's natural anatomy more closely, and reduce the likelihood of complications and revision surgeries. As a result, hip prostheses are available in a wide variety. Younger populations were the target market for hip implants with short stems. As a bone conservative surgery, it allows for more physiological stress transfer to the proximal femur by preserving the medial greater trochanter and maintaining a high neck retention [12]. The current designs do reduce stress shielding to a lesser degree than traditional implants [22], however, they have not been able to completely eliminate it [11].

Methods for utilizing and predicting data are covered in Machine Learning Techniques (MLTs). Techniques for this type of analysis are known to use complex algorithms that are designed to replicate model behaviour [13-16]. There are several successful applications of these models for prediction tasks [11] in many different fields, like industrial [18], electronic [13], space science [14] geology [15,16] and language [17], among many others.

The same techniques have also been successfully applied within the medical field, in addressing diverse clinical applications such as interpreting electrocardiograms, detecting dementia, cardiovascular diseases, or predicting the prognosis and survival rates for breast cancer or melanomas [18–23]. By adopting MLT into medical analysis, diagnostic accuracy is increased, costs are reduced, and human resources are saved [24,25]. To date, the effectiveness of MLTs for enhancing joint implant design has not been examined. In order to determine the relationship between stem performance and stem design for long stems, shape optimization algorithms different from MLTs have been used [26–30].

In this chapter we will concentrate on the nano materials and the roles they play in machine-learning based prosthesis. This will help us develop much more efficient systems in the future.

2.   Basics of joint replacement

Joints that are cartilaginous include hip and knee joints. Articular surfaces protect the joint's surface, allowing pain-free movement. By cushioning the joint, the cartilage facilitates bone movement. Many causes can lead to this surface wearing out; generally, the cause is not known. Pain occurs when the ends of the bones rub against one another when the cartilage wears down. [32] The term arthritis is used to describe numerous conditions in which the joint surfaces (cartilage) wear out.

An artificial device (prosthesis) is inserted into the joint during total joint replacement (TJR) to cure pain and restore movement to the joint. A typical orthopaedic implant today lasts 10 to 15 years. Knee and hip replacements last even longer [35]. A large part of the population is growing older and younger patients are requesting orthopaedic surgery. This means implants should last for at least 35 years.

Since the first primary joint replacement was performed more than a decade ago, the success rate has increased from 85 percent to 95 percent, A recent report indicated that, (i) total joint replacements (TJR) are receiving more revisions than in the past; (ii) Revision implant products grew by an estimated 12-15% annually during the past few years - Growth rates for primary implants are more than twice as fast [33-36].

Many challenging and contentious accounting issues continue to be debated among accountants and auditors around the world. These issues will require a unified approach based on a proper understanding of these standards.

A lot of research is being conducted throughout the world regarding the interface between implants and bones to determine how to improve osseointegration by functionalizing implant surfaces. The interaction between a material and its environment can be enhanced by specifically modifying its surface characteristics [42]. Chemical and mechanical treatments are often applied to commercially available devices' surfaces to achieve desirable features. There has been research studying osteoblasts' response to rough surfaces. The osteoblasts are found to exhibit less proliferation on rough surfaces and more differentiation [45]. The traditional method for designing endosseous devices is using this method. An innovative method for enhancing endosseous implant integration is to create biomimetic surfaces that function as adhesives and present cells with adherent factors. Adhesion of osteoblasts is mediated by at least two mechanisms: an interaction with integrin receptors on outer cell membranes is the most studied one, while another involves interactions between integrins and the heparin-binding sites of extra-cellular matrix proteins [9]. The "biochemical approach" in designing endosseous devices has led to the search for biomolecules that promote cell adhesion to the implant surface and enhance adhesion strength.

During hip resurfacing, metal is deposited on a metal surface and circulated in the bloodstream. A metallic hip replacement or hip resurfacing surgery patient's cobalt and chromium ions are detectable all over his or her body, and the possibility of negative side effects is not fully understood [18-19]. A recent publication raises the possibility of cancer-causing effects and hypersensitivity reactions in the medium-to-short term. The metal ions could be eliminated by using ceramic surfaces.

Therefore, resurfacing techniques that have thin implants need ceramic materials that are too thick, and this poses more design challenges. Knees also suffer from similar problems. A total knee replacement is not necessary for all patients. Unicondylar knee replacements may be effective in treating patients with arthritis that only affects part of their knee [45-50]. As a result of these replacements, the hospital stay can be shortened, the incision is

Emerging Applications of Nanomaterials
Materials Research Foundations 141 (2023) 353-371

Materials Research Forum LLC
https://doi.org/10.21741/9781644902295-14

smaller, and the patella does not need to be replaced. Still, the use of ceramics for this application has been limited.

2.1   Proposed design

A design is proposed for a knee joint assembly automating stance phase lock for transfemoral amputees [6]. This work used biomechanical simulation in order to develop a smart knee joint for use by children that can be controlled automatically according to children's stance. A limited sample size was used to determine the prototype's applicability to other types of prosthetics, reasons for amputation, and men and women of various ages and genders among other characteristics.

In order to detect clinical benefits and drawbacks associated with the developed knee prototype, it was important to test these factors using greater quantities. It was proposed to develop a biomechatronic prosthetic knee for transfemoral amputees [41]. The proposed study used accelerometer signals and analysis of event detection to identify gait but did not consider the determination of gait associated with prosthetic knee joints in amputees. The device needs continuous functionality, meaning that longer battery operating times and weight reduction strategies should be part of the design of the proposed prosthesis.

In a study presented, a microprocessor-controlled prosthetic knee was used [28]. Using a microprocessor-controlled prosthetic knee joint, Becher P.F. determined that their subject's balance and gait were changed more than with a mechanical knee joint. People with limited mobility were, however, unable to utilize the enhanced technical characteristics of a microprocessor-controlled Prosthetic Knee [28].

3.   Understanding role of machine learning in prosthesis

The goal of machine learning is to use data and algorithms to mimic the way humans learn, improving its accuracy over time. It lies at the intersection of artificial intelligence (AI) and computer science.

The algorithms of machine learning are powerful and learn to take decisions by hit and trial method, gradually improving at every step. Let us discuss the algorithms below.

Model Based Algorithms that are generally goal oriented and take decisions on the basis of trial and error comprise Deep Learning or Reinforcement Learning. Model Free Algorithms, on the other hand, are generally trial and error based and have a spectrum of deep learning algorithms. Real time applications, however, may not have much room for trial and error.

Therefore, the model used in such applications should be properly trained to account for unforeseen situations. Reinforcement learning is a different technique from other machine learning techniques because it optimizes reward functions while minimizing penalty functions. A knee joint for which both algorithms and reward functions are appropriately trained and configured so that it perfectly mimics a human's gait sounds like this would be the perfect solution.

Emerging Applications of Nanomaterials
Materials Research Foundations 141 (2023) 353-371

Materials Research Forum LLC
https://doi.org/10.21741/9781644902295-14

4. Phase lock knee joint design

Designing a phase lock knee joint to automate stance and stability for young amputees with arthritis. There has been some debate regarding transfemoral amputees [12]. This work used biomechanical simulation in order to develop a smart knee joint for use by children that can be controlled automatically according to children's stance. Based on the proposed framework, the prototype could be applied to other prosthetics, reasons for amputation, and ages and genders with a limited sample size. It was recommended that the therapeutic benefits of the prototyped knee be evaluated using greater quantities and that the drawbacks can be understood.

This paper studies the different nanomaterials used for biomechanical prosthetic knee for transfemoral amputees. Gait detection with a Prosthetic knee joint in amputees was not considered in the proposed study, which used accelerometer signals and event detection to analyse acceleration signals. For the proposed prosthesis to remain functional for a long time, it needs to be powered by batteries for longer operation and weigh less.

It evaluated the performance of an electric knee prostheses has been presented [2]. There were significant differences in the positions and gait of the users of a microprocessor-controlled knee compared to a mechanical knee. People with limited mobility were, however, unable to utilize the enhanced technical characteristics of a microprocessor-controlled Prosthetic Knee. The proposed controller makes use of three phase algorithms, such as slope estimation, ground-search, and impedance modulation [22]. Amputees can achieve stable stance quickly with prosthetic knee joints. The proposal controller was tested on a knee prosthesis based on the results of a series of stances. It was not taken into consideration because the tone of the amputee's balance and stability needed to be improved.

By adjusting the impedance control parameters by the controller for a prosthetic knee joint, we are able to achieve human gait with adaptive dynamic programming for transfemoral prostheses.

Analysing the gait phases of the amputees while walking and training the model online to follow the goals of each user, we were able to personalize user efficiency characteristics. By using dynamic programming, human kinematics could be achieved by changing the controller of the prosthesis [32]. Although the controller wasn't tested on a diverse range of amputees, it was based on certain criteria they ignored, such as the load on the ankle and the walking speed over uneven terrain.

5. Methodology

In the following section we will be looking at the methodology of our research and how we worked to build a knee joint prosthetic with the help of machine learning, optimizing it to a greater extent so that a person can walk on uneven surfaces too in the following section we will look various types of nano materials which could be used in development of an intelligent lower limb prosthesis which can withstand the stress and have elasticity so as to support and mimic like natural human leg.

A knee joint prosthetic is presented. Functional validation is performed on the proposed design. We sought to develop a robust, lightweight, cost-effective and durable prosthetic knee by analysing several factors. Even though the results improved, the advanced features still need to be optimized. Moreover, it was found that the gait was stable as it went up and down the stairs. However, no details were given on energy consumption or stability. The idea of wearing a prosthesis while walking on uneven surfaces is presented in a model for inclined plane walking. In the proposed model, the controller gains during inclined plane walk could be determined without tuning techniques. The plains optimization, however, required one-time manual effort to be performed offline.

Real-time computation and a set of validation rules made it computationally efficient. The sample size was small, however, for assessment. Sitting and ascending stairs, on the other hand, were not included in this study. There is a need to test it with more samples, and also with other activities like stair climbing, sitting, etc., as it was only designed for a single motion, that of walking at a steady speed [12].

It was proposed to design an Active Leg that adapts to the user's needs. This was achieved by using bipedal robotics. We were attempting to identify a biological mechanism that should allow an amputee to control his body [42]. A central pattern generator (CPG) was devised to provide an amputee with appropriate gait patterns. Initially, fuzzy logic was used to construct an amputee's gait patterns. Nevertheless, more research needs to be conducted on making prostheses that are less costly and more energy efficient.

6. Reinforcement learning algorithm

Reinforcement Learning (RL) is a branch of Machine Learning and since RL learns from the environment it is considered to be the most sophisticated learning in real time applications. And we're studying the materials which will completement the prosthesis. Reinforcement learning considers value functions to play an important role. Reinforcement learning programming functions as a mapping between situations and their actions, which is called policy functions. This category of algorithms is tuned for maximizing rewards. There is always a reward for the actions in this category. As a result, it is recommended for the agent to leverage all previous episodes of high reward and to explore new actions as well in order to make the best decision.

The goal remains the same, therefore, of achieving high rewards for any action, despite exploration and exploitation always finding a fix and looking like an impossible problem. Hence, there must be a balance between exploitation and exploration. In the limit of 0 -1, the randomness of the function could be increased by adding an entropy coefficient. Markov Decision Processes (MDP) are usually employed to implement RL systems; they comprise an action A, a state S, and a reward R along with the probability of transitioning from one state to another. The claim that MDP states depend only on previous states and actions, and they have nothing to learn from their past [17].

Materials Research Forum LLC
https://doi.org/10.21741/9781644902295-14

The use of Reinforcement learning for real-time applications has always been apprehensive because this technique is often called Trial and Error Learning. There have always been questions concerning Reinforcement Learning's practical application. Below are some of the challenges associated with Reinforcement Learning.

Training a model on a simulator is challenging, so studying real-time systems is quite difficult

Apart from safety, Reinforcement learning is also always governed by rewards and penalty. It might take an action in pursuit of reward which might prove to be harmful for the agent(program).

- Another reason to use Reinforcement learning in real-time applications is the use of high-dimensional modelling.
- The inequity in reward policy is a result of long delays in seeking out the optimal policy.
- It is possible to derive an optimal policy from the exploration vs. exploitation problem, but the agent is always lost in dilemma to either explore new policy of exploit old policies.
- Diving deep into personalizing a real-time system has remained an area of research for a long time.

It can be trained for some real-world applications to perform brilliantly even though the challenges described above may occur, owing to the distinct strength of Reinforcement Learning as well as the effectiveness of its samples if used wisely.

7. Hard on hard joints

Over the past 30 years, hard-on-hard ceramics have been used for Total Joint Arthroplasty, in an attempt to substitute the CoCr–UHMWPE-bearing couple.

The reduction of debris generation and fracture risk has received a great deal of research effort. Thus, some of the research into monolithic oxide ceramics is mainly focused on the use of alumina and zirconia, and as a consequence, only those materials have been commercialized until now. Alumina and zirconia are the main constituents of ceramic composites [20], with the manufacturers of these materials paying attention to their properties. The main focus has been on improving the processing and reliability of ceramics such as alumina and zirconia rather than developing new microstructural designs and searching for new materials.

In order to improve the properties of monolithic ceramic materials, the size of the alumina and zirconia grains are to be reduced. A variety of raw materials with nanostructured crystals are available today; however, in order to maintain the nanostructures, they have to be processed conventionally into dense compacts which creates serious problems [5].

Emerging Applications of Nanomaterials   Materials Research Forum LLC
Materials Research Foundations 141 (2023) 353-371 https://doi.org/10.21741/9781644902295-14

Nanoparticles are composed of the form, volume, the percentage of face-centred atoms or defects, and behaviours that might include the ionization potential of electrons or band gaps. Growth parameters such as times, rates, and temperatures are also characteristics that determine the characteristics. There are many dimensions for each data point due to all these characteristics. The researchers were able to pick out trends in 2D maps of nanoparticles by generating 2D maps encoded by one characteristic at a time and allowing the data to become more dimensional.

T-distributed stochastic neighbour embedding (t-SNE) and self-organisation maps (SOM) can reduce the dimensionality of these kinds of data into 2D maps. According to the SOM algorithm, maps are based on an interconnected grid of neurons. A value is randomly assigned to each neuron at the beginning. Machine learning algorithms aim to find the most similar value within the grid for each data point recorded by taking the differences between the values.

The value of the neuron is then updated for both the neuron closest to the optimal matching unit and the neuron near the optimal matching unit, in order to "weight" it based on the data matching. Similar to the t-SNE, however, its grid is weighted by probability rather than distance, so the map is less meaningful when seeing distances and directions.

In order to analyze micro-micro composites made with zirconia Y-TZP at position 3, a very different analysis is required, though the conclusion would be the same. For zirconia Y-TZP, if the doping agent is not modified, the operational reinforcement is directly related to the grain size.

Furthermore, this material's grain size range for biomedical applications is established by ISO standard 13356. There is an additional problem, as the martensitic transformation process, which strengthens the zirconia structure, is also responsible for the degradation of zirconia by hydrothermal processes [22, 23]. One of the main manufacturers of femoral heads in the world was forced to withdraw from the market because of this last mechanism. Micro-micro composites made of zirconia can perform better mechanically when alumina is added to them, but the ageing problem still remains.

The reduction of grain size in the nonorange (100 nm) does not only result in a non-improvement, but also severely damages the performance of monolithic alumina and zirconia materials.

8. Micro nanocomposites

It is thought that nano nano composites contain two phases: the matrix, which is comprised of crystals of micron size, and the nanoparticles, which are nanometre sized and located inter- and intragranular.

This residual stress in the matrix is caused by dislocations of the particles in nanocomposites. This is due to the differences in the surface energy and interface energy between the matrix and nanoparticles, in addition to the mismatch between thermal

Emerging Applications of Nanomaterials
Materials Research Foundations 141 (2023) 353-371

Materials Research Forum LLC
https://doi.org/10.21741/9781644902295-14

expansion and elastic properties. Alumina-zirconia nanocomposites were the first to reveal the surprising effects of nanomaterials and nanostructures.

Stress intensity factor directly determines the static or dynamic wear of an implant; this is a performance parameter that guarantees that there is no subcritical propagation of cracks or defects and, therefore, is one of the most critical factors when designing new ceramics for structural applications [36].

Nanostructured multiphase powders derived by colloidal routes have opened a new research avenue for alumina-based nanocomposites [37]. In an effort to develop a new class of dense nanocomposite materials, researchers have developed micro structured alumina matrices that are surrounded by very small zirconia particles as well as individual grains. The knee implant is coated with zirconia nitride for patients with suspected metal ion hypersensitivity.

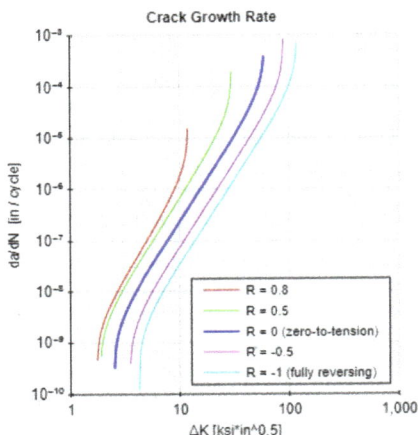

Figure 1. Stress Intensity Graph

Nanostructured nanocomposites have been shown to possess unique properties. Researchers have discovered that zirconia nanoparticles embedded within alumina crystals present a compressive residual stress field that retards crack front opening when applying external stress, preventing stress corrosion. Applied stress intensity factors with a lower residual stress intensity factor seem to have a greater relative effect. This can be seen from the Fig. 1 above, as the stress intensity factor increases, the crack growth rate also increases.

8.1 Nanocomposites

The phase transformation mechanism described in a paper by Benzai et. al. can also be used to explain why nano-structured materials exhibit excellent crack resistance [45].

There is no need to emphasize any nano-effect for the durability of these nanocomposites. Chevalier J. et. al. did point out, these nano-nano composites have, however, been observed to have higher KIC/KIO ratios than micro-nano composites discussed in [39] as those exhibit significant "nano" effects. In other words, transformation toughening and internal stresses that result from nanoparticles being encased inside other crystals do not affect both KIC and KIO values equally. Despite the fact that nano-nano composites have been described as interpenetrated microstructures, the amount of nanoparticles contained inside other crystals is quite small, so the effect of a network of dislocations does not affect the KIO value.

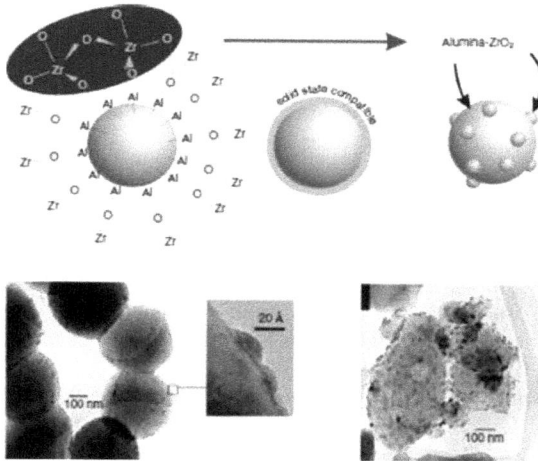

Figure 2. Nano-composite powders

In addition to this, Ce-TZP exhibits a high degree of plasticity, which contributes to the observed significant increase in toughness. In order to accomplish this, cooling after sintering relaxes the stresses. Only particles of zirconia nanoparticles present in alumina grains are able to produce the nano effect in this scenario. Ce-TZP compromises the toughening effects of transformation with the ability to induce internal stress that can affect the threshold for the stress intensity factor. By conventionally processing these nanocomposite powders, nano-nano composites with flexural strengths of 950-1000 MPa and fracture toughness of 12 MPa$\sqrt{m}$ are obtained.

8.2 Macro-micro-nanocomposites

As an example, revisions of total hip arthroplasty experience a 31 percent failure rate [33]. As an important element of calculating the minimum detectable defect size required for a structural life exceeding patient lifetime, quality-control procedures must be followed prior to the device entering service so as to detect the defect. Ceramic implants typically have defects in the range of ten to a hundred microns, whereas metal implants have defects in the range of hundreds of microns. There are two approaches to designing damage-resistance bioceramic microstructures: crack propagation versus crack initiation. A useful distinction between intrinsic and extrinsic fracture mechanisms is often useful [50]. There is an intrinsic mechanism that evolves ahead of crack tips in the alumina zirconia micro-nanocomposites described above (regardless of the crack's size) referred to as compression residual stress.

It remains largely unexplored in terms of extrinsic flaw state whether short crack growth resistances are important. Fabrication, machining, or finishing may result in damage to the component [49,50], heat sterilization (ageing) [31], or even surgery (handling of the prosthesis). Corrosion and degradation in vivo also do produce defects. When there are imperfections in a component, stress would build up, causing cracks to grow and the component to fracture [7]. Ceramic components that are fractured require revision surgeries because surgeons must carefully remove all the broken pieces.

Creating nanoscale Al_2O_3-ZrO_2-Nb composites offers the possibility of designing new materials with toughening mechanisms operating on a scale smaller than that of the matrix microstructure, as in natural bone, and enhances the material's inherent fracture properties. By creating structural synergy between the matrix and dispersoid phases, there will be an increased resistance to crack initiation and propagation, resulting in an increase in catastrophic failure resistance. The use of extrinsic mechanisms for crack growth resistance (R-curve) and flaw tolerance, such as the adhesion of elongated metallic particles, can improve flaw tolerance. So, in these cases, both resistance and strength increase.

In order to develop better crack growth resistance materials, it is necessary to understand the respective influences of nano-, micro-, and macro-scale mechanisms. In addition to them providing improved mechanical properties, macromolecular-nanosystem composite materials are now being used for new biomedical applications including spinal and knee components. They are also used for larger and thinner versions of existing components, such as ball heads and inserts.

9. Bioinspired materials hierarchy

TJR uses those materials that allow movement of the joint similar to that of a normal one. By teaming up with engineers and other scientists, orthopaedists are improving surgical materials and techniques rapidly. The metals stainless steel, chrome, and cobalt alloys, as well as titanium alloys, have been extensively used in dentistry, to fix fractures, and replace joints for many decades [22].

Osseointegration is a critical issue for metal implants. Recent advances in glass coating fabrication have found their way into the literature concerning bone implant interface enhancement [5-9]. Inorganic or organic materials could be combined with the glasses to tune resorption (reabsorbing) rates of coatings or create therapeutic coatings.

However, implant materials with a younger modulus than the bone result in negative effects due to a fundamental flaw: bioimplants with young moduli cause bone degradation, such as bone loss from stress shielding. Despite their advantages, plastic stems increase the amount of proximal interface stress. They may also lead to increased loosening rates due to interface debonding and micromotion, therefore offering a cure for bone resorption (reabsorbing). In conclusion, the change in bone morphology around hip stems is caused by implant flexibility and bone resorption (reabsorbing) around the stems, as well as stress shielding.

Ideally, scaffolds should have controlled pore sizes, shapes, and orientations, and should be produced reliably and economically [41]. Although porous materials do not have equal strengths, all of them have one thing in common that is also a limitation of these materials, the lack of strength that is quite inherent with porosity. Therefore, they generally work best in low-stress conditions, like breaking jaws or fractured skulls. Consequently, the problem of how to create a scaffold that is both porous and strong remains unsolved. Using freeze casting [32], polymeric or ceramic parts can be produced with complex shapes that are frequently overlooked.

The ceramic part is formed by pouring a ceramic slurry into a mould and freezing it. When the part is demoulded, the frozen solvent works as a temporary binder to hold the part together. During freeze drying, the solvent is sublimated under vacuum in order to prevent the drying stresses that may cause cracks and warping. In the process of sintered drying, the compacts are made stronger, stiffer, and contain the desired porosity. After freezing the scaffold acquires an anisotropic microstructure that is complex and often anisotropic. It is possible to direct crystal growth in a way that dictates the porosity orientation in the final material.

10. Use of nanomaterials in machine learning based prosthesis

In earlier section we discussed various types of nano materials. In this section we will typically discuss about the generation method and use of nano based neural material. Carbon-based materials, metal nanoparticles, metal-polymer combinations, hydrogels, and other nanomaterials are employed to make prosthetic nerve sensors. The usable lifespan of prosthetic limbs is impacted by the fact that nanoparticle-reinforced creep composites creep more readily than carbon nanoparticle materials. Carbon-based nanomaterials perform well in creep resistance. Bionic electronic exogenous nerve sensor may monitor physiological indices and feel bodily motions. It can also be investigated how electrophysiological signals, such as those from additional wearable devices that measure body temperature, breathing rate, heart rate, and blood pressure, are sent. The neural sensor [51] can be put to the skin's surface layer and utilised to transmit nerve signals within the limbs. When utilised to transmit nerve signals from the skin's surface layer, polyester thread made of carbon nanotubes, carbon-based materials are very stretchy and insensitive to temperature changes to deformation and has a high cycle life, quick responsiveness, great ruggedness, geometric customised cuttability and layered sensing.

There is 3D graphene-based piezoresistive sensors that operate well in electronic skin. Because of its ease of preparation and output, it is derived from electrospinning and 3D printing. Because of its simple mechanism, low power consumption, and easy signal collection, it is regarded as a potential flexible sensor.

Nickel-cobalt nanoparticles [52] obtained on the surface of the moon is also chemically functionalized multiwall carbon nanotubes and work as sensors with greater stability and sensitivity repeatability. When metal polymers and carbon nanomaterials are used together, the polymer component can change the metal's modulus, and the material itself can lessen the difference between soft tissues and electrodes and enhance the performance of prosthetic limbs. one of several choices for creating nerves is the newly developed hydrogel substance. It is biocompatible, biodegradable, and antibacterial as an artificial tissue or artificial limb that may be implanted with bioelectronics. It also provides protection from heat and cold, and it takes a while to recover. Stronger than other materials in terms of tensile, compressive, dynamic durability, and strain sensitivity.

Conclusions

Despite its intelligence and complexity, the human body is the most engineered structure created by God. The orthotist and prosthetist find it challenging to replace its lost anatomical structures and functions. AI and robotic advancements have given hope to millions of people with disabilities. Artificial intelligence is in its infancy when it comes to prosthetics and orthotics and is only widely used in a few fields. It is still early in the commercialization process for many AI projects. Many disabled persons are unable to afford these devices due to the high costs. In order for the highest quality and latest technology to reach a greater population of disabled people at affordable prices, government agencies, manufacturers, and funding agencies must step up and invest in this field.

Following a survey of the available work in the field of prosthetic knee joints and studying the benefits of reinforcement learning algorithms, we determined that the approach of developing a smart knee joint with Model Based outputs and Model Free algorithms is an effective strategy. In the future, we plan to develop an Artificial Knee with a natural gait like a human one through a novel approach.

Thus, we have made a study of all the nano materials that are used in the Machine Learning based knee joint prosthetics. This will also help with future development of better equipment with the help of the flaws listed, this research will pave a way for much better nano-technology development for operating in surgeries.

Nanotechnology has the ability to drastically improve people's lives. We have started by employing low-cost, lightweight solar polymers, making solar energy more broadly available. Nanoparticles can be used to clean up dangerous chemical spills as well as contaminants in the air. Nanomaterials, according to experts, might be employed in space exploration.

Materials Research Forum LLC
https://doi.org/10.21741/9781644902295-14

References

[1] R. Langer, D. A. Tirrell, Designing materials for biology and medicine, Nature, 428 (2004) 487-492. https://doi.org/10.1038/nature02388

[2] M. K. Anderson, Dreaming about nanomedicine. Wired Magazine, November, 2000.

[3] M. R. Gwinn, V. Vallyathan, Nanoparticles: Health effects-pros and cons, Environmental Health Perspectives 114 (2006) 1818-1825. https://doi.org/10.1289/ehp.8871

[4] I. Linkov, F. K. Satterstrom, L. M. Corey, Nanotoxicology and nanomedicine: Making hard decisions, Nanomedicine: NBM 4 (2008) 167-171. https://doi.org/10.1016/j.nano.2008.01.001

[5] K. K. Jain, Nanomedicine: Application of nanobiotechnology in medical practice, Med. Prin. Pract. 17 (2008) 89-101. https://doi.org/10.1159/000112961

[6] A. Cavalcanti, B. Shirinzadeh, M. J. Zhang, L. C. Kretly, Nanorobot hardware architecture for medical defense, Sensors 8 (2008) 2932-2958. https://doi.org/10.3390/s8052932

[7] J. M. Hootman, C. G. Helmick, Projections of US prevalence of arthritis and associated activity limitations, Arthritis Rheum. 54 (2006) 226-229. https://doi.org/10.1002/art.21562

[8] M. H. Huo, J. Parvizi, N. F. Gilbert, What's new in hip arthroplasty, JBJS 88 (2006) 2100-2113. https://doi.org/10.2106/JBJS.F.00595

[9] A. Bagno, A. Piovan, M. Dettin, A. Chiarion, P. Brun, R. Gambaretto, G. Fontana, C. Di Bello, G. Palù, I. Castagliuolo, Human osteoblast-like cell adhesion on titanium substrates covalently functionalized with synthetic peptides, Bone 40 (2007) 693-699. https://doi.org/10.1016/j.bone.2006.10.007

[10] A. W. Bridges, N. Singh, K. L. Burns, J. E. Babensee, L. A. Lyon, A. J. Garcia, Reduced acute inflammatory responses to microgel conformal coatings, Biomaterials 29 (2008) 4605-4615. https://doi.org/10.1016/j.biomaterials.2008.08.015

[11] H. Wagner, Surface replacement arthroplasty of the hip, Clin. Orthop. 134 (1978) 102-130. https://doi.org/10.1097/00003086-197807000-00014

[12] W. C. Head, Wagner surface replacement arthroplasty of the hip: Analysis of fourteen failures in forty-one hips, The Journal of Bone and Joint Surgery, American volume 63-A (1981) 420-427. https://doi.org/10.2106/00004623-198163030-00017

[13] W. N. Capello, T. M. Trancik, G. Misamore, R. Eaton, Analysis of revision surgery of resurfacing hip arthroplasty, Clin. Orthop. 170 (1982) 50-55. https://doi.org/10.1097/00003086-198210000-00007

[14] B. E. Bierbaum, R. Sweet, Complications of resurfacing arthroplasty, Orthop. Clin. North Am. 13 (1982) 761-775. https://doi.org/10.1016/S0030-5898(20)30234-0

Materials Research Forum LLC
https://doi.org/10.21741/9781644902295-14

[15] W. C. Head, The Wagner surface replacement arthroplasty, Orthop. Clin. North Am. 13 (1982) 789-797. https://doi.org/10.1016/S0030-5898(20)30236-4

[16] M. A. Freeman, G. W. Bradley, ICLH surface replacement of the hip: An analysis of the first 10 years, The Journal of Bone and Joint Surgery, British volume 65 (1983) 405-411. https://doi.org/10.1302/0301-620X.65B4.6409905

[17] J. Daniel, P. B. Pynset, D. J. W. McMinn, Survival analysis of metal-on-metal hip resurfacing in patients under the age of 55 years with osteoarthritis, The Journal of Bone and Joint Surgery, British volume 86-B (2004) 177-84. https://doi.org/10.1302/0301-620X.86B2.14600

[18] H. Pandit, S. Glyn-Jones, P. McLardy-Smith, R. Gundle, D. Whitwell, C. L. Gibbons, S. Ostlere, N. Athanasou, H. S. Gill, D. W. Murray, Pseudotumours associated with metal-on-metal hip resurfacings, J. Bone Joint Surg. Br. 90 (2008) 847-51. https://doi.org/10.1302/0301-620X.90B7.20213

[19] E. Dunstan, D. Ladon, P. Whittingham-Jones, R. Carrington, T. W. R. Briggs, Chromosomal aberrations in the peripheral blood of patients with metal-on-metal hip bearings, The Journal of Bone and Joint Surgery, American volume 90 (2008) 517-522. https://doi.org/10.2106/JBJS.F.01435

[20] M. N. Rahaman, B. S. Bal, J. P. Garino, M. Ries, A. Yao, Ceramics for prosthetic hip and knee joint replacement, J. Am. Ceram. Soc. 90 (2007) 1965-1988. https://doi.org/10.1111/j.1551-2916.2007.01725.x

[21] M. Nygren, Z. Shen, On the preparation of bio-, nano- and structural ceramics and composites by spark plasma sintering, Solid State Sciences, 5 (2003) 125-131. https://doi.org/10.1016/S1293-2558(02)00086-9

[22] M. Yoshimura, T. Noma, K. Kawabata, S. Somiya, Role of H_2O on the degradation Process of YTZP, J. Mater. Sci. Lett. 6 (1987) 465-467. https://doi.org/10.1007/BF01756800

[23] J. Chevalier, B. Cales, J. M. Drouin, Low-temperature ageing of Y-TZP ceramics, J. Am. Ceram. Soc. 82 (1999) 2150-2154. https://doi.org/10.1111/j.1151-2916.1999.tb02055.x

[24] R. C. Garvie, R. H. J. Hannink, R. T. Pascoe, Ceramic steel, Nature 258 (1975) 703. https://doi.org/10.1038/258703a0

[25] C. M. Sharkness, S. Hamburger, R. M. Moore, R. G. Kaczmarek, Prevalence of artificial hips in the United States, J. Long Term Effects Med. Implants 2 (1992) 1-8.

[26] Y. Murase, E. Kato, K. Daimon, Stability of ZrO_2 phases in ultrafine ZrO_2-Al_2O_3 mixtures, J. Am. Ceram. Soc. 69 (1986) 83-87. https://doi.org/10.1111/j.1151-2916.1986.tb04706.x

Emerging Applications of Nanomaterials   Materials Research Forum LLC
Materials Research Foundations 141 (2023) 353-371   https://doi.org/10.21741/9781644902295-14

[27] D. J. Green, Critical microstructures for microcracking in Al2O3-ZrO2 composites, J. Am. Ceram. Soc. 65 (1982) 610-614. https://doi.org/10.1111/j.1151-2916.1982.tb09939.x

[28] P. F. Becher, Transient thermal stress behaviour in ZrO2-toughenend Al2O3, J. Am. Ceram. Soc. 64 (1981) 37. https://doi.org/10.1111/j.1151-2916.1981.tb09555.x

[29] A. H. De Aza, J. Chevalier, G. Fantozzi, M. Schehl, R. Torrecillas, Crack growth resistance of alumina, zirconia and zirconia toughened alumina ceramics for joint prostheses, Biomaterials 23(2002) 937-945. https://doi.org/10.1016/S0142-9612(01)00206-X

[30] C. Pecharromán, J. F. Bartolomé, J. Requena, J. S. Moya, S. Deville, J. Chevalier, G. Fantozzi, R. Torrecillas, Percolative mechanism of aging in zirconia-containing ceramics for medical applications, Adv. Mater. 15 (2003) 507-511. https://doi.org/10.1002/adma.200390117

[31] S. Deville, J. Chevalier, G. Fantozzi, J. F. Bartolomé, J. Requena, J. S. Moya, R. Torrecillas, L. A. Díaz, Low temperature ageing of zirconia-toughened alumina ceramics and its implication in biomedical implants, J. Eur. Ceram. Soc. 23 (2003) 2975-2982. https://doi.org/10.1016/S0955-2219(03)00313-3

[32] S. Deville, J. Chevalier, C. Dauvergne, G. Fantozzi, J. F. Bartolomé, J. S. Moya, R. Torrecillas, Microstructural investigation of the aging behavior of (3Y-TZP)-Al2O3 composites, J. Am. Ceram. Soc. 88 (2005) 1273-1280. https://doi.org/10.1111/j.1551-2916.2005.00221.x

[33] T. Nakanishi, M. Sasaki, J. Ikeda, F. Miyaji, M. Kondo, Mechanical and phase stability of zirconia toughened alumina, Key Engineering Materials 330-332 II (2007) 1267-1270. https://doi.org/10.4028/www.scientific.net/KEM.330-332.1267

[34] G. Pezzotti, K. Yamada, S. Sakakura, R. P. Pitto, Raman spectroscopic analysis of advanced ceramic composite for hip prosthesis, J. Am. Ceram. Soc. 91 (2008) 1199-1206. https://doi.org/10.1111/j.1551-2916.2007.01507.x

[35] S. Choi, H. Awaji, Nanocomposites - a new material design concept, Science and Technology of Advanced Materials 6 (2005) 2-10. https://doi.org/10.1016/j.stam.2004.06.002

[36] A. H. De Aza, J. Chevalier, G. Fantozzi, M. Schehl, R. Torrecillas, Slow crack growth behaviour of zirconia toughened alumina ceramics processed by different methods, J. Am. Ceram. Soc. 86 (2003) 115-20. https://doi.org/10.1111/j.1151-2916.2003.tb03287.x

[37] M. Schehl, L. A. Díaz, R. Torrecillas, Alumina based nanocomposites from powder-alcoxide mixtures, Acta Materialia 50 (2002) 1125-1139. https://doi.org/10.1016/S1359-6454(01)00413-X

[38] S. Deville, J. Chevalier, G. Fantozzi, J. F. Bartolomé, J. Requena, J. S. Moya, R. Torrecillas, L. A. Díaz, Development of advanced zirconia-toughened alumina nanocomposites for orthopaedic applications, Key Engineering Materials 264-268 (2004) 2013-2016. https://doi.org/10.4028/www.scientific.net/KEM.264-268.2013

[39] J. Chevalier, S. Deville, G. Fantozzi, J. F. Bartolomé, C. Pecharromán, J. S. Moya, L. A. Díaz, R. Torrecillas, Nanostructured ceramic oxides with a slow crack growth resistance close to covalent materials, Nano Letters 5 (2005) 1297-1301. https://doi.org/10.1021/nl050492j

[40] J. Chevalier, A. H. De Aza, G. Fantozzi, M. Schehl, R. Torrecillas, Extending the lifetime of ceramic orthopaedic implants, Advanced Materials 12 (2000) 1619-1621. https://doi.org/10.1002/1521-4095(200011)12:21<1619::AID-ADMA1619>3.0.CO;2-O

[41] M. Nawa, S. Nakamoto, T. Sekino, K. Niihara, Tough and strong Ce-TZP/Alumina nanocomposites doped with titania, Ceram. Inter. 24 (1998) 497-506. https://doi.org/10.1016/S0272-8842(97)00048-5

[42] K. Tanaka, J. Tamura, K. Kawanabe, M. Nawa, M. Uchida, T. Kokubo, T. Nakamura, Phase stability after aging and its influence on Pin-on-Disk wear properties of Ce-TZP/Al2O3 nanocomposite and conventional Y-TZP, J. Biomed. Mater. Res. 67A (2003) 200-207. https://doi.org/10.1002/jbm.a.10006

[43] K. Tanaka, J. Tamura, K. Kawanabe, M. Nawa, M. Oka, M. Uchida, T. Kokubo, T. Nakamura, Ce-TZP/Al2O3 nanocomposites as a bearing material in total joint replacement, J. Biomed. Mater. Res. 63 (2002) 262-270. https://doi.org/10.1002/jbm.10182

[44] M. Uchida, H. M. Kim, T. Kokubo, M. Nawa, T. Asano, K. Tanaka, T. Nakamira, Apatite-forming ability of a Zirconia/Alumina nano-composite induced by chemical treatment, Inc. J. Biomed. Mater. Res. 60 (2002) 277-282. https://doi.org/10.1002/jbm.10071

[45] R. Benzaid, J. Chevalier, M. Saâdaoui, G. Fantozzi, M. Nawa, L. A. Diaz, R. Torrecillas, Fracture toughness, strength and slow crack growth in a ceria stabilized zirconia-alumina nanocomposite for medical applications, Biomaterials, 29 (2008) 3636-3641. https://doi.org/10.1016/j.biomaterials.2008.05.021

[46] R. Torrecillas, L. A. Díaz, Nanocomposites for biomedical applications, International Journal of Materials Research, 2008.

[47] H. G. Richter, G. Willmann, Realiability of ceramic components for total hip endoprostheses, British Ceramic Transactions 98 (1999) 29-34. https://doi.org/10.1179/bct.1999.98.1.29

[48] C. Piconi, G. Maccauro, L. Pilloni, W. Burger, F. Muratori, H. G. Richter, On the fracture of a zirconia ball head, J. Mater. Sci.: Mat. In Med. 17 (2006) 289-300. https://doi.org/10.1007/s10856-006-7316-0

[49] H. G. Richter, Fractography of bioceramics, Key Eng. Mat. 223 (2002) 157-180. https://doi.org/10.4028/www.scientific.net/KEM.223.157

[50] S. Deville, J. Chevalier, L. Gremillard, Influence of surface finish and residual stresses on the ageing sensitivity of biomedical grade zirconia, Biomaterials 27 (2006) 2186-2192. https://doi.org/10.1016/j.biomaterials.2005.11.021

[51] M. Cao, J. Su, S. Fan, H. Qiu, D. Su, L. Li, Wearable piezoresistive pressure sensors based on 3D graphene, Chemical Engineering Journal, 406 (2021) 126777. https://doi.org/10.1016/j.cej.2020.126777

[52] K. Arikan, H. Burhan, R. Bayat, F. Sen, Glucose nano biosensor with non-enzymatic excellent sensitivity prepared with nickel-cobalt nanocomposites on f-MWCNT, Chemosphere 291 (2022) 132720. https://doi.org/10.1016/j.chemosphere.2021.132720

Keyword Index

About the Editors

Prof. N.B. Singh, former Head Chemistry Department and Former Dean Faculty of Science, DDU Gorakhpur University, Gorakhpur is presently Emeritus professor at Department of Chemistry and Biochemistry and Research Development Cell, Sharda University, Greater Noida, UP. Dr. Singh received the most prestigious Alexander von Humboldt fellowship (Germany) in 1977 and has worked at different universities in Germany. His main areas of research are Solid state chemistry, materials science, cement chemistry, thermodynamics, and water purification. He has published more than 300 research articles, 8 books, and more than 50 book chapters. Prof. Singh is president of Indian Association for Solid State Chemists and Allied Scientists Jammu and received many awards. His name appeared twice in the list of 2% scientists of world declared by Stanford University, USA.

Dr. Md. Abu Bin Hasan Susan, a Professor of Chemistry at Dhaka University, Bangladesh, received his Ph.D. degree from Yokohama National University, Japan in 2000 as a MEXT scholar and was awarded VBL, JSPS, CREST, and Bridge postdoctoral fellowships. He has 176 articles including 18 book chapters. He collaborates with 14 renowned laboratories and visited eleven countries. He was nominated for awards from FACS, in 2005 and ISESCO, in 2010. He received the Dean's Award of the Faculty of Science, Dhaka University (2011), the UGC Awards of Bangladesh in 2011 and 2013, the United Group Paper Awards in 2016 and 2017, and the Bangladesh Academy of Sciences

Gold Medal in 2012. He received the Silver Medal of the Society of Promotion of Education and Science, India in 2019. He is an *Associate Editor* of Spectrum of Emerging Sciences, an *Editorial Board Member* of the Journal of Bangladesh Academy of Sciences and Universal Journal of Electrochemistry, and the *Journal of Scientific and Technical Research*. He served as the National Representative of Division I of IUPAC. He has been a Fellow (2018) of the Bangladesh Academy of Sciences and IUPAC (2016). His citations are 10,901, *h*-index- 30 and *i*-10 index- 79.

Dr. Ratiram Gomaji Chaudhary is presently working as Associate Professor and Head, Department of Chemistry, Seth Kesarimal Porwal College of Arts, Science and Commerce, Kamptee. His research areas are *Biogenic Synthesis*, *Phytosynthesis*, *Metal oxide/Graphene-based nanohybrids*, *Toxic Dyes Degradations* etc. He has been awarded two times with *'Rajiv Gandhi National Fellowship Award'* as JRF by UGC, New Delhi for pursuing MPhil and PhD degrees. He has published 3 books, 19 book chapters, and 102 regular articles, and 12 review articles in peer-reviewed SCI/Scopus indexing journals having Total Citations of 1351 with h_index 20. He is a recipient of several awards like 'Best Researcher Award', 'Young Scientist Award', 'Outstanding Reviewer Award by IOP-VAST', Mahatma Jyotirao Fule: State Level Best Researcher Award, etc. He worked as a Guest Editor for Scopus indexed journals *Materials Today: Proceeding*, Elsevier, and *Current Pharmaceutical Biotechnology*, *Current Pharmaceutical Design*, and *Current Nanosciences,* Bentham Science*.*